Lecture Notes in Artificial Intelligence 9110

Subseries of Lecture Notes in Computer Science

LNAI Series Editors

Randy Goebel
University of Alberta, Edmonton, Canada

Yuzuru Tanaka
Hokkaido University, Sapporo, Japan

Wolfgang Wahlster
DFKI and Saarland University, Saarbrücken, Germany

LNAI Founding Series Editor

Joerg Siekmann
DFKI and Saarland University, Saarbrücken, Germany

More information about this series at http://www.springer.com/series/1244

Tom Collins · David Meredith
Anja Volk (Eds.)

Mathematics and Computation in Music

5th International Conference, MCM 2015
London, UK, June 22–25, 2015
Proceedings

 Springer

Editors
Tom Collins
De Montfort University
Leicester
UK

Anja Volk
Utrecht University
Utrecht
The Netherlands

David Meredith
Aalborg University
Aalborg
Denmark

ISSN 0302-9743   ISSN 1611-3349 (electronic)
Lecture Notes in Artificial Intelligence
ISBN 978-3-319-20602-8   ISBN 978-3-319-20603-5 (eBook)
DOI 10.1007/978-3-319-20603-5

Library of Congress Control Number: 2015941498

LNCS Sublibrary: SL7 – Artificial Intelligence

Springer Cham Heidelberg New York Dordrecht London

Printed on acid-free paper

Springer International Publishing AG Switzerland is part of Springer Science+Business Media
(www.springer.com)

Preface

The 5th Biennial International Conference for Mathematics and Computation in Music (MCM 2015) took place June 22–25, 2015, at Queen Mary University of London, UK, co-hosted by the School of Electronic Engineering and Computer Science (Centre for Digital Music) and the School of Mathematical Sciences. As the flagship conference of the Society for Mathematics and Computation in Music (SMCM), MCM 2015 provided a dedicated platform for the communication and exchange of ideas among researchers in mathematics, informatics, music theory, composition, musicology, and related disciplines. It brought together researchers from around the world who combine mathematics or computation with music theory, music analysis, composition, and performance.

This year's program – full details at http://mcm2015.qmul.ac.uk – featured a number of distinguished keynote speakers, including Andrée Ehresmann (who spoke on contemporary mathematical approaches to creative systems), Emilia Gómez (who spoke on music technologies in classical orchestral music concerts), Gareth Loy (who spoke on steps toward a theory of musical interest), and Ge Wang (who spoke on the art of designing computer music), and a film (*From Circles to Hyperspheres, Part III*) by Gilles Baroin and Hugues Seress.

The program also emphasized performances of real music related to the theoretical discussions. A noontime concert, Blood and Tango, presented a new adaptation of Gareth Loy's *Blood from a Stone*, composed in 1992 for Max Matthew's electronic violin and performed at MCM 2015 by Laurel Pardue, followed by Susanne Beer and Ian Pressland's divergent interpretations of Piazzolla's *Le Grand Tango* for cello and piano, each accompanied by Elaine Chew. An evening concert, Geometries and Gestures (Gege), featured Tom Johnson's *Rational Melodies* performed on flute by Carlos Vaquero, and musical illustrations of chord geometries in Italian popular music performed by Moreno Andreatta (SMCM Vice President), and of hypergestures in free jazz improvisations by Guerino Mazzola (SMCM President).

The chapters in this book correspond to the papers and posters presented at the conference following a careful double-blind peer-review process. We received 64 submissions from 108 authors across 19 different countries. Each submission was assigned at least three reviewers for double-blind review. A paper was accepted only if a majority of its reviewers recommended it for acceptance. The format of each accepted paper (long or short paper) was also decided on the basis of the recommendations of the majority of the paper's reviewers. Submissions were accepted in two categories: long papers (with a limit of 12 pages) for oral presentation at the conference and short papers (with a limit of six pages) for poster presentation at the conference. A total of 24 long papers and 14 short papers were accepted following review.

Finally, we are grateful to the London Mathematical Society, the Institute of Musical Research, the Engineering and Physical Sciences Research Council, and the Society for Mathematics and Computation in Music for their generous support for, and promotion of, the conference.

May 2015

Oscar Bandtlow
Elaine Chew
David Meredith
Anja Volk
Tom Collins

Organization

Executive Committee

General Chairs

Oscar Bandtlow   Queen Mary University of London, UK
Elaine Chew   Queen Mary University of London, UK

Program Chairs

David Meredith   Aalborg University, Denmark
Anja Volk   Utrecht University, The Netherlands

Publication Chair

Tom Collins   De Montfort University, UK

Panels/Tutorials/Workshop Chair

Johanna Devaney   Ohio State University, USA

Communications Chairs

Janis Sokolovskis   Queen Mary University of London, UK
Luwei Yang   Queen Mary University of London, UK

Review Board

Joshua Albrecht   University of Mary Hardin-Baylor, USA
Anna Aljanaki   Utrecht University, The Netherlands
Emmanuel Amiot   Classes Préparatoire aux Grandes Ecoles, Perpignan, France
Christina   University of Athens, Greece
  Anagnostopoulou
Moreno Andreatta   IRCAM/CNRS/UPMC, France
Gerard Assayag   IRCAM/CNRS/UPMC, France
Oscar Bandtlow   Queen Mary University of London, UK
Brian Bemman   Aalborg University, Denmark
Emmanouil Benetos   City University London, UK
Louis Bigo   Universidad del País Vasco UPV/EHU, Spain
Jean Bresson   IRCAM/CNRS/UPMC, France
Chantal Buteau   Brock University, Canada
Clifton Callender   Florida State University, USA

Emilios Cambouropoulos	Aristotle University of Thessaloniki, Greece
Norman Carey	CUNY Graduate Center, USA
Elaine Chew	Queen Mary University of London, UK
Ching-Hua Chuan	University of North Florida, USA
David Clampitt	Ohio State University, USA
Darrell Conklin	Universidad del País Vasco UPV/EHU, Spain
Michael Cuthbert	Massachusetts Institute of Technology, USA
Johanna Devaney	Ohio State University, USA
Shlomo Dubnov	University of California, San Diego, USA
Tuomas Eerola	Durham University, UK
Thomas Fiore	University of Michigan-Dearborn, USA
Arthur Flexer	Austrian Research Institute for Artificial Intelligence, Austria
Harald Fripertinger	Karl-Franzens-Universität Graz, Austria
Mathieu Giraud	CNRS, France
Maarten Grachten	Austrian Research Institute for Artificial Intelligence, Austria
Yupeng Gu	Indiana University, Bloomington, USA
Dorien Herremans	University of Antwerp, Belgium
Keiji Hirata	Future University Hakodate, Japan
Anna Huang	Massachusetts Institute of Technology, USA
Özgür İzmirli	Connecticut College, USA
Philip Kirlin	Rhodes College, USA
Katerina Kosta	Queen Mary University of London, UK
Robin Laney	The Open University, UK
Olivier Lartillot	Aalborg University, Denmark
David Lewis	Goldsmiths, University of London, UK
Cristina Catherine Losada	University of Cincinnati, USA
Matija Marolt	University of Ljubljana, Slovenia
Alan Marsden	University of Lancaster, UK
Panayotis Mavromatis	NYU Steinhardt, USA
Guerino Mazzola	University of Minnesota, USA
David Meredith	Aalborg University, Denmark
Andrew Milne	University of Western Sydney, Australia
Thomas Noll	ESMuC Barcelona, Spain
François Pachet	Sony CSL, Paris, France
Marcus Pearce	Queen Mary University of London, UK
Robert Peck	Louisiana State Univesity, USA
Richard Plotkin	University at Buffalo, SUNY, USA
Ian Quinn	Yale University, USA
Richard Randall	Carnegie Mellon University, USA
Christopher Raphael	Indiana University, Bloomington, USA
Christophe Rhodes	Goldsmiths, University of London, UK
David Rizo	University of Alicante, Spain
Craig Sapp	Stanford University, USA

William Sethares	University of Wisconsin, USA
Alan Smaill	University of Edinburgh, UK
Bob Sturm	Queen Mary University of London, UK
David Temperley	University of Rochester, USA
Petri Toiviainen	University of Jyväskylä, Finland
Godfried Toussaint	McGill University, Canada
Peter van Kranenburg	Meertens Institute, The Netherlands
Gissel Velarde	Aalborg University, Denmark
Anja Volk	Utrecht University, The Netherlands
Gerhard Widmer	Johannes Kepler University Linz and Austrian Research Institute for Artificial Intelligence, Austria
Geraint Wiggins	Queen Mary University of London, UK
Jonathan Wild	McGill University, Canada
Daniel Wolff	City University London, UK
Jason Yust	Boston University, USA
Marek Žabka	Netherlands Institute for Advanced Study in the Humanities and Social Sciences, The Netherlands

Society for Mathematics and Computation in Music

President

| Guerino Mazzola | University of Minnesota, USA |

Vice President

| Moreno Andreatta | IRCAM/CNRS/UPMC, France |

Treasurer

| David Clampitt | Ohio State University, USA |

Secretary

| Johanna Devaney | Ohio State University, USA |

Journal of Mathematics and Music

Editors-in-Chief

| Thomas Fiore | University of Michigan-Dearborn, USA |
| Clifton Callender | Florida State University, USA |

Reviews Editor

| Jonathan Wild | McGill University, Canada |

Sponsoring Institutions

London Mathematical Society
Institute of Musical Research
Engineering and Physical Sciences Research Council
Society for Mathematics and Computation in Music
Queen Mary University of London
Centre for Digital Music
School of Mathematical Sciences
School of Electronic Engineering and Computer Science

Contents

Performance

Similarity and Contrast

Post-Tonal Music Analysis

Geometric Approaches

Deep Learning

Scales

Notation and Representation

A Structural Theory of Rhythm Notation Based on Tree Representations and Term Rewriting

Florent Jacquemard[1](✉), Pierre Donat-Bouillud[1,2], and Jean Bresson[1]

[1] UMR STMS: IRCAM-CNRS-UPMC and INRIA, Paris, France
florent.jacquemard@inria.fr, jean.bresson@ircam.fr
[2] ENS Rennes, Ker Lann Campus, Bruz, France
pierre.donat-bouillud@ens-rennes.fr

Abstract. We present a tree-based symbolic representation of rhythm notation suitable for processing with purely syntactic theoretical tools such as term rewriting systems or tree automata. Then we propose an equational theory, defined as a set of rewrite rules for transforming these representations. This theory is complete in the sense that from a given rhythm notation the rules permit to generate all notations of equivalent durations.

Introduction

Term Rewriting Systems (TRSs) [8] are well established formalisms for tree processing (transformation and reasoning). With solid theoretical foundations, they are used in a wide range of applications, to name a few: automated reasoning, natural language processing, foundations of Web data, etc. TRSs perform in-place transformations in trees by the replacement of patterns, as defined by oriented equations called rewrite rules. They are a classical model for symbolic computation, used for rule-based modeling, simulation and verification of complex systems or software (see e.g. the languages TOM[1] and Maude[2]). Tree Automata (TAs) [7] are finite state recognizers of trees which permit to characterize specific *types* of tree-structured data (*regular* tree languages). They are often used in conjunction with TRSs, acting as *filters* in the explorations of sets of trees computed by rewriting.

It is also common to use trees to represent hierarchical structures in symbolic music (see [15] for a survey). For instance, the GTTM [13] uses trees to analyse inner relations in musical pieces. Trees are also a natural representation of rhythms, where durations are expressed as a hierarchy of subdivisions. Computer-aided composition (CAC) environments such as Patchwork and OpenMusic [3,6] use structures called *rhythm trees* (RTs) for representing

This work is part of the EFFICACe project funded by the French National Research Agency (ANR-13-JS02-0004-01). A more complete version of this paper is available at https://hal.inria.fr/hal-01134096/.

[1] http://tom.loria.fr.
[2] http://maude.cs.illinois.edu.

© Springer International Publishing Switzerland 2015
T. Collins et al. (Eds.): MCM 2015, LNAI 9110, pp. 3–15, 2015.
DOI: 10.1007/978-3-319-20603-5_1

and programming rhythms [2]. Such hierarchical, notation-oriented approach (see also [15]) is complementary to the performance-oriented formats corresponding to the MIDI notes' onsets and offsets in standard computer music systems. It also provides a more structured representation of time than music notation formats such as MusicXML [9] or Guido [12], where durations are expressed with integer values. As highly structured representations, trees enable powerful manipulation and generation processes in the rhythmic domain (see for instance [11]), and enforce some structural constraints on duration sequences.

In this paper, we propose a tree-structured representation of rhythm suitable for defining a set of rewriting rules (*i.e.* oriented equations) preserving rhythms, while allowing simplifications of notation. This representation bridges CAC rhythm structures with formal tree-processing approaches, and enables a number of new manipulations and applications in both domains. In particular, rewriting rules can be seen as an axiomatization of rhythm notation, which can be applied to reasoning on equivalent notations in computer-aided music composition or analysis.

1   Preliminary Definitions

Let us assume given a countable set of variables $\mathcal{X}$, and a ranked signature Σ which is a finite set of symbols, each symbol being assigned a fixed arity. We denote as Σ_p the subset of Σ of symbols of arity p.

Trees. A Σ-labelled tree t (called *tree* for short in the rest of the paper) is either a single node, called *root* of t and denoted by $root(t)$, labeled with a variable $x \in \mathcal{X}$ or one constant symbol of $a_0 \in \Sigma_0$, or it is made of one node also denoted by $root(t)$ and labeled with a symbol $a \in \Sigma_n$ $(n > 0)$, and of an ordered sequence of n *direct subtrees* $t_1, \ldots, t_n$.

In the first case, the tree t is simply denoted x or a_0. In the second case, t is denoted $a(t_1, \ldots, t_n)$, $root(t)$ is called the *parent* of respectively $root(t_1), \ldots, root(t_n)$, and the latter are called *children* of $root(t)$. Moreover, for all i, $1 < i \leq n$, $root(t_{i-1})$ is called the previous *sibling* of $root(t_i)$. For $1 \leq i \leq p$, the previous *cousin* of $root(t_i)$ is either $root(t_{i-1})$ if $i > 1$, or the last children of the previous cousin of the parent $root(t)$ if $i = 1$ and if this node exists. In other terms, the previous cousin of a node ν in a tree t is the node immediately at the left of ν in t, at the same level. A node in a tree t with no children is called a *leaf* of t. In the following, we will consider the sequence of leaves of a tree t as enumerated by a depth-first-search (*dfs*) traversal.

Example 1. Some trees are depicted in Figs. 1 to 5. In the tree $2\big(n, 3(o, n, n)\big)$ of Fig. 2(b), the first leaf in *dfs* ordering (labeled with n) is the previous cousin of the node labeled by 3, and the second leaf in *dfs* ordering (labeled with o) is the previous sibling of the third leaf (labeled with n), which is in turn the previous sibling of the fourth leaf (also labeled with n). ◇

The set of trees built over Σ and $\mathcal{X}$ is denoted $\mathcal{T}(\Sigma, \mathcal{X})$, and the subset of trees without variables $\mathcal{T}(\Sigma)$. The definition domain of a tree $t \in \mathcal{T}(\Sigma, \mathcal{X})$, denoted

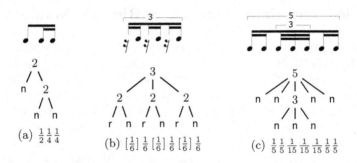

Fig. 1. Simple trees of $\mathcal{T}(\Sigma_{rn})$ with their corresponding rhythmic notations and values.

by $dom(t)$, is the set of nodes of t. The *size* $|t|$ of t is the cardinality of $dom(t)$. Given $\nu \in dom(t)$: $t(\nu) \in \Sigma \cup \mathcal{X}$ is the label of ν in t, $t|_\nu$ is the subtree of t at node ν, $t[t']_\nu$ is the tree obtained from t by replacement of $t|_\nu$ by t'. We define the *depth* of a single-node tree x or a_0 as 0 and the depth of $a(t_1, \ldots, t_n)$ as $1+$ the maximal depth of $t_1, \ldots, t_n$.

Pattern Matching. We call *pattern* over Σ a finite sequence of trees of $\mathcal{T}(\Sigma, \mathcal{X})$ of length $n \geq 1$, denoted as $t_1; \ldots; t_n$ (the symbol ; denotes the cousin relation). The *size* of a pattern is the sum of the sizes of its constituting trees.

A *substitution* is a mapping from variables of $\mathcal{X}$ into trees of $\mathcal{T}(\Sigma, \mathcal{X})$ with a finite domain. The application of substitutions is homomorphically extended from variables to trees and patterns: $\sigma\big(a(t_1, \ldots, t_p)\big) = a\big(\sigma(t_1), \ldots, \sigma(t_p)\big)$ and $\sigma\big(t_1; \ldots; t_n\big) = \sigma(t_1); \ldots; \sigma(t_n)$.

A tree t *matches* a pattern $t_1; \ldots; t_n$ at node ν with substitution σ if there exists a sequence of successive cousins $\nu_1, \ldots, \nu_n$ in $dom(t)$ such that $\nu_1 = \nu$ and $t|_{\nu_i} = \sigma(t_i)$ for all $1 \leq i \leq n$. When there exists such a sequence of cousins, we write $t[t'_1; \ldots; t'_n]_\nu$ for the iterated replacement $t[t'_1]_{\nu_1} \ldots [t'_n]_{\nu_n}$.

Term Rewriting. A *rewrite rule* is a pair of patterns of same length denoted $\ell_1; \ldots; \ell_n \rightarrow r_1; \ldots; r_n$ and a tree rewrite system (*TRS*) over Σ is a finite set of rewrite rules over Σ.

A tree $s \in \mathcal{T}(\Sigma, \mathcal{X})$ rewrites to $t \in \mathcal{T}(\Sigma, \mathcal{X})$ with a TRS $\mathcal{R}$ over Σ, denoted by $s \xrightarrow{\mathcal{R}} t$ ($\mathcal{R}$ may be omitted when clear from context) if there exists a rewrite rule $\ell_1; \ldots; \ell_n \rightarrow r_1; \ldots; r_n \in \mathcal{R}$, a node $\nu \in dom(s)$ and a substitution σ over Σ such that s matches $\ell_1; \ldots; \ell_n$ at ν with σ and $t = s[\sigma(r_1); \ldots; \sigma(r_n)]_\nu$. The reflexive and transitive closure of $\xrightarrow{\mathcal{R}}$ is denoted by $\xrightarrow{*}{\mathcal{R}}$, and the reflexive, symmetric and transitive closure by $\xleftrightarrow{*}{\mathcal{R}}$.

This definition strictly generalizes the standard notion of term rewriting [8], which corresponds to the particular case of rewrite rules with patterns of length one (*i.e.* trees). Our notion of rewriting cousin-patterns can be captured by extensions of rewriting such as the spatial programming language MGS [5].

2   Ranked Tree Representation of Rhythm Notation

We consider a particular signature Σ_{rn} for expressing rhythm notations. It contains the following symbols of arity zero (*constant symbols*): n (representing a note), r (rest), s (slur), d (dot), and o (for composition of durations, as explained below). Moreover, Σ_{rn} contains a subset $\mathbb{P}$ of symbols denoted as prime integers, each $p \in \mathbb{P}$ having arity p, and a copy $\bar{\mathbb{P}} = \{\bar{p} \mid p \in \mathbb{P}\}$, where $\bar{p}$ has also arity p. More precisely, $\mathbb{P}$ is assumed to contain a (small) prime integer $\max(\mathbb{P})$, assumed fixed throughout the paper, and all prime numbers smaller than $\max(\mathbb{P})$, *i.e.* $\mathbb{P} = \{2, 3, 5, \ldots, \max(\mathbb{P})\}$. Typically, $\max(\mathbb{P}) = 11$. The symbols of $\mathbb{P} \cup \bar{\mathbb{P}}$ will be used to build tuplets defined by equal subdivision of a duration.

2.1   Tree Semantics

Intuitively, a tree of $\mathcal{T}(\Sigma_{rn})$ represents a sequence of notes and rests, denoted by symbols n and r in the leaves, their duration being encoded in the structure of the tree. The symbols s and d are used to group the durations of successive leaves. The symbol o is used to group the durations of successive cousins, and possibly further subdivise the summed duration.

Formally, given a tree $t \in \mathcal{T}(\Sigma_{rn}, \mathcal{X})$, we associate recursively a *duration value*, denoted $dur_t(\nu)$, to each node $\nu \in dom(t)$ as follows:

- If $\nu = root(t)$, then $dur_t(\nu)$ is a number of beats $n \geq 1$ associated to t (t can represent e.g. one or several beats or a whole bar).
- Otherwise, let ν_0 be the parent of ν in t and let p be the arity of $t(\nu_0)$, $dur_t(\nu) = \frac{dur_t(\nu_0)}{p} + cdur_t(\nu)$, where $cdur_t(\nu) = dur_t(\nu')$ if ν has a previous cousin ν' such that $t(\nu') = o$, $cdur_t(\nu) = 0$ otherwise.

A tree $t \in \mathcal{T}(\Sigma_{rn}) \setminus \{o\}$ represents a sequence $val(t)$ of durations. Let $k \geq 1$ be the number of leaves of t not labeled by o and let $\nu_1 \ldots, \nu_k$ be the enumeration of these leaves in *dfs*. The *duration sequence* of t is defined as $ds(t) = \langle t(\nu_1), dur_t(\nu_1)\rangle, \ldots, \langle t(\nu_k), dur_t(\nu_k)\rangle$. Let $i_1, \ldots, i_{\ell+1}$ ($\ell \geq 0$) be an increasing sequence of indices defined as follows: $i_1 = 1$, $i_2, \ldots, i_\ell$ is the subsequence of indices of nodes in $\{\nu_2 \ldots, \nu_k\}$ labeled by n or r, $i_{\ell+1} = k + 1$.

The *rhythmic value* $val(t)$ of t is the sequence of pairs $u_1, \ldots, u_\ell$, where for each j, $1 \leq j \leq \ell$,

$$u_j = \langle t(\nu_{i_j}), \sum_{i=i_j}^{i=i_{j+1}-1} dur_t(\nu_i)\rangle$$

According to this definition, the first component of each pair u_j is either n or r, and the second component is the sum of the durations of next leaves labeled by s or d. For convenience, we shall omit the first components of pairs, and denote $val(t)$ as a sequence of durations, where the duration of rests r are written in brackets to distinguish them from durations of notes n.

Example 2. Fig. 1 displays some examples of trees with the corresponding notation and rhythm value.[3]   ◇

Example 3. The trees in Fig. 2 contain the symbol o for the addition of durations as defined in the above semantics. In the tree of Fig. 2(a), the duration values of the second leaf (labeled by o) and third leaf (labeled by n) are summed to express that the second note has a duration of $\frac{1}{2}$ beat. Note that these two leaves are cousin nodes. The idea is the same in Fig. 2(b) (here the second note has duration $\frac{1}{3}$). In Fig. 2(c), two duration values of $\frac{1}{4}$ are also summed, like in Fig. 2(a), but here, the obtained duration value of $\frac{1}{2}$ is further divided by 3. This is expressed by the 3 in the notation, which actually stands for 3 : 2 (3 *in the time of* 2). Similarly, in the bar Fig. 2(d), we have 5 quavers in the time of 3.   ◇

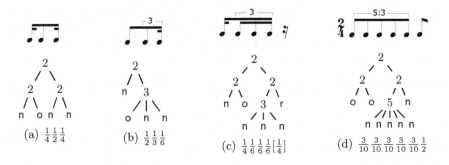

Fig. 2. Example of trees of $\mathcal{T}(\Sigma_{\mathrm{rn}})$: summation with symbol o.

Example 4. Examples of the interpretation of s and d in terms of notation are given in Fig. 3. In the tree Fig. 3(a), a note of duration $\frac{1}{2}$ is dotted, extending its duration to $\frac{3}{4}$. The rhythm value of Fig. 3(b) and (c) is the same as for Fig. 3(a), but the notation Fig. 3(a) is more recommended (see [10]).   ◇

We define as equivalent the trees representing the same actual rhythm.

Definition 1. *Two trees* $t_1, t_2 \in \mathcal{T}(\Sigma_{\mathrm{rn}})$ *are equivalent iff* $val(t_1) = val(t_2)$.

The tree equivalence relation is denoted $t_1 \equiv t_2$. This notion of equivalence makes it possible to characterize different notations of the same rhythm, like *e.g.* the trees (a), (b) and (c) in Fig. 3.

---

[3] In the examples and figures of this paper, when not specified otherwise with a time signature, we will consider that the duration associated to each of the trees is 1 beat (this duration can also be found by summing the indicated fractional durations).

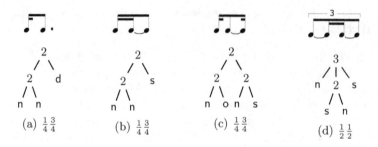

Fig. 3. Trees of $\mathcal{T}(\Sigma_{\mathsf{rn}})$ with slurs and dots.

2.2   Interpretation of Trees into Common Western Notation

The previous examples showed how the interpretation of trees of $\mathcal{T}(\Sigma_{\mathsf{rn}})$ as Common Western Notation is generally straightforward. We describe here some aspects that need particular treatments. We have already said a few words about the case of dots (symbol d), see Example 4. Let us present below another example about tuplets beaming using the symbols of $\bar{\mathbb{P}}$.

Example 5. Figure 4 presents different ways of beaming a sextuplet (see [10]). The three trees are *equivalent*. In Fig. 4(b) and (c) the division is respectively bipartite and tripartite. In Fig. 4(a), the division is unclear. The symbols $\bar{2}$ and $\bar{3} \in \bar{\mathbb{P}}$ at the top of the trees Fig. 4(b) and (c) indicate that there must be only one beam between subtrees (the value one corresponds to the depth of $\bar{2}$ and $\bar{3}$). The default rendering, in Fig. 4(a), with label $3 \in \mathbb{P}$, is that the number of beams between subtrees is the same as the number of beams in subtrees. In Fig. 5, similar variants are presented at the scale of a whole bar.   ◊

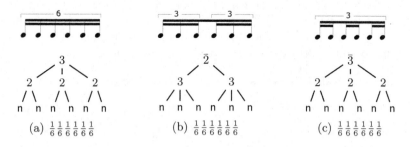

Fig. 4. Tuplet beaming with symbols of $\bar{\mathbb{P}}$ (one beat).

At this point, let us make a few remarks about the above tree semantics.

(1) In the tree of Fig. 3(a), the dot symbol d labels a leaf node of duration $\frac{1}{2}$, which comes after another leaf node labelled by n and of duration $\frac{1}{4}$. This

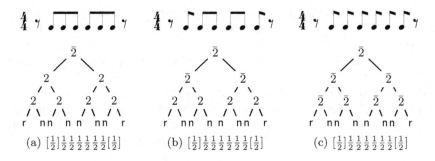

Fig. 5. Tuplet beaming with symbols of $\bar{\mathbb{P}}$ (one bar).

may seem counter-intuitive with respect to standard rhythm notation: we could expect the node of duration $\frac{1}{2}$ to be labeled by n and the node of duration $\frac{1}{4}$ to be labeled by d. However in our tree semantics, we have chosen to always represent rooted or tied notes by a n followed by some s or d, to avoid ambiguities.

(2) Labels d must be used with care for ensuring a correct interpretation into notation. Section 2.3 will discuss some constraints to be satisfied for this sake.

(3) The interpretation of the slur symbol s and the dot symbol d is the same regarding durations. This is also the case of $p \in \mathbb{P}$ and $\bar{p} \in \bar{\mathbb{P}}$. The symbols d and $\bar{p} \in \bar{P}$ have been introduced only to give notation-related information, and the choice of one symbol over the other equivalent will be only dictated by notation preferences (see also Sect. 2.3).

(4) The symbols of $\mathbb{P}$ and $\bar{\mathbb{P}}$ are somehow redundant with the tree structure, since every inner node is labeled with its degree. This technical facility however allowed us to define our trees in the algebra $\mathcal{T}(\Sigma_{rn})$ over a ranked signature Σ_{rn}.

2.3   Syntactical Restrictions and Tree Automata

In general, several notations can be associated to the same rhythmic value, and the preferences regarding the notation details may depend on varied factors like the metre, usage, or personal preferences of the author. For instance, we have seen in Example 4 (Fig. 3) that the dot symbol d produces the same rhythm value as the slur symbol s, but different notations. Also, separating innermost beams by using $\bar{p} \in \bar{\mathbb{P}}$ instead of $p \in \mathbb{P}$, like in Figs. 4(b) and (c) and Fig. 5, can be useful to reflect the correct subdivisions of a tuplet (following the metre) or indicate accentuations, see [10].

Some constraints regarding notation details can be expressed using tree automata (TAs) over Σ_{rn}. TAs in this case correspond to "style files" for rhythm notation.

The first and most important TA that we need in this context is the one characterizing trees with correct interpretation as notation. For instance, following the above remark (2), one can build a TA checking that the d symbols are well placed. This TA recognizes the set of trees of $T(\Sigma_{rn})$ where a symbol d can only occur in a pattern of the form $2(2(x, n), d)$ or $2(n, 2(d, x))$ is a regular language. We can also extend to double dots by also allowing patterns of the form $2(2(2(x, n), d), d)$ or $2(n, 2(d, 2(d, x)))$.

Moreover, the compositional properties of TAs make it possible to combine (by union, intersection, complementation) arbitrarily different notation constraints expressed as TAs. For instance, in a binary metre, there exists a TA that allows beaming as in Fig. 4(b) and forbids beaming as in Fig. 4(c).

3   Rewrite Rules

We define a set of rewrite rules on trees, which do not change the rhythmic values (*i.e.* the actual rhythm), but may change its notation. These rules can therefore be used to produce equivalent notations of a same rhythm. The set of rewrite rules over Σ_{rn} defined in this section will be called $\mathcal{R}_{rn}$.

3.1   Normalization Rules

The following rules reflect the semantical equivalence between dots and slurs (Example 4),

$$d \to s \tag{1}$$

and between symbols of $\bar{\mathbb{P}}$ and their counterpart of $\mathbb{P}$ (Example 5).

$$\bar{p}(x_1, \ldots, x_p) \to p(x_1, \ldots, x_p) \quad p \in \mathbb{P} \tag{2}$$

Addition of rests. Unlike notes, successive rests are always summed up implicitly. Following this principle, we can decide to merge subdivisions of rests, with standard rewrite rules of the form $2(r, r) \to r$, $3(r, r, r) \to r$, *etc.*, which are generalized into

$$p(\underbrace{r, \ldots, r}_{p}) \to r \quad p \in \mathbb{P} \tag{3}$$

The use of slurs is useless with rests, hence we have also this rule with cousin patterns of length 2

$$r; s \to r; r \tag{4}$$

Similarly, the following rule complies with the semantics of o,

$$o; r \to r; r \tag{5}$$

Normalization of s. According to the semantics of symbols of $\mathbb{P}$, we can simplify fully tied tuplets with standard rewrite rules: $2(s, s) \to s$, $3(s, s, s) \to s$, *etc.*, generalized into

$$p(s, \ldots, s) \to s \quad p \in \mathbb{P} \tag{6}$$

We have also $2(n, s) \to n$, $3(n, s, s) \to n$, *etc.*, generalized into

$$p(n, s, \ldots, s) \to n \quad p \in \mathbb{P} \tag{7}$$

Normalization of o. The following rule replaces o by s when possible.

$$o; s \rightarrow s; s \tag{8}$$

The semantics presented in Sect. 2.1 make it possible to sum up the durations corresponding to cousin nodes labeled by o, and then subdivide the duration obtained by this sum. Following this principle, we can sometimes simplify a pattern beginning with a sequence of o's, according to the value of the sum and the number of subdivision. The base case is the subdivision by 1, corresponding to the atomic note n

$$o; n \rightarrow n; s \tag{9}$$

For a subdivision by 2, we have the following rewrite rules with variables: $o; 2(x_1, x_2) \rightarrow x_1; x_2$, $o; o; o; 2(x_1, x_2) \rightarrow o; x_1; o; x_2$, and so on for each multiple of 2. For a subdivision by 3, we have $o; o; 3(x_1, x_2, x_3) \rightarrow x_1; x_2; x_3$ *etc.* The general form of the expected transformations is

$$\underbrace{o; \dots; o}_{kp-1}; p(x_1, \dots, x_p) \rightarrow \underbrace{o; \dots; o}_{k-1}; x_1; \underbrace{o; \dots; o}_{k-1}; x_2; \dots; \underbrace{o; \dots; o}_{k-1}; x_p \tag{10}$$

Equation (10) represents a non-bounded number of rules (one for each value of k). It can be simulated in a finite number of rewrite steps, using auxiliary symbols (which cannot be presented here due to space restrictions).

Example 6. Figure 6 presents a rewrite sequence from the tree in Fig. 3(c) into Fig. 3(b), and from Fig. 3(a) also into Fig. 3(b). The nodes of application of rewrite rules are marked by circles, and rewrite rules are indicated. This shows that using the above rewrite rules, we can explore the equivalent trees of Fig. 3. ◇

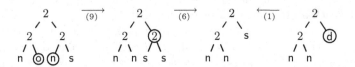

Fig. 6. Rewrite sequences starting from the trees of Figs. 3(c) and (a). The applied rewriting rule is between parenthesis.

3.2 Subdivision Equivalence

Finally, we propose standard rewrite rules for redefining subdivisions, such as

$$2(x_1, x_2) \rightarrow 3(2(o, o), 2(x_1, o), 2(o, x_2))$$
$$2(x_1, x_2) \rightarrow 5(2(o, o), 2(o, o), 2(x_1, o), 2(o, o), 2(o, x_2))$$
$$3(x_1, x_2, x_3) \rightarrow 2(3(o, x_1, o), 3(x_2, o, x_3)) \dots$$

The general form of these rules is

$$p(x_1, \ldots, x_p) \rightarrow p'\big(p(u_{1,1}, \ldots, u_{1,p}), \ldots, p(u_{p',1}, \ldots, u_{p',p})\big) \qquad (11)$$

where $p, p' \in \mathbb{P}$, $p \neq p'$, for all $1 \leq i \leq p'$, $1 \leq j \leq p$, $u_{i,j} \in \{o, x_1, \ldots, x_p\}$ and the sequence $u_{1,1}, \ldots, u_{1,p}, \ldots, u_{p',1}, \ldots, u_{p',p}$ has the form $\underbrace{o, \ldots, o}_{p'}, x_1, \ldots, \underbrace{o, \ldots, o}_{p'}, x_p$.

Example 7. Applying the above rules to the tree of Fig. 3(d), we obtain the rewrite sequence depicted in Fig. 7. The result is a simpler tree representing the same durations. ◊

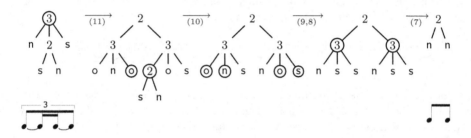

Fig. 7. Rewrite sequence starting from the tree in Fig. 3(d).

4   Properties

We show that the rewrite rules of $\mathcal{R}_{rn}$ are correct, in the sense that they preserve rhythmic values of trees, and complete, in the sense that given a tree t, it is possible to reach all trees equivalent to t using the rewriting rules of $\mathcal{R}_{rn}$.

For these properties to hold, we need to consider the following restrictions. A node ν in a tree $t \in \mathcal{T}(\Sigma_{rn})$ is called *dandling* if it is labeled by o and it is not the previous cousin of a node in $dom(t)$. A tree $t \in \mathcal{T}(\Sigma_{rn} \cup \Theta_{rn})$ is called o-*balanced* if for all successive cousins $\nu, \nu' \in dom(t)$ the multisets of labels on the two paths from ν and ν' up to the root of t are equal. Intuitively, it means that successive cousin nodes labeled by o represent the same duration. A tree $t \in \mathcal{T}(\Sigma_{rn})$ is *well-formed* iff it is o-*balanced* and without dandling nodes.

Example 8. All the trees of Fig. 3 are well-formed. The tree of Fig. 3(c) can be rewritten into the tree $2(2(n, o), n)$ by rule (7). The latter tree is however not well-formed because of the dandling o-node. ◊

Proposition 1. *For all well-formed trees* $t_1, t_2 \in \mathcal{T}(\Sigma_{rn})$, $t_1 \equiv t_2$ *iff* $t_1 \xleftrightarrow{*}_{\mathcal{R}_{rn}} t_2$.

Let us sketch the proof of Proposition 1. We show by a case analysis that for each rewrite rule $\ell_1; \ldots; \ell_n \to r_1; \ldots; r_n \in \mathcal{R}_{rn}$, and for each substitution σ grounding for the rule (*i.e.* such that $\sigma(\ell_1), \ldots, \sigma(\ell_n), \sigma(r_1), \ldots, \sigma(r_n)$ do not contain variables), $val\big(\sigma(\ell_1); \ldots; \sigma(\ell_n)\big) = val\big(\sigma(r_1); \ldots; \sigma(r_n)\big)$. The *if* direction of Proposition 1 then follows from a lifting of this result to the application of contexts, in order to consider rewriting at inner nodes (not only the root node). The proof of the *only if* direction works by structural induction on t_1 and t_2.

We can use the result of Proposition 1 in order to explore rhythm notations equivalent to a given tree, for instance for simplification like in Figs. 6 and 7. In our context, it is reasonable to assume a bounded depth for trees. By Kruskal lemma, it follows that the number of trees to consider is finite.

5   Conclusion

The choice of a representation determines the range of possible operations on a given musical structure, and thereby has a significant influence on compositional and analytical processes (see [11, 14] for examples in the domain of rhythm structures). In this paper we proposed a formal tree-structured representation for rhythm inspired by previous theoretical models for term rewriting. Based on this representation, tree rewriting can be seen as a means for transforming rhythms in composition or analysis processes. In a context of computer-aided composition for instance, this approach can make it possible to suggest to a user various notations of the same rhythmic value, with different complexities. Similarly, the rewrite sequence of Fig. 7 can be seen as a notation *simplification* for a given rhythm. An important problem in the *confluence* of the defined rewrite relation, *i.e.* whether different rewriting from a single tree will eventually converge to a unique canonical form. For a quantitative approach, it is possible to use standard complexity measures for trees (involving depth, number of symbols *etc.*). We can therefore imagine that this framework being used as a support for rhythm quantification processes [1] in computer-aided composition environments like OpenMusic.

The tree format that we are proposing has similarities with the Patchwork/OpenMusic *Rhythmic Tree* (RT) formalism [2].[4] Still, these two formats present a number of important differences. While RTs represent durations with integers labeling nodes (the subdivision ratios), our representation only uses the tree structure (*i.e.*, the labels in $\mathbb{P}$ are not formally needed) and labels in a finite (and small) set for leaves. This specificity makes the representation more amenable to purely syntactical processing, when RT processing needs arithmetic. Trees of $\mathcal{T}(\Sigma_{rn})$ and OM RTs are meant to be complementary, and some functions for converting trees back and forth between these two formats have been implemented. The definition domains of these functions are characterized by TA.

---

[4] A *rhythm tree* RT is defined as a pair $\langle d, S \rangle$ where $d \in \mathbb{N}$ is a duration and $S = s_0, \ldots, s_n$ is a sequence of subdivisions where for all $1 \le i \le n$, s_i is either a tree or a ratio. Formally, $s_i \in \mathbb{N}$ or s_i is a RT.

As mentioned in Sect. 2.3, it is also possible to use a TA to complement the rewriting rules and control the rhythm simplification processes by filtering out rewritten trees that do not correspond to actual notations (*e.g.* because of misplaced d), or/and restricting the search space to trees corresponding to acceptable or preferred notations. This approach is comparable to the use of *schemas* for XML data processing.

Note that the symbol n could be replaced by several symbols encoding pitches in order to represent actual melodies. Similar tree-based encodings have been used in [4] for the search of melodic similarities. Finally, other rewrite rules can be considered, including ones that do not preserve durations or rhythmic values. In this case, tree rewriting could constitute a novel creative approach to rhythm transformation in compositional applications. Another application could be the formalization of summarization of music by pruning trees like in [15].

References

1. Agon, C., Assayag, G., Fineberg, J., Rueda, C.: Kant: a critique of pure quantification. In: International Computer Music Conference Proceedings (ICMC), pp. 52–59 (1994)
2. Agon, C., Haddad, K., Assayag, G.: Representation and rendering of rhythm structures. In: Proceedings of 2nd International Conference on WEB Delivering of Music (CW 2002), pp. 109–113. IEEE Computer Society (2002)
3. Assayag, G., Rueda, C., Laurson, M., Agon, C., Delerue, O.: Computer assisted composition at IRCAM: PatchWork and OpenMusic. Comp. Music J. **23**(3), 59–72 (1999)
4. Bernabeu, J.F., Calera-Rubio, J., Iñesta, J.M., Rizo, D.: Melodic identification using probabilistic tree automata. J. New Music Res. **40**(2), 93–103 (2011)
5. Bigo, L., Spicher, A.: Self-assembly of musical representations in MGS. Int. J. Unconv. Comput. **10**(3), 219 (2014)
6. Bresson, J., Agon, C., Assayag, G.: OpenMusic: visual programming environment for music composition, analysis and research. In: Proceedings of the 19th ACM International Conference on MultiMedia, pp. 743–746. ACM (2011)
7. Comon, H., Dauchet, M., Gilleron, R., Jacquemard, F., Löding, C., Lugiez, D., Tison, S., Tommasi, M.: Tree automata techniques and applications (2007). http://tata.gforge.inria.fr
8. Dershowitz, N., Jouannaud, J.-P.: Rewrite systems. In: van Leeuwen, J. (ed.) Handbook of Theoretical Computer Science, vol. B, pp. 243–320. North-Holland, Amsterdam (1990)
9. Good, M.: Lessons from the adoption of MusicXML as an interchange standard. In: XML Conference (2006)
10. Gould, E.: Behind Bars: The Definitive Guide to Music Notation. Faber Music, London (2011)
11. Haddad, K.: Livre premier de motets: the concept of TimeBlocks in OpenMusic. In: Agon, C., Assayag, G., Bresson, J. (eds.) The OM Composer's Book 2. Delatour - Ircam, France (2008)
12. Hoos, H.H., Hamel, K.A., Renz, K., Jürgen, K.: Representing score-level music using the GUIDO music-notation format. Comput. Musicol. **12** (2001)

13. Lerdahl, F., Jackendoff, R.: A Generative Theory of Tonal Music. MIT Press, Cambridge (1983)
14. Malt, M.: Some considerations on Brian Ferneyhough's musical language through his use of CAC. In: Agon, C., Assayag, G., Bresson, J. (eds.) The OM Composer's Book 2. Delatour - Ircam, France (2008)
15. Rizo, D.: Symbolic music comparison with tree data structures. Ph.D. thesis, Universidad de Alicante, November 2010

Renotation from Optical Music Recognition

Liang Chen, Rong Jin, and Christopher Raphael[✉]

School of Informatics and Computing, Indiana University,
Bloomington 47408, USA
craphael@indiana.edu

Abstract. We describe the music renotation problem, in which one transforms a collection of recognized music notation primitives (e.g. note heads, stems, beams, flags, clefs, accidentals, etc.) into a different notation format, such as transposing the notation or displaying it in a rectangle or arbitrary size. We represent a limited degree of image understanding through a graph that connects pairs of symbols sharing layout constraints that must be respected during renotation. The layout problem is then formulated as the optimization of a convex objective function expressed as a sum of penalty terms, one for each edge in the graph. We demonstrate results by generating transposed parts from a recognized full score.

Keywords: Music renotation · Optical music recognition · Music notation layout

1 Introduction

Optical Music Recognition holds great promise for producing the symbolic music libraries that will usher music into the 21st century, allowing flexible display, automated transformation, search, alignment with audio, and many other exciting possibilities. Work in this area dates back to the 1960s [1–11], with a nice overview of important activity given by [12]. However, collective efforts have not yet produced systems ready for the grand challenge of creating large-scale definitive music libraries [13,15]. Viro's recent work on the IMSLP constitutes a promising current approach to large-scale OMR [14]. Simply put, the problem is very hard, lacking obvious recognition paradigms and performance metrics [13,27], while containing a thicket of special notational cases that make OMR difficult to formulate in a general and useful manner.

One of the many challenges, and one central to the effort discussed here, concerns the *representation* of OMR output. There are a number of fairly general music representations [16], such as MEI [17] and MusicXML [18], that provide sufficiently expressive formats for music encoding. As these representations are extensible, they can be modified to include additional information relevant for a particular perspective. However, there remains a significant gap between the natural results of OMR, which tend toward loosely structured collections of symbols, and the necessary symbol interpretations for encoding music data in these formats. Rhythm and voicing are among the issues that pose the greatest

© Springer International Publishing Switzerland 2015
T. Collins et al. (Eds.): MCM 2015, LNAI 9110, pp. 16–26, 2015.
DOI: 10.1007/978-3-319-20603-5_2

difficulty. Lurking in the background is the fact that OMR must try to capture the literal contents of the printed page, while music representations tend to take a more abstract view of music content. For instance, symbols or text that span several staves must be recognized as such in OMR, while this layout problem is not relevant for a symbolic encoding.

Rather than solving the OMR-to-encoding problem, here we explore the possibility of making OMR results useful with only a minimal understanding of the recognized symbols' meaning. In particular we address the *renotation* problem — perhaps the most important application of OMR. Renotation refers to a collection of problems that transform recognized notation into related formats, such as transposition, score-to-parts, and reformatting for arbitrary display sizes. In all cases, the output is music notation in an image format. In our approach, we seek only the level of notational understanding necessary to accomplish this task, which is considerably less than what is required by familiar symbolic representations. For instance, we do not need to understand the rhythmic meaning of the symbols. We bind our process to the original image, rather than a more abstract symbolic representation, in an attempt to leverage the many intelligent choices that were made during is construction. For instance we beam notes, choose stem and slur directions, etc. exactly as done in the original document, while we use the original symbol spacing as a guide to resolving symbol conflicts in our renotated document. However, we cannot renotate simply by cutting and pasting measures as they appear in the original source. For instance, the spacing considerations of parts are quite different than those for a score, where alignment of coincident events is crucial.

Our essential approach is to cast the renotation problem as an optimization expressed in terms of a graph that connects interdependent symbols. This view is much like the spring embedding approach to graph layout, so popular in recent years [19–22]. Our problem differs from generic graph layout in that many aspects of our desired layout are constrained, such as the vertical position of note heads, clefs and accidentals. This leads to an easier optimization task, where one can model with a convex objective functions whose global optimum is easy to identify and satisfies our layout objectives. A significant difference between our problem and graph layout is the origin of our graph edges, which come from known semantic relations between symbols. For instance we know that an accidental belongs to a note head, and thus introduce an edge that constrains their relative positions. Thus the edges of our graph come mostly from the *meaning* of the symbol, rather than their spatial proximity. This work bears some resemblance to the work of Renz [28], who takes a spring embedding view of one-dimensional music layout.

We will describe our basic approach and present a score-to-parts with example with transposition, using the *Nottorno* from Borodin's Second String Quartet.

1.1   Background: The Ceres System

Our *Ceres* OMR system [23–25] is named after the Roman goddess of the harvest, as we seek to harvest symbolic music libraries from the tens of thousands of images on the International Music Score Library Project (IMSLP) [26]. At present, *Ceres* is composed of two phases: recognition and correction. The recognition phase begins by identifying staff lines, then simultaneously grouping them into systems and identifying system measures. In the heart of the recognition process we identify the contents of each measure. Here we recognize both *composite* and *isolated* symbols. By composite symbols, we mean beamed groups and chords (including isolated notes), which are composed by grammatically constrained configurations of note heads, stems, beams, flags, and ledger lines, as well as the decorations that belong note heads and stems (accidentals, articulations, augmentation dots, etc.) The isolated symbols we currently seek include rests, clefs, slurs, "hairpin" crescendos, text dynamics, and various other symbols. We refer to both the isolated symbols and the building blocks of the composite symbols as *primitives.*

While OMR system comparisons are suspect due to the lack of accepted metrics and ground-truthed test data [13,27], *Ceres'* performance was competitive with what many regard as the currently-best system, *SharpEye* [1], a commercial system, in a recent limited test [23]. That said, all OMR researchers we know ackowledge that system performance varies greatly between music documents, while our evaluation was narrow relying on hand-marked ground truth. While we expect that considerable progress is still possible with the core recognition engine, it seems futile to pose the problem entirely in terms of recognition. Music notation contains a long and heavy tail of special cases and exceptions to general rules that must be handled somehow. The OMR researcher is continually faced with modeling scenarios where accounting for greater notational generality may lead to worse overall recognition performance. As an example, beamed groups can span several staves, several measures, and can have stems that go in both directions, however, allowing for these unusual cases inevitably results in false positive detections. The only reasonable modeling strategy chooses tradeoffs on the basis of overall recognition performance. Consequently, we should expect that our recognition results will contain errors, perhaps many.

Ceres complements its recognition engine with a user interface allowing a person to correct mistakes and address special cases not handled through recognition. The *Ceres* user interface, depicted in Fig. 1, facilitates a drag-and-drop process we call "tagging," in which the user views the original image with the recognized results superimposed, while editing the recognized notation at the primitive level. In choosing to operate on primitives, we disregard the grammatical structure of composite symbols recovered in the recognition phase, instead treating the results as a "bag of primitives." While we lose useful information in the process, there are significant advantages to this approach. First of all, we can present the user with a well-defined task requiring no knowledge of the inner-workings of our recognition engine: she must simply cover the image "ink" with appropriately chosen primitives. In addition, operating at the primitive level allows one to recover from partially correct recognition results that are awkward to handle at the composite symbol level. For instance, consider the situation of the last measure in Fig. 2, where the beamed groups have each been recognized as two separate beamed groups with opposite stem directions. Converting this

Fig. 1. *Ceres'* drag-and-drop user interface for correction primitives. Color figure with color coding of various symbol types can be seen at www.music.informatics.indiana. edu/papers/mcm15

Fig. 2. Working at the primitive level allows the user to easily recover from partially correct results, as depicted in the rightmost measure. Color figure showing recognized symbols can be seen at www.music.informatics.indiana.edu/papers/mcm15

result at the beamed group level requires that all of the hierarchical relations are correctly expressed through the editor; in contrast, editing at the primitive level only requires that one delete the two recognized beams while substituting a longer beam in their place.

2 Music Renotation

Music *renotation* seeks to transform recognized music notation into related formats, such as converting a score into parts, transposition, or reformatting notation to fill a window of arbitrary size. Each of these tasks requires *some* degree of understanding of the notated symbols' meaning. For instance, we must know which configurations of primitives constitute beamed groups and chords, since these must be rendered subject to collective constraints (stems must terminate at beams, note heads and flags must touch stems, multiple beams must be parallel, etc.). We also need to know which symbols *belong* to which measures, since they must "travel" with the measure as the notation is reformatted. We need to understand other ownership notions such as the way accidentals, augmentation dots, and articulations belong to note heads, as spacing and alignment constraints must implicitly represent this ownership. We need to understand time coincidence between notes, rests, and other symbols, as such relations are expressed, notationally, as vertical alignment. Finally, for polyphonic music it is helpful to understand some degree of voicing as the events within a voice require spacing that makes the results readable. It is worth noting that there is much that we do *not* need to understand, including details of rhythm, meaning of most text, dynamics, and other "mark up" symbols. In fact, when transposition is not involved we don't even need to understand the pitches of the notes. As the interpretation of music notation is a challenging problem, a basic tenet of our approach here is to accomplish renotation with the minimal degree of understanding possible.

2.1   The Symbol Graph

A simple way to represent these relations is by introducing a graph structure with the *connected* symbols as vertices, meaning those collections of primitives, including singletons, that form connected regions of image ink (e.g. beamed groups or chords without their non-touching *satellites,* or any other isolated symbol). We then connect these vertices with labeled edges representing the desired spatial relationships between the symbols they connect. Thus, the graph structure describes the pairwise interrelations between symbols needed to constrain the renotation process' modification of symbol locations and parameters. We briefly describe the construction of this graph, accomplished bottom up using the known locations and labels of the symbols through a process called *assembly*. In this process we first construct the subgraphs for each composite symbol, treating the connected portion as a single symbol, and connecting the *satellites* (accidentals, augmentation dots, articulations, etc.) to the body with edges. We then create the symbol graph on the entire collection of symbols by introducing additional edges. The edges we introduce are labeled as either "horizontal" or "vertical" meaning that they constraing the horizontal or verticial distance between symbols while being indifferent to the other dimension.

We assemble the primitives into beamed groups and isolated chords in a rule-based greedy manner. In essence, we regard the primitives as being either "plugs" or "sockets," and seek to hook up these connectors, choosing an order of decisions that avoids ambiguity. For instance, every non-whole note head must connect to a stem; every beam must connect to two stems; every flag must connect to a stem, etc. For brevity's sake we will omit the individual assembly steps, as the details are many. While such approaches tend to be vulnerable to early incorrect decisions, we have not encountered this difficulty in practice. It is worth noting that the recognition and tagging processes distinguish between various "look-alike" symbols, such as *staccato* marks and augmentation dots, or key signature accidentals and note modifier accidentals, thus simplifying the process.

Occasionally during this process we encounter "plugs" with no available "socket." Such cases are nearly always due to either mislabeled or unlabeled primitives. When this occurs, our interface displays the offending composite symbol in a special color to alert the user to the inconsistency. Thus, during the tagging process, the user is directed toward uninterpretable situations requiring further attention.

Fig. 3. Example of a symbol graph resulting from our assembly process.

Having constructed the composite symbols and connected their various satellites with edges, we then order the notes, rests, and clefs, within each staff, establishing horizontal connections between neighboring symbols, while adding vertical connections between apparently time-coincident members of this set. Clearly this process could benefit from a rhythmic or voice analysis, and we expect this will be necessary as we examine greater degrees of polyphony than exhibited in our current experiments. However, at present, we detect coincidence simply by thresholding differences in horizontal position. Finally vertical connections are established between the slurs, hairpins, and remaining symbols representing the appropriate coincidence relations.

We emerge from this assembly/tagging process with a graph structure that expresses the various ownership and hierarchical relations necessary to correctly interpret and use the symbols for renotation. Figure 3 gives an example of a symbol graph.

2.2   Music Renotation as Optimization

We formulate renotation as an optimization problem using the graph structure of the previous section. Here we denote the vertices of our graph by S (the connected symbols), while the drawing of each $s \in S$ is governed by a parameter vector $\theta(s)$. For instance, if s is a beamed group containing n notes, then $\theta(s)$ would represent the $(n + 2)$-tuple needed to fully specify its rendering: the horizontal positions of the note heads and the two "corners" of the primary beam. (The vertical positions of the note are fixed at the appropriate staff position). For any non-composite $s \in S$, $\theta(s)$ simply gives the location of the symbol. In many cases there is a sharing or "tying" of parameters, representing hard layout constraints. For instance, the vertical position of an accidental, s, should be the same as the note head it modifies, thus constraining its vertical coordinate. Thus the edge between a note head and its accidental refers to the flexible horizontal distance. Similarly, a *staccato* dot should be centered above the note head it belongs to, thereby constraining its horizontal coordinate. In such cases $\theta(s)$ would have only a single component representing the "free" parameters that are not determined by such constraints.

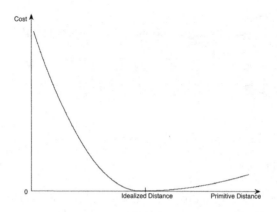

Fig. 4. The quadratic spline we use to represent the asymmetric penalty associated with an edge.

We denote by E the collection of edges in our symbol graph. Each edge, $e = (s_1, s_2) \in E$, $s_1, s_2 \in S$, has an associated affine function that reduces the edge parameters to a single quantity: $\lambda(e) = l_1^t \theta(s_1) + l_2^t \theta(s_2) + c$. In most cases the l_1^t, l_2^t vectors simply "choose" a single parameter of $\theta(s_1)$ or $\theta(s_2)$ though we occasionlly need the greater degree of generality the linear combination offers. We then write our objective function, H, as

$$H = \sum_{e \in E} Q_e(\lambda(e)) \tag{1}$$

where Q_e is a quadratic spline function as that depicted in Fig. 4.

The idea here is best illustrated in terms of an example. Consider the case of an accidental that modifies a note head. There is a desired horizontal distance between the two: other considerations aside we would prefer to have a fixed and known distance separating the two symbols. However we feel differently about moving the two symbols closer together and further apart. As the symbols become too close they crowd one another, and eventually touch, which is highly undesirable in music layout. However, the presence of other accidentals may require that we separate the symbols further than the ideal. As this is common with the notation of accidentals surrounding a chord, this choice should not come at a great cost. Thus when $\lambda(e) < 0$, $|\lambda(e)|$ measures the degree to which we are less than the ideal, while when $\lambda(e) > 0$, $\lambda(e)$ measures the amount we are greater than the ideal. Q_e captures our asymmetric penalty for these two situations. The situation described above applies more generally to the layout problem, thus all aspects of notation we wish to control are expressed as edges in the the graph, with corresponding penalty terms.

As a sum of nearly quadratic terms, the objective function, H, is easy to optimize and converges in just a few iterations of Newton's method. As H is strictly convex, this point of convergence is the global optimum. However, since the objective function modifies the parametrization of the notation, pairs of symbols can come into contact that have no edges penalizing their spacial relations — we don't know what symbols are in potential conflict when we construct the original graph. Thus, when such conflicts arise, we augment our graph with additional edges for each conflicting pair. The penalty term for each such edge causes the symbols to "repel" thus resolving the conflict. We iterate between optimizing our objective function and modifying the function to include new terms arising from newly-detected conflicts.

3   Results and Discussion

We applied our approach in a score-to-parts application on the 3rd movement, Notturno, of Borodin's Second String Quartet, with the Ernst Eulenburg edition from around 1920. In addition, the parts were trasposed from A major to B♭ major. The complete score can be seen at http://imslp.org/wiki/String_Quartet_No.2_(Borodin,_Aleksandr). In rendering our results we used the Bravura music font by Daniel Spreadbury available under the SIL open font license [30]. Perhaps the most challenging aspect of the score-to-parts problem is that scores require the alignment of time-coincident notes and rests between parts, thus creating spacing that would be completely unnatural for a single part. Thus the spacing needs to be modified considerably to create something that looks acceptable — this is essentially the main task of our algorithm.

There are some aspects of the score-to-parts problem that don't occur in our particular example, thus not addressed here. For instance, some scores, such as those for large ensemble, leave out staves of instruments that don't play on any given page. Thus the system must determine which instruments play which staves before staves can be assembled into complete parts. Also, sometimes scores will notate two or more instruments on a single staff, for instance, the bassoons and contrabassoon in orchestral score. Producing single-instrument parts in this case requires that the voices must be identified and separated. Both of these problems are examples of interest to us, though we do not implement solutions here.

Here we present results in "page" mode, in which we show the part as a single (tall) page, using line breaks and laying out the symbols so that the right bar lines of each staff align. We don't add page breaks, though this could be handled

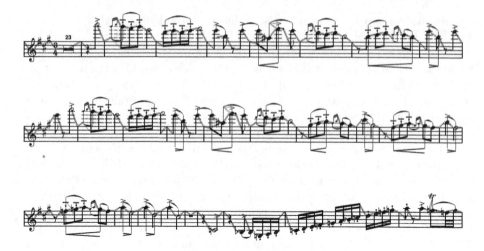

Fig. 5. A portion of the symbol graph generated for the 1st violin part of the Borodin 2nd String Quartet, 3rd movement. The red and green edges correspond to horizontal and vertical soft constraints (terms in Eq. 1). The blue edges are terms that appeared during the optimization due to unanticipated conflicts.

analogously to line breaks. We accomplish this by first partitioning the measures into lines using the standard Knuth dynamic programming approach [29]. As bar lines are treated like other stand-alone symbols, this simply amounts to fixing the horizontal location of each right bar line while optimizing over the remaining parameters. Thus each line of the page is regarded as a separate optimization problem.

A portion of the resulting symbol graph for the first violin part is shown in Fig. 5, while the complete set of resulting parts and graphs are available at www.music.informatics.indiana.edu/papers/mcm15. One can observe from the discussion surrounding Eq. 1 that our objective function, H, is a sum of one-dimensional penalty terms; these penalties consider either horizontal or vertical distances between symbols, though it is possible that both arise in some cases. In Fig. 5 these are shown as red (horizontal), green (vertical), and blue (conflict) edges drawn between symbols. One will also see that there are a number of edges that connect adjacent notes in a beamed group. While these are technically edges that go from a vertex to itself, this requires no change in our formulation of the problem, reflecting a desired relation, e.g. note spacing, between the parameters of a beamed group.

We perform transposition simply by moving the staff position of each note by a fixed number of steps, changing the key signature, and respelling all accidentals, viewing them as either +1,0, or −1 modifiers. For instance, an E♯ in the key of D major moves the E up by +1, and would thus appear as F double sharp in the key of E major or B♮ in the of A♭. For the Borodin we find this creates some rather unusual spellings, such as the section rendered in B double flat major

(9 flats!) at the start of the fifth line of the violin part. Of course, this would normally be spelled as A major, though this would require harmonic analysis to detect, and certainly constitutes a rare case.

A notation maven may find a fair bit to criticize about the resulting layout. We view our effort as more of a proof of concept, rather than recipe for ideal notation. We have not considered many aspects of layout usually included in serious notation systems, such as spacing that reflects note length. Rather, it has been our intent to show that OMR results can be utilized effectively by addressing only a minimal portion of the symbol interpretation problem. We continue to explore the symbol graph as a possible alternative to more expressive music representations such as MEI and MusicXML — capturing less about the notational structure and its meaning, but easier to derive automatically from OMR. Analogous approaches are promising for other OMR applications requiring symbol interpretation, such as music playback.

References

1. Jones, G., Ong, B., Bruno, I., Ng, K.: Optical music imaging: music document digitisation, recognition, evaluation, and restoration. In: Interactive Multimedia Music Technologies, pp. 50–79. IGI Global, Information Science Reference (2008)
2. Ng, K.C., Boyle, R.D.: Recognition and reconstruction of primitives in music scores. Image Vis. Comput. **14**(1), 39–46 (1996)
3. Bitteur, H.: Audiveris (2014). https://audiveris.kenai.com/
4. Fujinaga, I.: Adaptive optical music recognition. Ph.D. Thesis, McGill University, Montreal (1997)
5. Pugin, L., Burgoyne, J.A., Fujinaga, I.: MAP adaptation to improve optical music recognition of early music documents using hidden markov models. In: Proceedings of International Symposium on Music, Information Retrieval, pp. 513–516 (2007)
6. Choudhury, G.S., DiLauro, T., Droettboom, M., Fujinaga, I., Harrington, B., MacMillan, K.: Optical music recognition system within a large-scale digitization project. In: Proceedingsof International Symposium on Music Information Retrieval (2000)
7. Fahmy, H., Blostein, D.: A graph-rewriting paradigm for discrete relaxation: application to sheet-music recognition. Int. J. Pattern Recogn. Artif. Intell. **12**(6), 763–799 (1988)
8. Rossant, F., Bloch, I.: Robust and adaptive OMR system including fuzzy modeling, fusion of musical rules, and possible error detection. EURASIP J. Appl. Signal Process. **2007**(1), 160 (2007)
9. Couasnon, B., Retif, B.: Using a grammar for a reliable full score recognition system. In: Proceedings of International Computer Music Conference, pp. 187–194 (1995)
10. Carter, N.P.: Conversion of the Haydn symphonies into electronic form using automatic score recognition: a pilot study. In: Proceedings of SPIE, pp. 279–90 (1994)
11. Bainbridge, D., Bell, T.: The challenge of optical music recognition. Comput. Humanit. **35**, 95–121 (2001)
12. Fujinaga, I.: Optical music recognition bibliography (2000). http://www.music.mcgill.ca/ich/research/omr/omrbib.html

13. Rebelo, A., Capela, G., Cardoso, J.S.: Optical recognition of music symbols. Int. J. Doc. Anal. Recogn. **13**, 19–31 (2009)
14. Viro, V.: Peachnote: music score search and analysis platform. In: Proceedings of the International Society for Music Information Retrieval Conference (ISMIR), pp. 359–362 (2011)
15. Blostein, D., Baird, H.S.: A critical survey of music image analysis. In: Baird, H.S., Bunke, H., Yamamoto, K. (eds.) Structured Document Image Analysis, pp. 405–434. Springer, Berlin (1992)
16. Selfridge-Field, E.: Beyond MIDI : The Handbook of Musical Codes. MIT Press, Cambridge (1997)
17. Hankinson, A., Roland, P., Fujinaga, I.: The music encoding initiative as a document encoding framework. In: 12th International Society for Music Information Retrieval Conference, pp. 293–298 (2011)
18. Good, M.: MusicXML for notation and analysis. Comput. Musicol. **12**, 113–124 (2001)
19. Kabourov, S.: Spring embedders and force directed graph drawing algorithms (2012). CoRR. abs/1201.3011
20. Fruchterman, T., Reingold, M.: Graph drawing by force-directed placement. Softw. Prac. Exp. (Wiley) **21**(11), 1129–1164 (1991)
21. Eades, P.: A heuristic for graph drawing. Congr. Numer. **42**(11), 149160 (1984)
22. Kamada, T., Kawai, S.: An algorithm for drawing general undirected graphs. Inf. Process. Lett. (Elsevier) **31**(1), 715 (1989)
23. Raphael, C., Jin, R.: Optical music recognition on the international music score library project. Document Recognition and Retrieval XXI (2014)
24. Jin, R., Raphael, C.: Interpreting rhythm in optical music recognition. In: Proceedings of International Symposium on Music, Information Retrieval, pp. 151–156 (2012)
25. Raphael, C., Wang, J.: New approaches to optical music recognition. In: Proceedings of International Symposium on Music, Information Retrieval, pp. 305–310 (2011)
26. Guo, E.: The IMSLP, petrucci music library (2014). http://imslp.org/
27. Byrd, D., Schindele, M.: Prospects for improving optical music recognition with multiple recognizers. In: Proceedings of International Symposium on Music, Information Retrieval, pp. 41–46 (2006)
28. Renz, K.: Algorithms and data structure for a music notation system based on GUIDO notation. Ph.D. Dissertation, Technischen Universität Darmstadt Proceedings of International Symposium on Music Information Retrieval (2002)
29. Knuth, D., Plass, M.: Breaking paragraphs into lines. Softw. Prac. Exp. **11**, 1119–1184 (1981)
30. Spreadbury, D.: http://www.smufl.org/fonts/

Music Generation

Foundations for Reliable and Flexible Interactive Multimedia Scores

Jaime Arias[1]($\boxtimes$), Myriam Desainte-Catherine[1],
Carlos Olarte[2], and Camilo Rueda[3]

[1] Université de Bordeaux, LaBRI, UMR 5800, 33400 Talence, France
jarias@labri.fr
[2] ECT, Universidade Federal do Rio Grande do Norte, Natal, Brazil
[3] DECC, Pontificia Universidad Javeriana Cali, Cali, Colombia

Abstract. Interactive Scores (IS) is a formalism for composing and performing interactive multimedia scores with several applications in video games, live performance installations, and virtual museums. The composer defines the temporal organization of the score by asserting temporal relations (TRs) between temporal objects (TOs). At execution time, the performer may modify the start/stop times of the TOs by triggering interaction points and the system guarantees that all the TRs are satisfied. Implementations of IS and formal models of their behavior have already been proposed, but these do not provide usable means to reason about their properties. In this paper we introduce REACTIVEIS, a programming language that fully captures the temporal structure of IS during both composition and execution. For that, we propose a semantics based on tree-like structures representing the execution state of the score at each point in time. The semantics captures the hierarchical aspects of IS and provides an intuitive representation of their execution. We also endow REACTIVEIS with a logical semantics based on linear logic, thus widening the reasoning techniques available for IS. We show that REACTIVEIS is general enough to capture the full behavior of IS and it also provides declarative ways to increase the expressivity of IS with, for instance, conditional statements and loops.

1 Introduction

Preliminaries. Interactive multimedia (e.g., live-performance arts) refers to computer-based design systems consisting of multimedia content that interacts with the performer's actions and other external events. Multimedia content is structured in a spatial and temporal order according to the author's requirements. The potential high complexity of these systems requires adequate specification languages for the complete description and verification of scenarios.

Interactive Scores (IS) [6] is a formalism for composing and performing interactive multimedia scores where the performer has the possibility to influence the execution of the score. This means the composer allows the performer to modify, during execution, the temporal organization of the score by adding *interaction*

T. Collins et al. (Eds.): MCM 2015, LNAI 9110, pp. 29–41, 2015.
DOI: 10.1007/978-3-319-20603-5_3

points (IPs). Hence, the performer enjoys a certain freedom in choosing the time of interaction (or whether it takes place) leaving the system the task of maintaining the temporal constraints of the score. The IS model thus combines two temporal paradigms used in current multimedia tools [6]: *time-line* and *time-flow*. The former is represented at composition time when the composer defines multimedia processes by their start and end times, as well as by temporal relations between them. The time-flow paradigm is represented by the time at which the processes are actually executed.

Let us describe a simple example to introduce the terminology we shall use. In IS, boxes represent *temporal objects* (TOs) whose temporal organization is defined by asserting *temporal relations* (TRs) that those objects must obey. TRs define temporal (quantitative) and logical (qualitative) relations between TOs. More precisely, there are two qualitative relations that are defined between boxes: *precedence* and *posteriority*. Hence, TRs are enhanced with quantitative constraints by giving a range of possible durations in $[0, \infty]$. Consider for instance the IS on the left of Fig. 1 which specifies the atmosphere of a cloud forest in a theatrical installation. The composer defines the score S in which the box A controls a machine that generates white smoke; the box B controls a group of fans that evenly distributes the smoke; all the boxes in the box C are performed once the smoke has been well distributed (defined by TRs r_3 and r_4); box E controls a set of lights in order to represent a sudden beam of light; finally, box D plays the sound of the howling of a wolf whose starting time depends on a performer's action (a mouse click).

TOs are classified into *textures* and *structures*. Textures represent the execution in time of a given multimedia process (e.g., changing the brightness of a light) while structures (i.e., the hierarchical organization of the score) represent the execution of a group of TOs with their own temporal organization. In our example, texture A has a duration of 2 time-units (TU) and it starts at TU 1 (relation r_1); structure C starts after 5 TU of stopping A (r_3) and after 3 TU of stopping B (r_4) and stops when textures D and E have finished (r_7, r_8); texture D starts when the message *"/mouse 1"* arrives between 2 and 5 TU after starting C (r_5); Finally, the score S finishes when C has finished (r_9).

In all executions of the score, the start time and duration of textures A, B and E do not change. Such TOs are seen as *static control points* that must be handled by the system without interaction with the environment. The start time of texture D, however, depends on triggering an IP. The starting of D modifies the duration of structures C and S. Hence, those TOs are controlled by *dynamic control points* that depend on the interaction with the environment. We also note that texture D starts automatically after 5 TU of starting C if the IP is not triggered. This *default action* guarantees that TRs are satisfied during execution.

The IS model is implemented in I-SCORE (http://i-score.org), a tool that offers two different stages or times: *composition* and *performance*. In the former, composers place TOs on a horizontal time-line. Then, they add IPs and connect TRs between the TOs in order to define temporal properties. During the execution stage, the performer can dynamically trigger the IPs while the static

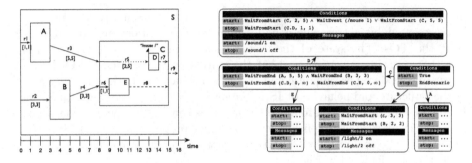

Fig. 1. Example of an interactive score and its program tree.

control points are triggered by the system. Since multimedia processes and IPs are handled by external applications, I-SCORE uses multimedia protocols like OSC in order to send/receive the messages defined by the composer.

Scores in I-SCORE are executed by the *ECO machine* [9] which is responsible of (1) triggering the static control points; (2) controlling the triggering of the dynamic control points; and (3) maintaining the temporal organization of the score. This machine relies on a Hierarchical Time Stream Petri Net (HTSPN) [12] to represent and execute the partially ordered set of events. Therefore, each time a score is written or modified, it must be translated into a HTSPN to be executed.

Motivation and Contributions. Some applications of IS such as video games, live performance installations, and virtual museum visits [1] demand two features that I-SCORE as well as its execution model (HTSPN) do not currently support: (1) the use of more flexible control structures such as conditionals and loops [5]; and (2) scores must be verified before being played since they can be seen as critical systems where raise conditions (abnormal behaviors) should not happen. As an example of (1), consider a score where the composer may define an IP that decides between executing the TO A or B. As for (2), consider the situation where a given texture is never played due to inconsistent start/end conditions or a multimedia resource that receives an unexpected number of messages that it cannot handle concurrently. Dealing with (1) in I-SCORE (i.e., on a horizontal time-line) would be hard and it would require a complete redesign of the HTSPN execution model. Moreover, due to the fact that there is one language to specify the score and another, completely different, to define the execution model, it does not seem trivial to define effective reasoning techniques to deal with (2).

In this paper we define REACTIVEIS, a programming language that takes advantage and extends the full capacity of temporal organization during the composition and execution of IS. The syntax of REACTIVEIS allows composers to define arbitrary hierarchies of processes and conditional commands –(1) above–. We endow the language with an operational semantics based on labelled trees that we claim to be simpler and more flexible than the current execution model in HTSPN. These structures allow to model the hierarchical aspect of IS and provide an intuitive representation of their execution. Roughly, the program is represented by a tree whose nodes define the conditions needed to stop/ start the TOs. The state of the system is a proper subtree of the program tree that contains information about the start/stop times of each TO. Hence, trees are considered as semantic and syntactic formal objects that are very close to the structure and behavior of IS. More interestingly, they can be defined and handled by means of a well-founded theory. This simple yet powerful characterization of IS allowed us to quickly develop an interpreter of REACTIVEIS written in OCAML. The tool produces a graphical representation of the execution of the IS as the one depicted in Fig. 2.

In order to deal with (2) above, we give a declarative interpretation of REACTIVEIS programs as formulas in intuitionistic linear logic (ILL) [7] with subexponentials [4]. We show that such interpretation is adequate: derivations in the logic correspond to traces of the program and vice-versa. Then, we can use all the meta-theory of ILL to reason about IS. In particular, we can verify whether an IS is free of raise conditions. Moreover, we can rely on the recent developments on the specification of temporal and spatial modalities in ILL (see [10]) to declaratively enrich REACTIVEIS with new constructs. For instance, it would be possible to define IS whose hierarchy may change dynamically by allowing TOs to *move* into another TO according to the stimulus from the environment.

REACTIVEIS thus offers the following advantages wrt to its predecessor I-SCORE: (1) it offers an intuitive yet precise description of the behavior of IS; (2) the tree-based semantics gives a more concrete guidance to the implementer on how a score should be executed without dealing with the HTSPN model; (3) it is a first step towards a model for defining non-linear behavior (e.g., conditional statements) in IS; (4) the ILL characterization sets the basis for developing techniques and tools for the verification and analysis of IS.

Organization. Section 2 develops the theory of REACTIVEIS: syntax, semantics and its properties (Sects. 2.1, 2.2, 2.3). Section 2.4 is dedicated to the logical interpretation of programs and the kind of properties that can be verified. Section 2.5 discusses the ideas on how to extend the IS model to handle more flexible structures. Section 3 concludes the paper. Due to space restrictions, some auxiliary definitions and results appear in the extended version of this paper [3].

2   ReactiveIS: A Language for Specifying IS

In this section we introduce the syntax, semantics and logic characterization of REACTIVEIS. We start with the constructors already available in I-SCORE and later, in Sect. 2.5, we introduce the mechanisms for conditional statements.

Syntax. REACTIVEIS programs are built from the following syntax:

$\langle score \rangle ::= \langle structure \rangle$

$\langle texture \rangle ::= \texttt{texture}(\langle params \rangle \ \langle msg \rangle \ \langle msg \rangle)$

$\langle structure \rangle ::= \texttt{structure}(\langle params \rangle \ \langle TO\text{-}list \rangle)$

$\langle params \rangle ::= \langle name \rangle \ \langle condition \rangle \ \langle condition \rangle$

$\langle TO\text{-}event \rangle ::= \texttt{start} \ \langle name \rangle \mid \texttt{end} \ \langle name \rangle$

$\langle condition \rangle ::= \texttt{wait}(\langle TO\text{-}event \rangle \ \langle min \rangle \ \langle max \rangle)$
$\mid \quad \texttt{event} \ \langle msg \rangle$
$\mid \quad (\langle condition \rangle \wedge \langle condition \rangle)$
$\mid \quad (\langle condition \rangle \vee \langle condition \rangle)$

Recall that a *structure* is a TO used to define the hierarchical organization of the score and a *texture* represents the execution of a given multimedia process by an external application. Hence, a *score* is a *structure* that represents the execution of a set of TOs (i.e., structures and textures). A structure is comprised of a set of parameters (explained below) and a (possibly empty) list of other TOs (TO-list). A texture requires, besides the parameters, two messages used to start and stop the external process. These messages are the output of the system and so they have to be sent to some other application by means of multimedia protocols such as OSC.

The syntactic unit *params* specifies a *name* (an identifier) for the TOs and also the starting and stopping *conditions*. Such conditions represent the TRs between TOs and define the temporal organization of the score.

Conditions in REACTIVEIS can be: (1) `wait` conditions that define a delay from the start or from the end of a TO (*TO-event*). Delays are defined as a range between 0 and ∞, thus allowing flexibility in temporal specifications; (2) an `event` condition represents the triggering of a specific event by the environment. Such events are messages (*msg*), for instance *"/mouse 1"*, sent by the performer during execution (at IPs). Such messages represent the inputs of the system. More complex conditions can be written by using conjunctions and disjunctions.

As an example, consider the definition for the structure C in Fig. 1:

```
1   Structure C = {
2     start.c = (Wait(End(A),5,5) & Wait(End(B),3,3));
3     stop.c  = (Wait(End(D),0,INF) & Wait(End(E),0,INF));
4     Texture D = {
5       start.c = ((Wait(Start(C),2,5) & Event("/mouse 1")) | Wait(Start(C),5,5));
6       stop.c  = Wait(Start(D),1,1);
7       start.msg = "/sound/1 on";  stop.msg = "/sound/1 off";
8     }; ... };
```

Attributes `start.c` and `stop.c` represent, respectively, the start and stop conditions of the TO. The `Wait` condition receives three arguments: an event representing the start/end of a TO, its minimum and its maximum duration (that can be infinite, denoted `INF`). Condition `Event` receives a particular OSC message that will be sent by the performer (e.g., *"/mouse 1"*). If 5 time-units have elapsed after starting C and such message has not yet arrived, D will automatically start (due to the disjunction in the starting condition). Attributes `start.msg` and `stop.msg` specify the messages that must be sent to external multimedia processes.

2.1   Conditions and Program Representation

In this section we give a tree-based representation of REACTIVEIS programs and we formalize the idea of *conditions*. Such definitions will be later used to describe the operational semantics of the language.

Conditions are built from a *Condition System* (CS) which is a first-order signature Σ that contains the distinguished predicates `WaitFromStart`, `WaitFromEnd`, `EndScenario` and `WaitEvent`. We also assume a (decidable) first-order theory Δ over Σ for dealing with deductions such as $x > 40 \models x > 0$. We shall use $\mathcal{C}$ to denote the set of conditions (formulas) built from Σ and the grammar: $F, G, \ldots := \texttt{true} \mid A \mid F \wedge G \mid F \vee G$, i.e., conditions can be atomic formulas (e.g., predicates) or conjunctions/disjunctions of formulas.

A program in REACTIVEIS is defined as a *labelled tree* whose nodes represent the TOs of the score. We will sometimes abuse notation and refer to TOs simply as nodes. Each node is associated with the conditions for starting and stopping the TO, and the corresponding messages.

Definition 1 (Program Tree). *Let $\mathcal{N}$ be a countable set of nodes, $\mathcal{B}$ the set of labels representing the names of* TO*s, and $\mathcal{M}$ the set of messages. A program tree is a labelled tree $P = \langle N, E, \ell, m, r \rangle$ where: $N \subseteq \mathcal{N}$ is the set of nodes; $E \subseteq N \times \mathcal{B} \times N$ is the set of edges; $\ell : N \to \mathcal{C} \times \mathcal{C}$ is a total function representing the start/end conditions; $m : N \rightharpoonup \mathcal{M} \times \mathcal{M}$ is a partial function representing the messages for starting/stopping an external application; and $r \in N$ is the root of the tree. Given $n \in N$, we shall use $c_s(n)$ and $c_e(n)$ (resp. $m_s(n)$ and $m_e(n)$) to to denote the starting/stopping conditions (resp. messages) for n.*

For a given tree T, the nodes, the edges, and the root node of T are denoted by $V(T)$, $E(T)$ and $root(T)$, respectively. We write $s \xrightarrow{a} t$ to represent an a-labeled edge from s (the source) to t (the target). As usual, sequences of labels $\alpha = a_0.a_1 \ldots a_n$ represent a path from the root r to a given node u in T. We use the empty sequence ε to represent the root of T. For a path p in T, $target_T(p)$ is its ending node.

The right part of Fig. 1 shows a fragment of the program tree for our running example. The predicates `WaitFromStart(p,t1,t2)` and `WaitFrom End(p,t1,t2)` hold when the time elapsed since the start and the end, respectively, of the target node of the path `p` is within the interval `[t1,t2]`. `WaitEvent(e)` waits for the external message `e`. Observe that the root node has no wait condition for starting (`true`) and it finishes when all its children have finished (`EndScenario`).

2.2 State Tree and Tree Operations

An execution state of a REACTIVEIS program is also represented as a *labelled tree* that identifies the TOs currently being executed and the ones that have already stopped. Each node in the tree has associated the times on which the TO started and stopped. If a TO has not been stopped yet, we use as stop time the special symbol $\perp \notin \mathbb{Z}_+$. We shall use $\mathbb{Z}_\perp$ to denote $\mathbb{Z}_+ \cup \{\perp\}$.

Definition 2 (State Tree). *A state tree is a labelled tree $S = \langle N, E, \ell, r \rangle$ where N, E and r are as in Definition 1 and $\ell : N \to \mathbb{Z}_+ \times \mathbb{Z}_\perp$ is a total function giving, for each node, its starting and ending times. Functions $t_s : N \to \mathbb{Z}_+$ and $t_e : N \to \mathbb{Z}_\perp$ give the starting and stopping time of a node, respectively.*

S is a *valid* state for a program tree P if S is homomorphic to P, i.e., there exists $f : V(S) \to V(P)$ that preservers the structure: $f(root(S)) = root(P)$ and $s \xrightarrow{a} t \in E(S)$ iff $f(s) \xrightarrow{a} f(t) \in E(P)$ (see Fig. 2(a)).

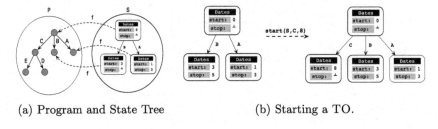

(a) Program and State Tree   (b) Starting a TO.

Fig. 2. Basic operations on state trees.

Now we define two operations on state trees, stopping and starting a TO:

Stopping a TO. When a node n is stopped, its stop time, and the stop time of its (non already stopped) children, must be updated with the current time of execution. Formally, $stop(S, p, t) \triangleq \langle N, E, \ell \Leftarrow \{n \mapsto (t_s(n), t) \mid n \in k\}, r \rangle$ where $k = \{ n \mid n \in D(target_S(p)) \land t_e(n) = \bot \}$. "$\Leftarrow$" means *relational overriding*, i.e., $\ell \Leftarrow R \triangleq R \cup \{ x \mapsto y \mid x \mapsto y \in \ell \land x \notin dom(R)\}$; $D(v)$ denotes the set containing v and its descendants in S; p is a path in S; and $t \in \mathbb{Z}_+$ is the current time of execution.

Starting a TO. This operation causes that a new b-labelled edge be added to the current state tree. This edge points to a new node having the current time as its start time and an undefined stop time. More precisely, for a non-empty path p, let $up(p)$ be the sequence of labels without the last label, and $last(p)$ be the last label of the sequence. For a path p in S and a time $t \in \mathbb{Z}_+$, starting a TO is defined as $start(S, p, t) \triangleq \langle N \cup \{n_1\}, E \cup \{n \xrightarrow{b} n_1\}, \ell \cup \{n_1 \mapsto (t, \bot)\}, r \rangle$ where $n_1 \notin N, n = target_S(up(p)), b = last(p)$. Figure 2(b) shows the start of C.

2.3 Operational Semantics

The semantics of ReactiveIS considers two kind of reduction relations, $\longrightarrow$ and $\Longrightarrow$, parametric on the program tree P (see Fig. 3). Recall that the input of the program is a set of messages produced by the environment and the output is the set of messages the program must produce during a time-unit. Hence, the *observable* transition $\langle S, t \rangle \overset{I,O}{\Longrightarrow}_P \langle S', t + 1 \rangle$ means that at time t, the state tree S on input I reduces in one *time unit* to S' and outputs O. The observable transitions are obtained from finite sequences of internal transitions. Such *internal* transitions represent how the state S is gradually updated by starting/stopping TOs. It is important to notice that the changes in the state of the score are only visible at the end of the time-unit, i.e., it is assumed that internal transitions cannot be directly observed.

    The internal transition $\langle St, O \rangle^{I,t}_S \longrightarrow_P \langle St', O' \rangle^{I,t}_S$ means that, given that the input in the current time-unit is I and the initial state is S, the state St moves to St' possibly adding new messages to the set O leading to O'. Let us give some intuition. We use $L(S)$ to denote the set of all paths in S including

$$R_{\mathsf{START}} \frac{p \in \mathtt{canStart}(S, P) \quad \langle P, S, I, t\rangle \models c_s(n)}{\langle St, O\rangle_S^{I,t} \longrightarrow_P \langle start(St, p, t), O \cup \{m_s(n)\}\rangle_S^{I,t}} \quad \text{where } n = target_P(p)$$

$$R_{\mathsf{STOP}} \frac{p \in \mathtt{canStop}(S) \quad \langle P, S, I, t\rangle \models c_e(n)}{\langle St, O\rangle_S^{I,t} \longrightarrow_P \langle stop(St, p, t), O \cup \{m_e(n)\}\rangle_S^{I,t}} \quad \text{where } n = target_P(p)$$

$$R_{\mathsf{TIME}} \frac{\langle S, \emptyset\rangle_S^{I,t} \longrightarrow_P^* \langle S', O\rangle_S^{I,t} \not\longrightarrow_P}{\langle S, t\rangle \overset{I,O}{\Longrightarrow}_P \langle S', t+1\rangle}$$

Fig. 3. Rules for the internal reduction $\longrightarrow$ and the observable reduction $\Longrightarrow$.

ε. We define $p_{alive}(S) = \{p \mid p \in L(S) \wedge t_e(target_S(p)) = \bot\}$, i.e., the set of TOs that are currently running. Moreover, let $Children(p)$ be the set of paths of a program tree P from the root node to the children of the ending node of p (i.e., $target_P(p)$). Since a TO can only start if its parent is running and it has not stopped yet, we can compute the set of TOs that can start by defining $canStart(P, S) \triangleq \{p \mid p_{parent} \in p_{alive}(S) \wedge p \in Children(p_{parent})\} \setminus L(S)$.

The rule R_{Start} says that a TO is executed only if (a) it has not yet been executed and (b) its start condition is satisfied. Premise (a) is ensured with the aid of the set $canStart(S, P)$ explained before. Premise (b) is asserted by means of the relation $\langle P, S, I, t\rangle \models F$ that intuitively means that the current state satisfies the condition F. The precise definition of $\models$ is in [3].

The rule R_{STOP} dictates that a TO is stopped only if (a) it is currently being executed and (b) its end condition is satisfied. Premise (b) is similar as in the previous rule. Premise (a) is ensured with the aid of the set $canStop(S) \triangleq \{p \mid p \in L(S) \wedge t_e(target_S(p)) = \bot\}$ that contains the nodes in the state tree whose end time is not defined.

The only non-determinism of REACTIVEIS programs is due to the signals provided by the environment. Then, we can prove that the observable relation is indeed a function (the proof is in [3]).

Theorem 1 (Determinism). *For all state S and input I, if $\langle S, t\rangle \overset{I,O_1}{\Longrightarrow}_P \langle S_1', t'\rangle$ and $\langle S, t\rangle \overset{I,O_2}{\Longrightarrow}_P \langle S_2', t'\rangle$ then $O_1 = O_2$ and $S_1' = S_2'$.*

2.4 Logical Characterization of IS

An appealing feature of REACTIVEIS is that it allows a logic characterization as formulas in intuitionistic linear logic (ILL) [7] with subexponentials [4] (SELL). The technical details of this semantics can be found at [3]. In the following, we shall give some intuitions to understand what kind of properties we are able to verify about IS.

The formula $!^a F$ in SELL means that F is marked with a given modality a. The index a is taken from a poset $\langle I, \preceq\rangle$ (the subexponential signature) and it can be interpreted as a spatial location or a time-unit [10]. Here, we shall mark the formulas with subexponentials of the form $t.x$ where t represents the current time-unit and "x" can be:

$!^{t.i}F$: F is an input from the environment, e.g., $!^{t.i}\text{evt}(\text{mouse1})$

$!^{t.o}F$: F is an observable action, e.g., $!^{t.o}\text{msg}(m)$ means that the starting/ stopping message m was added.

$!^{t.s.p}F$: F represents information about the state, e.g., $!^{t.s.A}\text{box}(-,-)$ means that A has not been started yet. We use "$-$" instead of "$\perp$", as in the previous section, since "$\perp$" is a logical symbol in ILL.

The advantage of using subexponentials is that we can neatly split the logical context in a sequent. In our particular case, the context is split into different time-units and each time-unit stores information about inputs from the environment $(t.i)$, observable actions $(t.o)$ and information about the state of the system $(t.s)$. To better understand this idea, consider the following derivation:

$$\frac{!^{4.i}\text{evt}(e2), !^{4.i}\text{evt}(e3) \longrightarrow !^{4.i}\text{evt}(e3)}{!^{3.i}\text{evt}(e1), !^{4.i}\text{evt}(e2), !^{4.i}\text{evt}(e3), !^{4.s.A}\text{box}(5,7) \longrightarrow !^{4.i}\text{evt}(e3)} !_R$$

Roughly, we are trying to prove that the event $e3$ occurred in the time-unit 4. The introduction rule for $!$ ($!_R$, called promotion rule) forces to delete (weaken) from the context all the formulas with subexponentials not related to $4.i$. Then, we cannot use the information available on time-unit 3 (i.e., $!^{3.i}\text{evt}(e1)$) nor the information about the state of the system (i.e., $!^{4.s.A}\text{box}(5,7)$).

The encoding of each TO in a REACTIVEIS program gives rise to three formulas. Namely, ctr to control when to start/stop the TO and str and stp to handle the action of starting/stopping the TO. Let us consider the texture D in our running example whose starting condition depends on the starting of C and an event from the environment. We define the control formula as follows:

$$\text{ctr(D,t)} \stackrel{\text{def}}{=} !^{t.s.D}\text{P\_STOP} \multimap \text{stop-imm}(D,t) \& !^{t.s.D}\text{P\_RUN} \multimap \text{decide}(D,t) \&$$
$$!^{t.s.D}\text{P\_IDLE} \multimap \forall n, m.(!^{t.s.D}\text{box}(n,m) \multimap !^{(t+1).s.D}\text{box}(n,m))$$

Intuitively, D can only proceed if its parent C has already added to the context one of the predicates P_STOP, P_RUN or P_IDLE notifying its current state (stopped, currently running or idle –already stopped or not started–). We recall that $F \multimap G$ represents linear implication. The additive conjunction & allows us to choose between three possible choices: stop immediately (stop-imm), decide to start or stop (decide) or continue in the same state. The definition of these formulas can be found in [3].

Let us consider the formulas needed to start the execution of D:

$$\text{str(D,t)} \stackrel{\text{def}}{=} (\text{condition} \multimap \text{start}(D,t)) \& (\text{default} \multimap !^{(t+1).s.D}\text{box}(-,-))$$
$$\text{start}(D,t) \stackrel{\text{def}}{=} !^{t.s.D}\text{box}(t,-) \otimes !^{(t+1).s.D}\text{box}(t,-) \otimes !^{t.o}\text{msg}(m_s(D))$$

The formula condition is obtained by translating the starting condition of D into a SELL formula. The formula default (see [3]) says that the starting conditions cannot be satisfied in the current time-unit and then, the state of D remains the same. Note that the formula $\mathtt{start}(D,t)$ adds to the context the information needed to deduce that D started at time-unit t and it also adds to the context $t.o$ the starting message of D. The formulas controlling the stopping of D can be defined similarly.

With the aid of these formulas and some auxiliary definitions in [3], we can define an encoding $\llbracket \cdot \rrbracket$ mapping REACTIVEIS programs into SELL formulas. Moreover, by relying on a focused [2] proof system for SELL, we can show that operational steps correspond to derivations in SELL and vice-versa (see theorem below). Focusing is a discipline on proofs to reduce the non-determinism during proof search. Hence, focused proofs can be interpreted as the normal form proofs for proof search. Roughly, once we choose to work on a formula (i.e., we focus on it), we do not have more choices that decompose it completely. Hence, in the end of the focused phase, we observe that the logical derivation mimicked exactly the operational steps of the encoded program.

Theorem 2 (Adequacy). *Let P be a REACTIVEIS program. Then, $\langle S,t \rangle \overset{I,O}{\Longrightarrow}_P \langle S', t+1 \rangle$ iff the sequent $\llbracket P \rrbracket, \llbracket S \rrbracket_t, \llbracket I \rrbracket_t \longrightarrow \llbracket S' \rrbracket_{t+1} \otimes \llbracket O \rrbracket_t$ is provable in SELL.*

The previous theorem opens the possibility of reasoning about IS by using well established techniques in proof theory. Moreover, all the tools developed for SELL [10] can be applied to verify properties of scores. For that, we consider SELL sequents of the form $\llbracket P \rrbracket, \llbracket S_{\mathtt{init}} \rrbracket_0, \mathtt{env} \longrightarrow \mathcal{G}$ where P is the program (the score); $S_{\mathtt{init}} = \langle \{r\}, \emptyset, \{r \mapsto (0, \bot)\}, r \rangle$ is the initial configuration of the score; $\mathtt{env}$ encodes any possible input from the environment (i.e., a disjunction of all the possible inputs from the environment); and $\mathcal{G}$ encodes the property to be verified (i.e., the goal).

Let us give some examples of $\mathcal{G}$-formulas. Consider the case where we want to verify that two TOs A and B must be executed concurrently. Then we can set $\mathcal{G} = \Cup t.\exists n, m.!^{t.s.A}\mathtt{box}(n, -) \otimes !^{t.s.B}\mathtt{box}(m, -)$ meaning that there exists a time-unit t ($\Cup t$) such that in that time-unit A and B have already started and not stopped. As another example, consider the fact that regardless the inputs from the environment, the TO B cannot be currently playing if A has finished its execution. In that case, we have $\mathcal{G} = \Cap t.\exists n_1, n_2.!^{t.s.A}\mathtt{box}(n_1, n_2) \multimap \exists n'_1, n'_2.!^{t.s.B}\mathtt{box}(n'_1, n'_2)$ meaning that for all time-unit t ($\Cap t$), if A has already stopped, then it must be the case that B has already stopped too. Now assume that there is a precedence relation between A and B, i.e., B cannot start if A is currently playing. In this case, $\mathcal{G} = \Cap t.(\exists n.!^{t.s.A}\mathtt{box}(n, -)) \multimap !^{t.s.B}\mathtt{box}(-, -)$. Finally, one may be interested in proving that there exists at least one execution path such that a given TO A is executed. This can be formalized by proving the property $\mathcal{G} = \Cup t.\exists n, m.!^{t.s.A}\mathtt{box}(n, m)$.

2.5 Extending the IS Model

Having a formal model for IS opens the possibility to propose new programming constructs and reason about its behavior. For instance, it turns out that the notion of conditions as logical formulas and its realization in the operational semantics ($\langle P, S, I, t \rangle \models F$) are general enough to define conditional statements in REACTIVEIS. For that, let us extend the syntax of REACTIVEIS to consider conditions of the form $\langle conditions \rangle ::= \ldots \mid event\ (e(n)\ op\ \langle value \rangle)$ where e is an event with a carried value n and op is a relational operator (e.g., $\leq$, $=$, etc.). Let $mouse$ be a parametric event with two possible values, 1 and 2. By defining the starting condition of a TO A as $\texttt{event}(\texttt{mouse}(n) = 1)$, and the condition of B as $\texttt{event}(\texttt{mouse}(n) = 2)$, we can define a score that waits until the $mouse$ event is detected. Then, it decides whether to execute A or B. Since the conditions on the carried value n are mutually exclusive, it would be more natural to write if c then A else B to specify such behavior. Hence, IPs can now be used to express non-linear behaviors: once an event is detected, the evaluation of the expression "$n\ op\ value$" will determine the flow of the score.

Interestingly, the logical semantics of REACTIVEIS will allow the composer to verify that, regardless the path taken by the performer, the desired properties of the score hold. These techniques can also be used to avoid mistakes during composition. For instance, consider the situation where the starting condition of another TO C depends on both the starting of A and B. In this case, the logical semantics will detect that C will be never played.

Loops can be also obtained in a similar fashion. However, special attention must be paid to avoid two copies of the same TO during execution. For that it suffices to consider that the repetition of a TO A is restricted to: (1) A has already stopped and (2) the performer must send an event to control whether A has to be played again (i.e., loops are guarded by IPs). The semantics and the logic characterization require a minimal change to reset the starting and stopping times of A in order to enable its next execution.

3   Concluding Remarks and Related Work

We have introduced REACTIVEIS, a new programming language for the composition and performing of interactive scores. We defined an operational semantics for REACTIVEIS based on labelled trees which is simpler and more intuitive that the HTSPN model of I-SCORE. We also endowed REACTIVEIS with a declarative semantics based on intuitionistic linear logic with subexponentials, thus allowing us to prove the correct behavior of scores. REACTIVEIS then strives at setting the foundations for reasoning about IS and, more importantly, to extend its functionality in a declarative way.

The idea of a tree-based semantics is influenced by the works in [8] and [14]. In [8] a semantics for ORC, a language to specify programs to orchestrate the invocation of sites that are subject to constraints on their execution, is presented. Such semantics considers trees annotated with information about the values of the program variables and the times at which they are assigned. In [14] a graph model to represent the semantics of video (VIDEOGRAPH) is presented. In such model the nodes represent events and they are linked to each other based on their containment and temporal relationships. On the other side, our logical characterization of REACTIVEIS is based on the ideas in [10] where subexponentials in linear logic were used to give logical semantics to concurrent programming languages featuring modalities. In [11,13] the authors propose a semantics for IS based on process calculi. However, no practical techniques were proposed for the verification of the score as the logical characterization presented here. Moreover, the models proposed in [11,13] cannot be straightforwardly extended to deal with non-linear behavior.

Our next step is to formalize, for instance in Coq, the semantics presented here in order to generate a verified interpreter. Since the use of SELL formulas for specifying score's properties may be cumbersome for non-experts, we plan to develop a front-end, e.g., a small assertion language, to express such properties.

References

1. Allombert, A., Marczak, R., Desainte-Catherine, M., Baltazar, P.: GarnierLaurent: virage: designing an interactive intermedia sequencer from users requirements and theoretical background. In: International Computer Music Conference (2010)
2. Andreoli, J.M.: Logic programming with focusing proofs in linear logic. J. Log. Comput. **2**(3), 297–347 (1992)
3. Arias, J., Desainte-Catherine, M., Olarte, C., Rueda, C.: Foundations for reliable and flexible interactive multimedia scores. Technical Report, LaBRI, University of Bordeaux, March 2015
4. Danos, V., Joinet, J., Schellinx, H.: The structure of exponentials: uncovering the dynamics of linear logic proofs. In: Mundici, D., Gottlob, G., Leitsch, A. (eds.) KGC 1993. LNCS, vol. 713, pp. 159–171. Springer, Heidelberg (1993)
5. De la Hogue, T., Baltazar, P., Desainte-Catherine, M., Chao, J., Bossut, C.: OSSIA: open scenario system for interactive applications. In: Journées d'Informatique Musicale, pp. 78–84. Bourges (2014)
6. Desainte-Catherine, M., Allombert, A., Assayag, G.: Towards a hybrid temporal paradigm for musical composition and performance: the case of musical interpretation. Comput. Music J. **37**(2), 61–72 (2013)
7. Girard, J.: Linear logic. Theor. Comput. Sci. **50**, 1–102 (1987)
8. Hoare, T., Menzel, G., Misra, J.: A tree semantics of an orchestration language. In: Broy, M., Gruenbauer, J., Harel, D., Hoare, T. (eds.) Engineering Theories of Software Intensive Systems. NATO Science Series II: Mathematics, Physics and Chemistry, vol. 195, pp. 331–350. Springer, Netherlands (2005)
9. Marczak, R., Desainte-Catherine, M., Allombert, A.: Real-time temporal control of musical processes. In: Proceedings of the Third International Conferences on Advances in Multimedia, MMEDIA 2011, pp. 12–17 (2011)
10. Nigam, V., Olarte, C., Pimentel, E.: A general proof system for modalities in concurrent constraint programming. In: D'Argenio, P.R., Melgratti, H. (eds.) CONCUR 2013 – Concurrency Theory. LNCS, vol. 8052, pp. 410–424. Springer, Heidelberg (2013)
11. Olarte, C., Rueda, C.: A declarative language for dynamic multimedia interaction systems. In: Chew, E., Childs, A., Chuan, C.-H. (eds.) MCM 2009. CCIS, vol. 38, pp. 218–227. Springer, Heidelberg (2009)
12. Sénac, P., de Saqui-Sannes, P., Willrich, R.: Hierarchical time stream petri net: a model for hypermedia systems. In: DeMichelis, G., Díaz, M. (eds.) ICATPN 1995. LNCS, vol. 935, pp. 451–470. Springer, Heidelberg (1995)

13. Toro, M., Desainte-Catherine, M., Rueda, C.: Formal semantics for interactive music scores: a framework to design, specify properties and execute interactive scenarios. J. Math. Music **8**(1), 93–112 (2014)
14. Tran, D.A., Hua, K.A., Vu, K.: VideoGraph: a graphical object-based model for representing and querying video data. In: Laender, A.H.F., Liddle, S.W., Storey, V.C. (eds.) ER 2000. LNCS, vol. 1920, pp. 383–396. Springer, Heidelberg (2000)

Genetic Algorithms Based on the Principles of *Grundgestalt* and Developing Variation

Carlos de Lemos Almada[✉]

Federal University of Rio de Janeiro, Janeiro, Brazil
calmada@globo.com

Abstract. This article describes a specific aspect of a broad research project which aims at realization of analytical and compositional approaches theoretically based on the principles of *Grundgestalt* and developing variation, both elaborated by Arnold Schoenberg. The present study introduces a group of four complementary and sequential modules modeled as genetic algorithms (forming the geneMus complex), employed for the systematical production of variants from a basic musical cell. The description of the interaction and functioning of the four modules is followed by examples of their application in the whole process.

Keywords: Grundgestalt · Developing variation · Genetic algorithms · Musical composition

1 Introduction

This article is part of a research focused on systematical studies based on the principles of developing variation and *Grundgestalt* under analytical and compositional perspectives. Both principles were elaborated by Arnold Schoenberg (1874–1951) and are associated to a sort of organic musical creation, in other terms, to a gradual production of a large set of motive-forms and themes from a small group of basic elements (the *Grundgestalt*) by recursive use of transformational processes, which corresponds to developing variation techniques. The present paper describes a set of four sequential and complementary computational modules, modeled as a genetic algorithm complex, forming a system created for organic production of variations from a basic musical structure.

2 Theoretical Grounds

Among Schoenberg's innumerable contributions to musical theory, the principles of *Grundgestalt* and developing variation are probably the most important ones, and maybe, the most influential considering the strong interest they recently arose in the academic fields of musical analysis and composition.[1] Both concepts are originated from an organic conception of musical creation that characterized

---

[1] See, for example, [1–3], among others.

© Springer International Publishing Switzerland 2015
T. Collins et al. (Eds.): MCM 2015, LNAI 9110, pp. 42–51, 2015.
DOI: 10.1007/978-3-319-20603-5_4

part of the work of Austro-German classical and romantic composers, namely the lineage formed by Mozart, Beethoven and especially, Brahms. For Schoenberg, Brahms' employment of variation techniques was highly sophisticated and represented a remarkable characteristic of his "progressive" compositional personality.[2] In fact, an important part of Schoenberg style formation is due to a strong influence from Brahmsian mastery in the derivative treatment [5].

Essentially, the principle of *Grundgestalt* (normally translated as "basic format") corresponds to a set of basic musical elements from which - at least in the idealized case - a composer could extract all the necessary material (even the most contrasting ideas) for his/her work. The construction of a musical piece should then grow as an organism, in a gradual process of progressive transformation, which is ultimately associated to the principle of developing variation, or else, variation over variation with the consequent production of extensive generations of derived forms.

3   The Gr-System and the geneMus Complex

Since 2011, Carlos Almada has been developing a research project whose main objective is to produce systematical studies based on both Schoenbergian principles, considering two perspectives: analytical and compositional. The first approach resulted in the elaboration of an analytical model, which was applied to the examination of organically-constructed musical pieces. Some analyzes were made [6–8], contributing to the consolidation of special terminology, symbology and graphic analytical tools. As a basic research's assumption, it is considered that the derivative operations can hypothetically occur in two levels, one abstract and the other concrete, consequently, resulting in two kinds of developing variation procedures, respectively, of first and second orders. The first order's developing variation (DV1) is responsible for the production of basic variants from abstract referential forms (intervallic or rhythmic sequences, for instance), by application of transformational operations (like inversion, augmentation, etc.). In turn, the second order's developing variation (DV2) occurs in the concrete level, affecting "real" and more complex musical structures (built from the concatenation of the basic variants formed in the DV1 process), also producing variants by application of similar kinds of operations.

In 2013, based on the analytical model's elements, in a reverse engineering process, Almada created the Gr-System ("Gr" for *Grundgestalt*), aiming at the use of intensive variation in musical composition. The main motivation for this initiative was to investigate if the processes for production of variants from a basic musical cell could be systematized, covering the widest possible spectrum of possibilities. Taking for granted that such a task could only be realized with the help

---

[2] In order to emphasize the importance of Brahms' derivative practice, Schoenberg presented in 1933 in Frankfurt Broadcast a radio conference entitled *Brahms the progressive*, celebrating his birth's centennial. Schoenberg provocatively chose this title as a kind of answer to the common sense that considered Brahms as a conservative composer in comparison to Wagner. That conference was years later transformed in a homonymous essay [4].

of computational means, Almada developed a genetic algorithm complex, named geneMus (gM), formed by four sequential and complementary modules, written and implemented in MATLAB. Unlike conventional genetic algorithms, which normally generate their offspring from the moment they are put in action, constrained by pre-established conditions, the functioning of the modules depends on an intense interaction with an user/composer, who controls the whole system and chooses, at each stage of the process, the fittest options according to his/her particular musical intentions.

Figure 1 presents a general overview of the functioning of gM's four modules (they are properly described in the following sections).

Besides the employment of elements from the formal systems [9,10] and genetic algorithms [11] theories, the basic structure of the system (and, consequently, of gM) is deeply anchored on principles, concepts and terminology

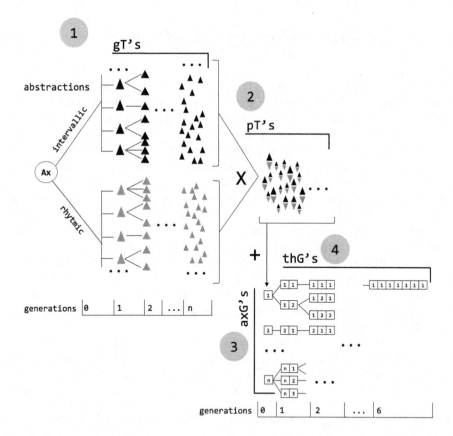

Fig. 1. Scheme of functioning of geneMus' four modules: (1) from the basic cell (Ax) are produced separate abstracted variants (gT's); (2) their recombination form concrete unities (pT's); (3) these are concatenated to form more complex structures (axG's); (4) that serve as base for further derivation, with the production of concrete variants (thG's)

from genetics and evolutionary biology [10,12,13], among others. Some aspects of this kind of influence can be observed throughout the text.

3.1   Module 1: Production of Abstracted Variants (geno-Theorems)

The first module is responsible for the production of variant unities (geno-Theorems, or gT's) from a basic cell, the *Grundgestalt* (renamed as the axiom of the system). By definition, an axiom is a brief monophonic musical segment, as exemplified in Fig. 2, in musical notation.

Fig. 2. A hypothetical axiom

The axiom is read as a MIDI file. Its musical information is then automatically transcribed into integers, being distributed into so-called chromosomes I (intervallic sequence), R (rhythmic sequence) and M (melodic sequence), formatted as vectors. The chromosomes ultimately represent abstractions of the axiom's musical data. These are the conventions adopted for their numeric notation: (a) for chromosome I, the integers represent the number of semitones contained in their intervals (1 = minor second), and the +/− signs correspond to their directions (up or down); (b) for chromosome R, integers represent temporal durations (1 = 16th note), and the signs +/−, respectively, if they correspond to sound or rest; (c) for chromosome M, the numbers are conventional MIDI pitches (60 = middle C). Taking the axiom of Fig. 2 as an example, its chromosomes are the following: I <+4+3+5>, R <+4+2+2+8>, and M <60 64 67 72>.

An important element of the system is the *coefficient of similarity* (Cs), a real number from 0 up to 1 that measures the "parenthood" degree of a given variant in relation to the referential form from which it is derived. By convention, the Cs of the axiom (more precisely, of its chromosomes) is maximal (= 1). Special algorithms for comparison of melodic or rhythmic contours calculate the loss of similarity between a given form and its immediate offshoot, which results in a Cs inversely proportional to the structural transformation caused in the variant. These algorithms are essentially based on vectorial subtraction of the numeric transcriptions of the respective involved forms, with addition/subtraction of weigths considering the relative position of each element within the vectors. This fact is associated to the intuition that the closer to the beginning of a melodic sequence an element (interval or duration) is positioned, the greater its influence on the melody's identity. Consequently, a vectorial difference with low values in initial entries will receive a bigger weigth than an opposite case, since it supposedly represents a couple of more similar vectors (and, by extension, abstract musical forms).

Table 1. Formal descriptions of gTs derived from axiom of Fig. 2: order numbers of production, referential forms and chromosomes of origin, applied operations, resultant gT's, coefficients of similarity and generation numbers

Order	Ref. form	Chrom.	Operation	Res. gT	Cs	Gen.
1	$< +4 + 3 + 5 >$	I	Intervalar expansion $(+/- 1$ semitone$)$	$< +5 + 4 + 6 >$	.82	1
2	$< +4 + 3 + 5 >$	I	Inversion and retrogradation	$< -5 - 3 - 4 >$	.43	1
3	$< +4 + 2 + 2 + 8 >$	R	Diminution	$< +2 + 1 + 1 + 4 >$	.94	1
4	$< +4 + 2 + 2 + 8 >$	R	Rotation	$< +2 + 8 + 4 + 2 >$	.33	1

The production of geno-Theorems (gT's) may proceed independently, considering either chromosomes I or R.[3] Having chosen one of them for initiating the production (in this case, the DV1 process) the user selects an operation to apply at the referential form (i.e., the chosen chromosome), automatically yielding a gT.[4] If the result is approved by the user it is selected, becoming a potential new referential form for production of gT's in a second generation. The process must be then replicated indefinitely. In other words, the user performs an artificial selection (in a Darwinian sense) - like a farmer chooses the bigger rabbits for reproduction -, taking only the results that will be fruitful according to his/her compositional intentions.

As an example of application of this module, be the following gT's (M-gT's and R-gT's, related to their respective referential chromosomes) as results of a first generation of variants from the axiom of Fig. 2 (see Fig. 3 and Table 1).

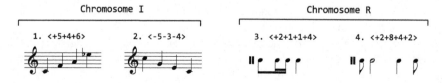

Fig. 3. A possible first generation of gT's derived from axiom of Fig. 2

These elements become referential forms for the production of further generations of geno-Theorems (Fig. 4 and Table 2).

---

[3] The chromosome M is not employed for variation, being only a transcription of chromosome I. It is necessary in order to provide to the system a melodic result with a compatible cardinality in relation to chromosome R (obviously, a sequence of intervals has always one element less than a sequence of corresponding durations).

[4] The geno-Theorems may be of two kinds, depending on the respective chromosome: R-gT if referred to chromosome R (or one of this "descendant"), or M-gT, in the alternative case.

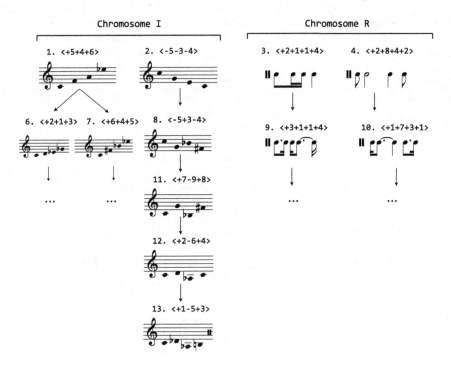

Fig. 4. Some gT's of 2nd to 5th generations, derived from gT's of Fig. 3

3.2   Module 2: Production of Concrete Variants (pheno-Theorems)

In this module the separate M-gT's and R-gT's yield in the first module are crossed over, resulting into concrete musical structures, the pheno-Theorems (or pT's). Intending to avoid "overpopulation", the user (adopting again an artificial selection strategy) may choose to apply some filters (or fitness functions) at the obtained breed, eliminating then undesirable or ill-formed results (as, for instance, redundant pT's, or those ones that extend some pre-determined pitch range).

Figure 5 shows some possible pheno-Theorems, based on the abstracted forms presented in Fig. 4.

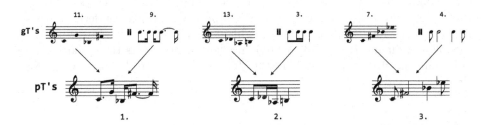

Fig. 5. Three possible pT's constructed from gT's of Fig. 4

Table 2. Formal descriptions of gTs derived from axiom of Fig. 2, considering generations 2 to 5: order numbers of production, referential forms and chromosomes of origin, applied operations, resultant gT's, coefficients of similarity and generation numbers

Order	Ref. form	Chrom.	Operation	Res. gT	Cs	Gen.
6	$< +5 + 4 + 6 >$	I	Intervalar contraction (+/− 3 semitone)	$< +2 + 1 + 3 >$	.72	2
7	$< +5 + 4 + 6 >$	I	Permutation	$< +6 + 4 + 5 >$	.79	2
8	$< -5 - 3 - 4 >$	I	Selective inversion (just one interval is inverted)	$< -5 + 3 - 4 >$	.61	2
9	$< +2 + 1 + 1 + 4 >$	R	Selective puntuaction (just one duration is added in 50 %)	$< +3 + 1 + 1 + 4 >$	.45	2
10	$< +2 + 8 + 4 + 2 >$	R	Rhythmic contraction (+/− 1 16th note)	$< +1 + 7 + 3 + 1 >$	.22	2
11	$< -5 + 3 - 4 >$	I	Intervalar complementation	$< +7 - 9 + 8 >$	.52	3
12	$< +7 - 9 + 8 >$	I	Multiplication (*2, mod12)	$< +2 - 6 + 4 >$	.57	4
13	$< +2 - 6 + 4 >$	I	Intervalar contraction (+/− 1 semitone)	$< +1 - 5 + 3 >$	.38	5
...	....	...	...	$< ... >$		...

3.3   Module 3: Production of Axiomatic-Groups

In this stage the collection of produced pT's become the basis for the construction of more complex musical structures, named as axiomatic groups (axG's). The axG's result from the concatenation of two or more pT's (that function as building blocks in this process), conditioned by decisions taken from some questions proposed to the user, about melodic transpositions, metrical displacements and suppression of notes or rests. Once selected (again, according to the user's own compositional criteria), an axG receives an identifier label, expressed as a 1×7 matrix, named as *Gödel*-vector (Gv).[5] The seven positions of this vector correspond to a sequence of generations (0 to 6),[6] with their content (integers) representing the chronological order in which a given form is created. Being by definition a referential form, an axG is considered as a "zero-generation" structure, in other terms, a sort of "patriarch" of potential lineages of derived forms (to be created in the fourth module). Therefore, its Gv is always expressed in the following manner: $< n\ 0\ 0\ 0\ 0\ 0\ 0 >$, where n represents the chronological order number of its appearance (i.e., "1" for the first axG to be formed, "2" for the second, and so on).

---

[5] For a detailed description of this vector and of a closing related element, the *Gödel*-address, see [14].

[6] The number of six generations was arbitrarily determined and can, of course, be expanded in the future.

Figure 6 presents a possible axG formed by concatenation of the three pT's of Fig. 5.

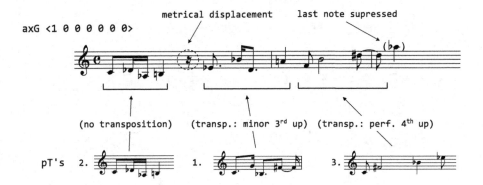

Fig. 6. A possible axG formed by the pT's of Fig. 5

3.4   Module 4: Production of Theorem-Groups

In module 4 the previously yield axG's become referential forms for the production of theorem-Groups lineages, i.e., concrete variant forms, corresponding to the already mentioned process of second order's developing variation.

Analogously to what happens in the first module, the thG's result from application of transformational operations at a selected axG. There are two types of operations in this module: (a) general: in this case, the whole axG content - in this case considered as a melodic-rhythmic unity - is affected by the algorithm (ex: inversion, retrogradation, permutation, etc.); (b) mutational: only a random-selected element is transformed by the algorithm (e.g., by intervallic expansion of one semitone). The use of the mutational operations in this stage is intended to

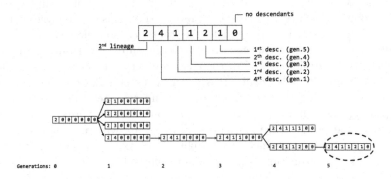

Fig. 7. Genealogical description of a hypothetical thG

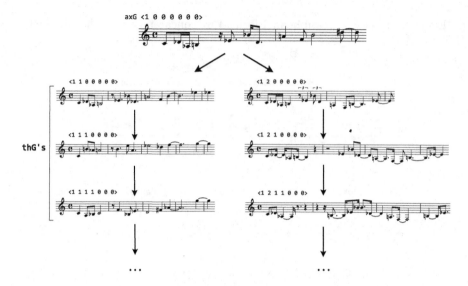

Fig. 8. Some possible derivations from axG of Fig. 6

someway replicate the biological processes of micromutations, thereby allowing gradual divergences through randomly small-scale transformations.

A resulting well-formed thG is classified according to its provenance (i.e., the patriarch from which it derives), and its genealogical description is automatically informed, properly expressed by its *Gödel*-vector. Figure 7 presents an example of this kind of identification.

Figure 8 shows some possible thG's (encompassing three generations) resulted from axG of Fig. 6, illustrating the process of second order's developing variation.

4   Concluding Remarks

Having already been applied to composition of some musical pieces (with varied instrumentations, extensions, genres, etc.), the geneMus complex can be considered as a robust auxiliary tool for the composer, since it is able to systematically provide a large quantity of derived forms (that may be selected according to particular constructive intentions) from a basic unity, contributing (at least, considering the material aspect) to the economy and coherence of the whole.[7]

Further studies will be realized in order to extend the current derivative processes (considering the four modules) beyond the user's decisions, by also encompassing possibilities for "natural" selection. In other words, it will be necessary to enable the system to produce "alone" (i.e., without a user's interference) lineages of variants, which can be achieved by expansion and improvement of the current fitness functions and filters implemented in some of the four stages.

---

[7] Recently, the four gM's modules were formatted as computational applications, being available for download in www.musmat.org.

References

1. Frisch, W.: Brahms and the Principle of Developing Variation. University of California Press, Los Angeles (1984)
2. Burts, D.: An application of the Grundgestalt concept to the first and second Sonatas for Clarinet and Piano Op. 120 no. 1 and no. 2, by J. Brahms. Tampa. Dissertation (Masters in Music). University of South Florida (2004)
3. Embry, J.: The role of organicism in the original and revised versions of Brahmss Piano Trio In B Major, Op. 8, Mvt. I: A comparison by means of Grundgestalt analysis. Amherst. Dissertation (Masters in Music). University of Massachusetts (2007)
4. Schoenberg, A.: Brahms the progressive. In: Style and Idea: Selected Writings of Arnold Schoenberg, pp. 398–444. Faber and Faber, London (1984)
5. Frish, W.: The Early Works of Arnold Schoenberg (1893–1908). University of California Press, Los Angeles (1993)
6. Almada, C.: Simbologia e hereditariedade na formacão de uma Grundgestalt: a primeira das Quatro Cancões Op.2 de Berg. Per Musi, 27, 75–88 (2013)
7. Almada, C.: Derivacão tematica a partir da Grundgestalt da Sonata para Piano op.1, de Alban Berg. In: II Encontro Internacional de Teoria e Analise Musical. São Paulo: UNESP-USP-UNICAMP (2011)
8. Almada, C.: A variacão progressiva aplicada na geracão de ideias tematicas. In: II Simposio Internacional de Musicologia. Rio de Janeiro: UFRJ, pp. 79–90 (2011)
9. von Bertanlanffy, L.: General Systems Theory: Foundations, Development Applications. George Braziller, New York (1976)
10. Hofstadter, D.R.: Gödel, Escher, Bach: An Eternal Golden Braid. Basic Books, New York (1999)
11. Prusinkiewsky, P., Lindenmayer, A.: The Algorithmic Beauty of Plants. Springer, New York (1990)
12. Dawkins, R.: The Selfish Gene. Oxford University Press, Oxford (2006)
13. Dawkins, R.: The Blind Watchmaker. Penguin Books, London (1991)
14. Almada, C.: Gödel-vector and Gödel-address as tools for genealogical determination of genetically-produced musical variants. In: International Congress on Music and Mathematics. Puerto Vallarta: University of Guadalajara (Mexico 2014, in press)

Describing Global Musical Structures by Integer Programming on Musical Patterns

Tsubasa Tanaka[1]($\boxtimes$) and Koichi Fujii[2]

[1] Institut de Recherche et Coordination Acoustique/Musique, Paris, France
tsubasa.tanaka@ircam.fr
[2] NTT DATA Mathematical Systems Inc., Tokyo, Japan
fujii@msi.co.jp

Abstract. Music can be regarded as sequences of localized patterns, such as chords, rhythmic patterns, and melodic patterns. In the study of music generation, how to generate sequences that are musically adequate is an important issue. In particular, generating sequences by controlling the relationships between local patterns and global structures is a difficult and open problem. Whereas grammatical approaches, which examine global structures, can be used to analyze how a piece is constructed, they are not necessarily designed to generate new pieces by controlling the characteristics of global structures, such as the redundancy of a sequence or the statistical distribution of specific patterns. To achieve this, we must overcome the difficulty of solving computationally complex problems. To deal with this problem, we take an integer-programming-based approach and show that some important characteristics of global structures can be described only by linear equalities and inequalities, which are suitable for integer programming.

Keywords: Musical patterns · Global structure · Hierarchy · Redundancy · Integer programming

1 Introduction

Music can be regarded as sequences of localized musical patterns such as chords, rhythmic patterns, and melodic patterns. In the study of music generation, it is important to know the characteristics of these sequences. For example, a Markov model is used to learn transition probabilities of localized musical elements in existing pieces or real-time performances [1], and pieces that imitate the original styles are expected to be generated from the model.

However, learning local characteristics is not sufficient to understand or generate music. Global musical structures or musical forms are necessary to be considered. Contrary to Markov models, grammatical approaches such as generative theory of tonal music [2] (GTTM, hereafter), are used to analyze the global structures of musical pieces. In GTTM, a musical piece is abstracted step by step by discarding less important elements and the whole piece is understood as a hierarchical tree structure. For example, Hamanaka et al. have implemented

© Springer International Publishing Switzerland 2015
T. Collins et al. (Eds.): MCM 2015, LNAI 9110, pp. 52–63, 2015.
DOI: 10.1007/978-3-319-20603-5_5

GTTM on a computer and analyzed musical pieces [3]. However, this model does not have a strategy for composing new pieces. We should note that the grammatical models are not designed to generate new pieces with specifying their characteristics of global structures such as the redundancy of the sequence and the statistical distribution of specific patterns that actually appear in the generated pieces.

Therefore, the objective of this paper is to propose a model that can give specifications of characteristics of global musical structures in the context of music generation. We expect the proposed model to be applied to help users generate new pieces of music and/or obtain desired musical structures.

Compared to the problem of analysis, the problem of generation has a difficulty of a combinatorial explosion of possibilities. When we compose a new piece or sequence, we have to choose a sequence from all the possible combinations of patterns, whose number increases exponentially depending on the number of basic patterns. To deal with such combinatorial problems, we propose an integer-programming-based approach. In order to apply the integer programming, the structural specifications have to be described by linear equalities and inequalities. This is the main challenge for our study.

Integer programming is a framework to solve linear programming problems whose variables are restricted to integer variables or 0–1 variables [4]. Although it shares something in common with constraint programming, which has been often used in the realm of music, one of the advantages of integer programming is that it has an efficient algorithm to find the optimum solution by updating the estimations of the lower and upper bounds of the optimum solution based on the linear programming relaxation technique. Integer programming has been applied to various problems. In our previous study [5], musical motif analysis was formulated as a set partitioning problem, which is a well-known integer programming problem. Thanks to the recent improvements of integer programming solvers such as Numerical Optimizer [6], more and more practical problems have been solved within a reasonable time. We expect that integer programming may also play an important role in the generation of music.

Other possible generation methods than integer programming and constraint programming include the use of metaheuristics. In the study of solving counterpoint automatically [7], counter melodies were generated based on the local search whose objective function is defined as how well the generated counter melody satisfies the rules of counterpoint. In such a search algorithm, there is no guarantee to find the optimum solution. Especially, in the case where there are conflicts between many rules, some rules might be violated. This is not preferable for us because we think that the structural specification should be strictly respected. Therefore, we take an integer-programming approach, which we think is suitable to find strict solutions for the discrete optimization problems.

There are several limitations about what we can describe in this paper. Three main limitations are as follows: (1) we do not describe the objective function and only focus on the constraints to be satisfied strictly (however, we are planning to introduce the objective function in the future work). (2) We do not propose a comprehensive formulation that covers various situations. We would rather explain our point of view based on a concrete examples of typical musical structure. (3) We only focus on the aspect of formulation and do not proceed to the steps of generation and evaluation.

The rest of this paper is organized as follows: In Sect. 2, we show the relationship between localized musical patterns and hierarchical global structure referring to the chord progression of a piece that has typical phrase structures. In Sect. 3, we describe how to implement this relationship by linear equalities and inequalities. In Sects. 4 and 5, we describe the similar relationships for the structures of rhythmic pattern sequence and intervallic pattern sequence by linear constraints. Then, these are combined to formulate the constraints for generating a melody. Section 6 gives some concluding remarks.

2   Hierarchy and Redundancy

From the micro-level point of view, composing a piece of music can be regarded as determining how the sequences of localized musical patterns are arranged. If such sequences are seen from the macro-level point of view, they can be regarded as the musical forms.

Then, a question arises: what are the global characteristics that a musical sequence should have to construct a global musical form and to generate well-organized music? In this section, we study the piece Op. 101 No. 74 by Ferdinand Beyer to consider this problem. The chord sequence of this piece is clearly related to the phrase structures and the musical form, and it will provide a good clue for this problem.

The chord sequence of this piece is TTSTTTDTSTSTTTDTDTDTDTDT, using T (tonic), S (subdominant), and D (dominant). Each chord corresponds to one measure. To see the phrase structures more clearly, let's combine each two bars, four bars, and eight bars. Then these four levels can be represented as the following sequences:

- 1st level: $A_1A_1A_2A_1A_1A_1A_3A_1A_2A_1A_2A_1A_1A_1A_3A_1A_3A_1A_3A_1A_3A_1A_3A_1$ (3/24)
- 2nd level: $B_1B_2B_1B_3B_2B_2B_1B_3B_3B_3B_3B_3$ (3/12)
- 3rd level: $C_1C_2C_3C_2C_4C_4$ (4/6)
- 4th level: $D_1D_2D_3$ (3/3)

Here, the same indices indicate the same patterns. For example, $A_1 = $ T, $A_2 = $ S and $A_3 = $ D. There are relationships between the consecutive levels such as $B_1 = A_1A_1$, $C_1 = B_1B_2$, $D_1 = C_1C_2$, and so on (The notations like "A_1A_1" represent the concatenations of the patterns and do not indicate multiplications). The numerators of the fractions indicated between parentheses after each sequence indicate the numbers of the variety of patterns that exist in the sequences in each level, and the denominators indicate the lengths of each level. We call the inverse numbers of these fractions redundancies[1]. Figure 1 visualizes this hierarchical

---

[1] The study [8] focuses on the redundancy of musical sequence, and models the musical style by referring to the Lempel-Ziv compression algorithm. We extend this perspective and pay attention to the redundancies of multiple levels simultaneously.

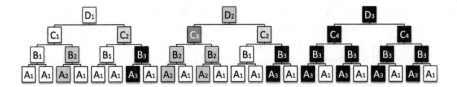

Fig. 1. Hierarchical structure of chord sequence of Beyer No. 74.

structure. Because the fourth level has three different elements, we see that this piece consists of three different parts. When observing the relationship between the third level and fourth level, we find that the latter parts of D_1 and D_2 are the same and D_3 consists of a repetition of C_4.

We can observe that there are many repetitions of the same patterns and that the lower the level, the higher the redundancy. In addition, T is more than twice as frequent as D, and D is twice as frequent as S. These biases of frequency are also notable characteristics of this sequence. Thus, the global structure of this piece can be regarded to be, at least, constrained by hierarchy, redundancies, and frequencies of patterns. Without repetitions in lower levels, the music will become unmemorable and lose the attention of human listeners. Moreover, repetitions in higher levels are related to known musical forms such as A-B-A ternary form or rondo form. Thus, frequency and redundancy are important fearures for describing musical structures.

Focusing on each level, state transition diagrams can be depicted per level (Fig. 2). From the global point of view, state transitions of multiple levels should be considered simultaneously. In these diagrams, the lowest level (first level) represents the rules of chord progression proper to this piece. Although the transition from S to D is possible in ordinary rules of harmony, this transition is not used in this piece.

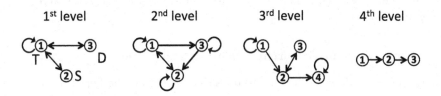

Fig. 2. Transition diagram of chord patterns in each level.

3   Implementation of Constraints on Chord Progression

It is difficult to find new sequences that have specific redundancies in respective levels, because the constraints for different levels may cause conflicts during the process of searching for the solutions. Also, the naive method of enumerating possible sequences one by one and checking whether or not they satisfy specific redundancies in each level won't be realistic because the number of possible sequences increases exponentially with the length of the sequence based on the number of the states in the lowest level.

To deal with this problem, we take an integer-programming-based approach, which has good search algorithms to practically solve computationally complex problems. In order to use integer programming, the characteristics of the sequence should be expressed by linear constraints (equalities and inequalities). In this section, we demonstrate that some important characteristics of the sequence observed in the Beyer's piece can be expressed by linear constraints.[2]

Let A, B, C, D be the set of patterns that can appear in respective levels. The number of elements in each level is denoted by K_1, K_2, K_3, and K_4. For explanation, we set these possible patterns as the patterns that actually appear in the Beyer's piece (i.e., $A = \{A_1, A_2, \cdots A_3\}$, $B = \{B_1, B_2, \cdots, B_3\}$, $\cdots$.).[3] Let a_t be a variable that corresponds to the t-th element in the sequence of variables for the patterns of the first level, b_t for those of the second level, c_t for those of the third level, and d_t for those of the fourth level. For example, in the original piece, $a_1 = A_1$, $a_2 = A_1$, $a_3 = A_2$, $a_4 = A_1$, $\cdots$, $a_{24} = A_1$, $b_1 = B_1$, $b_2 = B_2$, $b_3 = B_1$, $b_4 = B_2$, $\cdots$, $b_{12} = B_3$, and so on.

The followings are important characteristics of the original sequence explained in Sect. 2:

1. Hierarchy of phrase structures (relationships between neighboring levels): $b_t = a_{2t-1}a_{2t}(1 \leq t \leq 12)$, $c_t = b_{2t-1}b_{2t}(1 \leq t \leq 6)$, $d_t = c_{2t-1}c_{2t}(1 \leq t \leq 3)$.
2. Variety of patterns that appear in each level: the first level has three patterns, the second has three patterns, the third has four patterns, and the fourth has three patterns.
3. Frequency of each state in each level (in the first level, these are the frequency of each chord).
4. State transition rules (in the first level, these are the chord progression rules).

In this section, we show how to formalize these rules by variables and linear constraints on the variables.

3.1   Constraints of Hierarchical Phrase Structures

Let $x_{t,i}$, $y_{t,i}$, $z_{t,i}$, and $w_{t,i}$ be 0-1 variables that represent whether or not the t-th element of the sequence for each level is the pattern i (For example, the pattern

---

[2] In this paper, in order to avoid the explanation from being complicated, we only treat the stereotype examples of musical pieces whose groupings are always combinations of two consecutive elements in every level. However, in practice, such structures should vary depending on the specifications of the pieces that the user wants to create. For example, we can think of the case where the number of combination in the groupings are different between the levels. We can also think of the case where the consecutive patterns can be overlapped as is mentioned in [9]. How to formulate such cases is an important future issue.

[3] In practice, it is not necessary to stick to existing pieces.

i indicates A_i, in the case of the first level). $x_{t,i}$, $y_{t,i}$, $z_{t,i}$, and $w_{t,i}$ correspond to the first, second, third, and fourth levels respectively. For example, $x_{t,i} = 1$ if $a_t = A_i$, and otherwise $x_{t,i} = 0$. $y_{t,i} = 1$ if $b_t = B_i$ and otherwise $y_{t,i} = 0$. $z_{t,i}$ and $w_{t,i}$ are also defined similarly. a_t, b_t, c_t, and d_t take one of the values in A, B, C, and D, respectively. The numbers of elements of A, B, C, and D are denoted by K_1, K_2, K_3, and K_4, respectively. Because a_t, b_t, c_t, and d_t take only one value at a location t, respectively (here, location t means t-th element in the sequence of each level), they satisfy the following constraints:

$$\sum_{i=1}^{K_1} x_{t,i} = 1 \ (\forall t, 1 \leq t \leq 24), \quad \sum_{i=1}^{K_2} y_{t,i} = 1 \ (\forall t, 1 \leq t \leq 12),$$

$$\sum_{i=1}^{K_3} z_{t,i} = 1 \ (\forall t, 1 \leq t \leq 6), \quad \sum_{i=1}^{K_4} w_{t,i} = 1 \ (\forall t, 1 \leq t \leq 3). \tag{1}$$

Now the hierarchy can be expressed by constraints on $x_{t,i}$, $y_{t,i}$, $z_{t,i}$, and $w_{t,i}$. For example, if $B_i = A_{j_1} A_{j_2}$, the statement "$b_t = B_i$ (i.e. $y_{t,i} = 1$)" must be equivalent to "$a_{2t-1} = A_{j_1}$ and $a_{2t} = A_{j_2}$ (i.e., $x_{2t-1,j_1} = 1$ and $x_{2t,j_2} = 1$)", and this equivalence can be expressed by the three constraints: $y_{t,i} \leq x_{2t-1,j_1}$, $y_{t,i} \leq x_{2t,j_2}$, and $x_{2t-1,j_1} + x_{2t,j_2} - 1 \leq y_{t,i}$. Therefore, the constraints that correspond to the whole hierarchy are described as follows:

for all (i, j_1, j_2) that satisfy $B_i = A_{j_1} A_{j_2} (1 \leq i \leq K_1)$,

$$y_{t,i} \leq x_{2t-1,j_1}, \ y_{t,i} \leq x_{2t,j_2}, \ x_{2t-1,j_1} + x_{2t,j_2} - 1 \leq y_{t,i} \ (1 \leq t \leq 12), \tag{2}$$

for all (i, j_1, j_2) that satisfy $C_i = B_{j_1} B_{j_2} (1 \leq i \leq K_2)$,

$$z_{t,i} \leq y_{2t-1,j_1}, \ z_{t,i} \leq t_{2t,j_2}, \ y_{2t-1,j_1} + y_{2t,j_2} - 1 \leq z_{t,i} \ (1 \leq t \leq 6), \tag{3}$$

for all (i, j_1, j_2) that satisfy $D_i = C_{j_1} C_{j_2} (1 \leq i \leq K_3)$,

$$w_{t,i} \leq z_{2t-1,j_1}, \ w_{t,i} \leq z_{2t,j_2}, \ z_{2t-1,j_1} + z_{2t,j_2} - 1 \leq w_{t,i} \ (1 \leq t \leq 3). \tag{4}$$

Finally, the statement that every element in the last level is different can be expressed by the following constraints:

$$w_{1,i} + w_{2,i} + w_{3,i} \leq 1 \ (\forall i, 1 \leq i \leq K_4). \tag{5}$$

3.2   Constraints on Frequencies of Each Patterns

The constraints to control the frequency (or the range of frequency) of each state can be described by the following inequalities using 0–1 variables α_i, β_i, γ_i, and δ_i for each level:

$$L_{1,i}\alpha_i \leq \sum_{t=1}^{24} x_{t,i} \leq H_{1,i}\alpha_i \ (\forall i, 1 \leq i \leq K_1), \tag{6}$$

$$L_{2,i}\beta_i \leq \sum_{t=1}^{12} y_{t,i} \leq H_{2,i}\beta_i \ (\forall i, 1 \leq i \leq K_2), \tag{7}$$

$$L_{3,i}\gamma_i \leq \sum_{t=1}^{6} z_{t,i} \leq H_{3,i}\gamma_i \ (\forall i, 1 \leq i \leq K_3), \tag{8}$$

$$L_{4,i}\delta_i \leq \sum_{t=1}^{3} w_{t,i} \leq H_{4,i}\delta_i \ (\forall i, 1 \leq i \leq K_4), \tag{9}$$

where $L_{1,i}$, $L_{2,i}$, $L_{3,i}$, $L_{4,i}$, $H_{1,i}$, $H_{2,i}$, $H_{3,i}$, and $H_{4,i}$ are the constants[4] for lower and upper bounds in the case that the i-th element of each level appears. The statement "$\alpha_i = 0$" is equivalent to "$x_{t,i} = 0$ for all $t(1 \leq t \leq 24)$". This means that α_i represents whether or not the state a_i appears in the sequence. β_i, γ_i, and δ_i also have such meanings in their own levels.

Also, relative differences of frequencies that T is more than twice or twice as frequent as D and D is twice as frequent as S can be expressed by the following constraints:

$$\sum_{t=1}^{24} x_{t,1} \geq 2 \sum_{t=1}^{24} x_{t,3}, \quad \sum_{t=1}^{24} x_{t,3} = 2 \sum_{t=1}^{24} x_{t,2} \tag{10}$$

3.3 Constraints on Varieties of Patterns

Using the variables α_i, β_i, γ_i, and δ_i, the number of variety of patterns in each level can be described. For example, $\sum_i \alpha_i$ indicates the number of variety of patterns in the first level. Similarly, the statement that the numbers of variety of patterns in respective levels are 3, 3, 4, and 3 can be represented as:

$$\sum_{i=1}^{K_1} \alpha_i = 3, \quad \sum_{i=1}^{K_2} \beta_i = 3, \quad \sum_{i=1}^{K_3} \gamma_i = 4, \quad \sum_{i=1}^{K_4} \delta_i = 3. \tag{11}$$

3.4 Constraints on State Transitions

The possibility of state transitions in the first level can be controlled by posing the following inequality for all of the combinations of (t, i, j) whose transition from i to j at the location t is prohibited:

$$x_{t,i} + x_{t+1,j} \leq 1. \tag{12}$$

The same is true in other levels. One way to determine the allowed transitions is to prohibit the transitions that do not occur in the original piece.

Thus, linear constraints can describe both global structures and local state transitions.

---

[4] For example, they can be set depending on user's preference or statistics of the original piece. If $L_{j,i} \leq 1$ and $H_{j,i}$ is larger than or equal to the length of the sequence, these equations give no limitation to the number of each pattern that appears in the sequence.

4   Constraints on Rhythmic Patterns

In this section, the hierarchical structure of rhythmic patterns is illustrated in a similar way to the hierarchical structure of chords. We use a Beethoven's famous melody *Ode to Joy* as a model example (Fig. 3).

Fig. 3. Beethoven's melody *Ode to Joy.*

Adopting a beat as a unit length, the different elements of the first level are $A_1 = [f]$, $A_2 = [f\sim]$, $A_3 = [\sim e, e]$, $A_4 = [\sim f]$, and $A_5 = [e, e]$, where e, f, and s represent 8th note, 4th note, 2nd note, respectively, and "." and "$\sim$" represent a "dot" and a "tie," respectively. Each A_i represents one beat and does not necessarily correspond to the actual notations (the "ties" and "dots" actually correspond to the ties or prolongations of a note beyond the beat). The second level consists of $B_1 = [f, f]$, $B_2 = [f., e]$, $B_3 = [s]$, $B_4 = [f, e, e]$, $B_5 = [f, f\sim]$, and $B_6 = [\sim f, f]$. The entire hierarchy is as follows:

- a_t (1st level): $||A_1A_1A_1A_1|A_1A_1A_1A_1|A_1A_1A_1A_1|A_2A_3A_2A_4|| \cdots$ (5/64)
- b_t (2nd level): $||B_1B_1|B_1B_1|B_1B_1|B_2B_3|| \cdots$ (6/32)
- c_t (3rd level): $||C_1|C_1|C_1|C_2||C_1|C_1|C_1|C_2||C_1|C_3|C_3|C_4||C_5|C_1|c_1|c_2||$ (5/16)
- d_t (4th level): $||D_1D_2||D_1D_2||D_3D_4||D_5D_1||$ (5/8)
- e_t (5th level): $||E_1||E_1||E_2||E_3||$ (3/4)
- f_t (6th level): $||F_1 F_2||$ (2/2)

where, "|" is a bar line and "||" is a four-bar partition. This hierarchy can be described by linear constraints in a similar way to the previous section. However, this case has more levels than the previous case. In general, the number of possible patterns increases drastically when looking at a higher level. If the number of levels is too large, the number of variables for the level may become too large.

Therefore, we introduce an alternative way to describe the constraints for redundancies of higher levels using the variables for the first level $x_{t,i}(1 \le i \le K_1)$ (without using $y_{t,i}$, $z_{t,i}$, and $w_{t,i}$).[5]

---

[5] However, at the current moment, we do not know an alternative way to implement the constraints for state transitions and frequencies of each pattern in the high levels.

4.1 Alternative Way to Control Redundancies of High Levels

Let us consider a 0–1 variable $s_{n,t}$ whose value is 1 if and only if an element in a level n first appear at the location t. Then, $\sum_t s_{n,t}$ represents how many different elements appear in level n. This offers an alternative way to specify the redundancies. Our purpose here is to construct the variables $s_{n,t}$ that have such meaning. In order to do so, we introduce other subsidiary variables q_{n,t_1,t_2} and $r_{n,i,u1,u2}$.

q_{n,t_1,t_2} is a 0–1 variable that represents whether the sections t_1 and t_2 ($t_1 < t_2$) in level n are the same or not (we call each content of the time span of an element such as $b_t, c_t, \cdots$ a section). $q_{n,t_1,t_2} = 1$ means that all of corresponding elements a_{u_1} and a_{u_2} in the sections t_1 and t_2 in level n are the same. Therefore, it is necessary to introduce the constraints that make the statement "$q_{n,t_1,t_2} = 1 \iff x_{u_1,i} = x_{u_2,i}$ ($\forall(i, u_1, u_2)$ s.t. $i \in [1, K_n]$, $(u_1, u_2) \in D(n, t_1, t_2)$)" true, where $D(n, t_1, t_2)$ represents the range of the combinations (u_1, u_2) where a_{u_1} and a_{u_2} are all of the pairs of corresponding elements in the sections t_1 and t_2 in level n. This statement can be replaced by the statement "$r_{n,i,u_1,u_2} = 1$ ($\forall(i, u_1, u_2)$ s.t. $i \in [1, K_n]$, $(u_1, u_2) \in D(n, t_1, t_2)$) $\iff q_{n,t_1,t_2} = 1$," where the 0–1 variable r_{n,i,u_1,u_2} means whether $x_{u_1,i} = x_{u_2,i}$ or not.

Here, the statement "$r_{n,i,u_1,u_2} = 1 \iff x_{u_1,i} = x_{u_2,i}$," can be expressed by the following constraints:

$$r_{n,i,u_1,u_2} \leq 1 + x_{u_1,i} - x_{u_2,i}, \tag{13}$$

$$r_{n,i,u_1,u_2} \leq 1 - x_{u_1,i} + x_{u_2,i}, \tag{14}$$

$$1 - x_{u_1,i} - x_{u_2,i} \leq r_{n,i,u_1,u_2}, \tag{15}$$

$$-1 + x_{u_1,i} + x_{u_2,i} \leq r_{n,i,u_1,u_2}. \tag{16}$$

Also, the statement "$r_{n,i,u_1,u_2} = 1$ ($\forall(i, u_1, u_2)$ s.t. $i \in [1, K_n]$, $(u_1, u_2) \in D(n, t_1, t_2)$) $\iff q_{n,t_1,t_2} = 1$," can be expressed by the following constraints:

$$1 - \sum_{i \in [1, K_n]} \sum_{(u_1, u_2) \in D(n, t_1, t_2)} (1 - r_{n,i,u1,u2}) \leq q_{n,t_1,t_2}, \tag{17}$$

$$q_{n,t_1,t_2} \leq r_{n,i,u_1,u_2} \quad (\forall(i, u_1, u_2) \text{ s.t. } i \in K_n, (u1, u2) \in D(n, t_1, t_2)). \tag{18}$$

Let $v_{n,t}$ be $\sum_{t_1 < t} q_{n,t_1,t}$, then $v_{n,t}$ represents how many the same sections as t there are before the section t in level n. $v_{n,t} = 0$ means that the content of section t first appears. Therefore, the statement "$v_{n,t} = 0 \iff s_{n,t} = 1$" must be true to let $s_{n,t}$ have the proper meaning. This statement can be expressed by a constraint:

$$1 - s_{n,t} \leq v_{n,t} \leq M \cdot (1 - s_{n,t}), \tag{19}$$

where M is a sufficiently large constant number. Under the constraints above, we can control the redundancy of each level n by a constraint:

$$\sum_t s_{n,t} = Constant. \tag{20}$$

5   Constraints on Melodic Patterns

In this section, hierarchical structure of the melodic patterns, which has information of pitches and rhythms, is illustrated using the same melody *Ode to Joy*. Although pitch information and rhythmic information in melodic patterns are not completely independent of each other, the hierarchical structure of pitch patterns and rhythmic patterns are different in general. Therefore, we treat these two hierarchical structures separately. After that, we consider compatibility between them and combine them. Considering that the same pitch patterns can occur in different transpositions on the scale, we treat pitch information as the series of intervals. In Sect. 5.1, we introduce constraints on structure of intervallic patterns. In Sect. 5.2, constraints for controlling pitch range is introduced. Then, in Sect. 5.3, we describe how to combine the rhythmic structure and the intervallic structure to generate melodies.

5.1   Constraints on Intervallic Patterns

There are nine different one-beat intervallic patterns that appear in the piece *Ode to Joy*. The intervals are denoted by the interval numbers on the scale - 1, based on the diatonic scale[6]. These are $[0]$, $[1]$, $[-1]$, $[\sim]$, $[-1,0]$, $[-2]$, $[1,-1]$, $[-4]$, $[5]$, which are denoted by $A_1 \sim A_9$, respectively. These are series of intervals that start from the interval between the first pitch and the second pitch of the beat to the interval between the last pitch to the first pitch of the next beat. $[\sim]$ means the interval 0 by a tie or a prolongation of a pitch to the next pattern. Though there is no beat after the last beat, the last beat obviously corresponds to the last beats of the bar 4 and bar 8. Therefore, we regard the last beat as the same intervallic pattern as the last beats of the bar 4 and bar 8, namely $[1]$. The sequences of each level is as follows:

- a_t (1st level): $||A_1 A_2 A_2 A_1 | A_3 A_3 A_3 A_3 | A_1 A_2 A_2 A_1 | A_4 A_5 A_4 A_1|| \cdots$ (9/64)
- b_t (2nd level): $||B_1 B_2 | B_3 B_3 | B_1 B_2 | B_4 B_5|| \cdots$ (10/32)
- c_t (3rd level): $||C_1|C_2|C_1|C_3||C_1|C_2|C_4|C_3||C_5|C_6|C_7|C_8||C_1|C_2|C_1|C_3||$ (8/16)
- d_t (4th level): $||D_1 D_2||D_1 D_3||D_4 D_5||D_1 D_2||$ (5/8)
- e_t (5th level): $||E_1||E_2||E_3||E_1||$ (3/4)
- f_t (6th level): $||F_1\ F_2||$ (2/2)

Comparing, for example, the sequence of the fifth level of the rhythmic patterns $||E_1||E_1||E_2||E_3||$ and the sequence of the intervallic patterns ($||E_1||E_2||E_3||E_1||$), we see that the first two elements are the same and the first and last elements are not the same in the former sequence. On the other hand, the first two elements are not the same and the first and the last elements are the same in the latter sequence. This difference adequately represents that although the first, second, and fourth 4-bars phrases are almost the same, the first and second have the same

---

[6] Here, we can also represent the pitches and the intervals based on the chromatic scale. However, we use the scale degrees and the interval numbers based on the diatonic scale because that is more efficient.

rhythms but slightly different intervals and the first and last 4-bars phrases have the same intervals but slightly different rhythms. This is the benefit of treating the rhythmic structure and the intarvallic structure separately.

5.2   Constraints for Pitch Range

The intervallic structures described in the previous section do not have the limitation of the pitch range. This problem occurs because the intervallic patterns are based on relative intervals instead of absolute pitches. Therefore, in this subsection, we introduce extra variables and constraints for bounding the pitch range.

Let p_t be the first pitch of the pattern at the location t of the sequence. The pitch is counted on the scale, where the starting pitch p_1 is set as 0. In the case of *Ode to Joy*, C $= -2$, D $= -1$, E $= 0$, etc. The scale of the piece is $\{-5, -4, -3, -2, -1, 0, 1, 2\}$. Let $Intvl(A_i)$ be the total interval of the intervallic pattern A_i (e.g., $Intvl(A_7) = 0$, because the total interval of $A_7(\, = [1, -1])$ is 0 $(= 1 - 1)$). Then, p_t is equal to the accumulation of the total intervals from the first pattern to the $(t-1)$-th pattern as in the following equation:

$$p_t = \sum_{t_1 < t} Intvl(a_{t_1}) = \sum_{t_1 < t} \sum_i Intvl(A_i) \cdot x_{t_1, i} \tag{21}$$

Let $Int(i, j)$ be the total interval from the beginning of A_i to the end of jth interval. If $a_t = A_i$, the pitch after the jth interval of A_i in a_t is $p_t + Int(i, j)$. If the next constraint:

$$LB_t \leq p_t + Int(i, j) x_{t,i} \leq UB_t \tag{22}$$

is satisfied for all (t, i, j), the upper bound UB_t and lower bound LB_t for pitches in t-th element of the sequence can be set.

5.3   Compatibility of Rhythmic Pattern and Intervallic Pattern

Although the sequences of rhythmic patterns and intervallic patterns have been independently introduced in the previous subsections, it is necessary to combine these two sequences to complete a melody, which contains both rhythms and intervals. To combine both of the sequences, compatibility between rhythmic patterns and intervallic patterns must be assured. For example, [s] and [−1,0] are not compatible because the number of elements differs ([s] indicates that there is only one note in this rhythmic pattern, and [−1,0] indicates that there are two notes in the intervallic pattern).

Therefore, we introduce constraints to assure that both sequences can coexist and propose a formulation to generate both sequences simultaneously. Let's discriminate variables and constants for intervallic patterns from those of rhythmic patterns by adding a dash on the shoulder of the variable names. The compatibility between both sequences depends on whether a_t and a'_t are compatible in every location t or not. Therefore, the compatibility can be expressed by the following inequality for all combination A_{i_1} and A'_{i_2} that are not compatible:

$$x_{t,i_1} + x'_{t,i_2} \leq 1. \tag{23}$$

If we find a solution for the problem that is a combination of the problems for rhythmic patterns and intervallic patterns with these inequalities, we will be able to obtain a complete melody.

6   Conclusion

In this paper, we proposed a formulation to generate sequences of musical patterns controlling some global structures of the sequences especially focusing on hierarchy and degree of redundancy in each level. We showed that such structures can be expressed only by linear constraints, which are necessary to apply integer programming. Future tasks include describing global structures more comprehensively, implementation of the constraints and actually generating new pieces, defining constraints on the relationships between melody and chords, and defining the constraints for polyphonic relationships.

Acknowledgments. This work was supported by JSPS Postdoctoral Fellowships for Research Abroad.

References

1. Pachet, F.: The continuator: musical interaction with style. J. New Music Res. **32**(3), 333–341 (2003)
2. Lerdahl, F., Jackendoff, R.: A Generative Theory of Tonal Music. MIT Press, Cambridge (1983)
3. Hamanaka, M., et al.: Implementing "A generative theory of tonal music". J. New Music Res. **35**(4), 249–277 (2006)
4. Nemhauser, G.L., Wolsey, L.A.: Integer and Combinatorial Optimization. Wiley, New York (1988)
5. Tanaka, T., Fujii, K.: Melodic pattern segmentation of polyphonic music as a set partitioning problem. In: Proceedings of International Congress on Music and Mathematics (to be published)
6. http://msi.co.jp/nuopt/
7. Herremans, D., Srensen, K.: A variable neighbourhood search algorithm to generate first species counterpoint musical scores. Working Paper, University of Antwerp Faculty of Applied Economics Operations Research Group ANT/OR (2011)
8. Lartillot, O., et al.: Automatic Modeling of Musical Style. 8èmes Journées d'Informatique Musicale **2001**, 113–119 (2001)
9. Mazzola, G., et al.: The Topos of Music: Geometric Logic of Concepts, Theory, and Performance. Birkhäuser, Basel (2002)

Improved Iterative Random Walk
for Four-Part Harmonization

Raymond Whorley[1]([envelope]) and Darrell Conklin[2,3]

[1] Independent Researcher, Southampton, UK
map01rw@alumni.gold.ac.uk
[2] Department of Computer Science and Artificial Intelligence,
University of the Basque Country UPV/EHU, San Sebastián, Spain
darrell.conklin@ehu.es
[3] IKERBASQUE, Basque Foundation for Science, University of the Basque Country,
Bilbao, Spain

Abstract. Music generated by random walk from a statistical model generally fails to capture the style and quality of music in its training corpus. In the context of melody harmonization, this paper first demonstrates that violations of general rules of harmony are clearly related to cross-entropy. The paper then describes and evaluates an improved random walk method which efficiently samples low cross-entropy harmonizations. Applying the method, the relationship between cross-entropy and harmonization quality becomes even more apparent. These results will impact on future work in music generation from statistical models.

Keywords: Harmonization · Random walk · Probability threshold · Cross-entropy · Multiple viewpoint systems · Statistical model

1 Introduction

Music generation can be viewed as a process of drawing samples from statistical models learned from a style [1, 2]. The most popular sampling method, known as *random walk*, can be applied to any type of context model, that is, a model which predicts continuations of event sequences based on their context. Since this paper deals with four-part harmonization, the events under consideration are chords.

Random walk, while able to provide plausible local sequence continuations, even in real time, rarely produces satisfactory complete chord sequences. One reason for this is that the sampling of a low probability chord often leads to subsequent low probability chords which break the coherence of the musical sequence. To address this problem, random walk can be used in an *iterative* fashion, restarted multiple times while noting the high probability solutions found so far; however this procedure tends to produce samples close to the mean of the sequence probability distribution. The further one wants to reach into the high probability tail of the distribution, the less likely it is that iterative random walk will produce such sequences within a practical number of samples. In this paper

© Springer International Publishing Switzerland 2015
T. Collins et al. (Eds.): MCM 2015, LNAI 9110, pp. 64–70, 2015.
DOI: 10.1007/978-3-319-20603-5_6

we propose and evaluate an iterative random walk procedure which maintains the benefits of random walk, but also functions as an optimization method which rapidly samples high probability sequences from a statistical model.

2   Methods

The statistical model used in this paper for melody harmonization comprises three *multiple viewpoint systems* (see [3] and [4]); that is, one for each of three basic attributes being generated in a chord: note duration, note continuation (allowing the modelling of passing notes, and so on) and note pitch. The attributes are generated for alto, tenor and bass together (chord by chord), given the soprano part. Multiple viewpoint systems produce, for a contextual sequence, a distribution X mapping all possible events to a probability. To generate a harmonization by random walk, a chord x with probability $P(X = x)$ is sampled, and the procedure repeated until the end of the melody is reached. The model is capable of assigning an overall probability to all possible harmonizations of a given melody. It is convenient to work with logarithms rather than probability values, however, so the *cross-entropy* of a piece is used: the negative mean log probability of events in the sequence (see e.g. [3] for more details).

The *probability threshold* method [4] is a modification of random walk, where predictions with a probability lower than some fraction of the highest probability event in the distribution are ignored during sampling. More precisely, consider the distribution X constructed for an event in a sequence. A new probability distribution X' is created by removing all events from X that fall below a specified fraction $t \in [0, 1]$ of the highest probability event in X:

$$P(X' = x) = \begin{cases} 0 & P(X = x)/m < t \\ P(X = x)/z & \text{otherwise} \end{cases} \tag{1}$$

where $m = \max_{x \in X} P(X = x)$, and z is a normalizing constant for the new random variable X'. Although sampling is performed using distribution X', probabilities in the original distribution X are used to compute sequence cross-entropy. Regarding the extrema of t: when $t = 0$, the probability threshold method behaves exactly as standard random walk, because no events are removed from the distribution X. When $t = 1$, only one solution is possible: the solution containing the highest probability continuation at each sequence position.

3   Results

3.1   Relationship Between Harmonic Correctness and Cross-Entropy

The cross-entropy of samples produced by random walk is much higher than that of the 100 hymn tune harmonizations [5] in our corpus; in musical terms, the samples contain many chord progressions and note lengths which are atypical of the corpus. Our hypothesis was, therefore, that in general the lower the cross-entropy, the better the solution. Since iterative random walk produces solutions

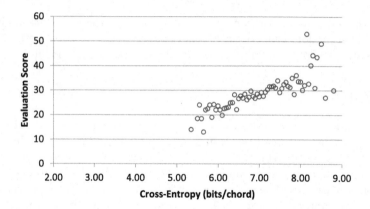

Fig. 1. Plot showing how evaluation score varies with cross-entropy (1024 samples, divided into bins of 0.05 bits/chord) for random walk. Note that lower scores mean better compliance with the rules of harmony checked, and that lower regions of the cross-entropy range are not visited by random walk.

with a wide range of cross-entropies, it was possible to objectively investigate this hypothesis. A set of 1024 harmonizations of the tune *Das walt' Gott Vater* ([5], hymn no. 36) were generated and computationally checked for violations of some general rules of harmony (as found in, e.g., [6]). We looked for part overlaps between adjacent chords; parallel fifths and octaves; and unusually large leaps within a part, with and without a note intervening (see [7] for a similar evaluation method). An evaluation score equal to the total number of rule violations was plotted against cross-entropy (see Fig. 1), leading to the conclusion that there is an objective improvement in the harmony as cross-entropy decreases.

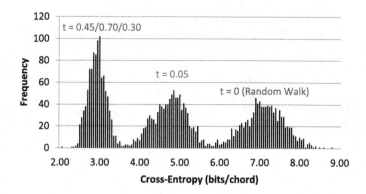

Fig. 2. Histogram (1024 samples per distribution divided into bins of 0.05 bits/chord) showing $t = 0.45/0.70/0.30$ (optimized for duration/continuation/pitch respectively), $t = 0.05$ and $t = 0$ (standard random walk) sample distributions.

3.2   Reduction and Minimization of Cross-Entropy

Having established a correlation between cross-entropy and harmonization quality for random walk, we wished to find out if the trend continued at cross-
entropies lower than the minimum found so far (5.33 bits/chord). To do this,
we developed a novel procedure for minimizing cross-entropy. Earlier work [4]
demonstrated the utility of probability thresholds in generating individual lower
cross-entropy solutions; indeed, $t = 1$ produces a single solution with a cross-
entropy of 2.68 bits/chord and an evaluation score of 3. This method is combined here with iterative random walk. Figure 2 shows that a t value of 0.05
dramatically reduces cross-entropy (cf. standard random walk), producing a minimum cross-entropy of 3.21 bits/chord. We have generated solutions with lower
cross-entropies than 2.68 bits/chord by employing t in the range 0.15 to 0.95;
but even lower low cross-entropy solutions can be found by optimizing three
different values of t for each musical attribute (i.e., duration, continuation and
pitch). Optimization was achieved by varying t for each attribute in turn, for the
generation of 1024 harmonizations of *Das walt' Gott Vater*, such that a minimal
cross-entropy sample was found (2.04 bits/chord). Figure 2 also shows the sample

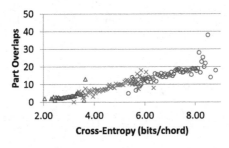

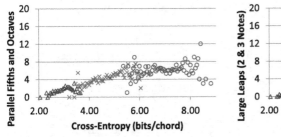

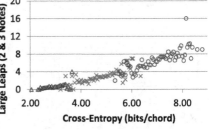

Fig. 3. Plots showing how the number of part overlaps, parallel fifths and octaves, and
large leaps (between adjacent notes and with a note intervening) varies with cross-
entropy (1024 samples, divided into bins of 0.05 bits/chord) for $t = 0.45/0.70/0.30$
(optimized for duration/continuation/pitch respectively), $t = 0.05$ and $t = 0$ (standard
random walk).

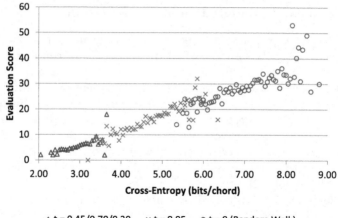

Fig. 4. Plot showing how evaluation score varies with cross-entropy (1024 samples, divided into bins of 0.05 bits/chord) for $t = 0.45/0.70/0.30$ (optimized for duration/continuation/pitch respectively), $t = 0.05$ and $t = 0$ (standard random walk).

distribution produced by optimized thresholds. The solution space is still very large, even with the more restrictive t values, indicating that diverse solutions are explored even after optimization.

Fig. 5. Musical scores of the lowest cross-entropy samples produced by $t = 0$ (standard random walk, top, 5.33 bits/chord) and $t = 0.45/0.70/0.30$ (optimized for duration/continuation/pitch respectively, bottom, 2.04 bits/chord).

To see if the objective improvement in harmony with decreasing cross-entropy continues down to the lower end of the cross-entropy scale, we subjected solutions obtained using optimized thresholds and $t = 0.05$ to the automatic rule violation checks outlined in Sect. 3.1. The results clearly show that the improvement in harmony continues all the way down to the lowest cross-entropies (see Figs. 3 and 4).

3.3 Example Harmonizations

Figure 5 shows musical scores of the lowest cross-entropy samples produced by random walk (5.33 bits/chord) and optimized thresholds (2.04 bits/chord). Note that although the minimal cross-entropy sample has by far the better harmony, it is not entirely without rule violations; for example, there are part overlaps in the third bar. It must also be pointed out that whereas this harmonization contains only three passing notes, J.S. Bach's original ([5], hymn no. 36) is more complex.

4   Conclusions and Future Work

We conclude that harmony generated from statistical models generally improves as cross-entropy decreases; that the application of iterative random walk in conjunction with probability thresholds is able to find very low cross-entropy solutions; and consequently that the solution space produced by the use of optimized thresholds is vastly superior to that produced by standard iterative random walk. This technique can be used to find low cross-entropy solutions to any sequence (not just music) within far fewer samples (and much less time) than would be required by standard iterative random walk. In future work, we shall endeavour to show that these results are more generally applicable, by consolidating and validating them with results from sampling involving more melodies.

Acknowledgments. This research is partially supported by the Lrn2Cre8 project which is funded by the Future and Emerging Technologies (FET) programme within the Seventh Framework Programme for Research of the European Commission, under FET grant number 610859. Special thanks to Kerstin Neubarth for valuable comments on the manuscript.

References

1. Conklin, D.: Music generation from statistical models. In: Proceedings of the AISB 2003 Symposium on Artificial Intelligence and Creativity in the Arts and Sciences, pp. 30–35. Aberystwyth, Wales (2003)
2. Herremans, D., Sörensen, K., Conklin, D.: Sampling the extrema from statistical models of music with variable neighbourhood search. In: Proceedings of the Sound and Music Computing Conference, pp. 1096–1103. Athens, Greece (2014)
3. Conklin, D., Witten, I.H.: Multiple viewpoint systems for music prediction. J. New Music Res. **24**(1), 51–73 (1995)

4. Whorley, R.P., Wiggins, G.A., Rhodes, C., Pearce, M.T.: Multiple viewpoint systems: time complexity and the construction of domains for complex musical viewpoints in the harmonization problem. J. New Music Res. **42**(3), 237–266 (2013)
5. Vaughan Williams, R. (ed.): The English Hymnal. Oxford University Press, London (1933)
6. Piston, W.: Harmony. Victor Gollancz, London (1976)
7. Suzuki, S., Kitahara, T.: Four-part harmonization using Bayesian networks: pros and cons of introducing chord nodes. J. New Music Res. **43**(3), 331–353 (2014)

Patterns

Location Constraints for Repetition-Based Segmentation of Melodies

Marcelo E. Rodríguez-López[(⊠)] and Anja Volk

Department of Information and Computing Sciences, Utrecht University,
Utrecht, The Netherlands
{m.e.rodriguezlopez,a.volk}@uu.nl

Abstract. Repetition-based modelling of melody segmentation relies on identifying and selecting repetitions of melodic fragments. At present, automatic repetition identification results in an overwhelmingly large number of repetitions, requiring the application of constraints for selecting relevant repetitions. This paper proposes constraints based on the locations of repetitions, extending existing approaches on constraints based on repetition length and frequency, and the temporal overlap between repetitions. To test our constraints, we incorporate them in a state-of-the-art repetition-based segmentation model. The original and constraint-extended versions of the model are used to segment 400 (symbolically encoded) folk melodies. Results show the constraint-extended version of the model achieves a statistically significant 14 % average improvement over the model's original version.

Keywords: Melody segmentation · Similarity matrix · Symbolic music processing

1 Introduction

Music segmentation is a fundamental listening ability, which can be described as "an auditory analog of the partitioning of the visual field into objects, parts of objects, and parts of objects" [6, p. 36]. Computational models of segmentation attempt to mimic this listening ability. Computational modelling of segmentation is important for fields like Music Information Research (for tasks such as automatic music archiving, retrieval, and visualisation), Computational Musicology (for automatic or human-assisted music analysis), and Music Cognition (to test segmentation theories).

In computational modelling of segmentation the task is most often to locate the time points separating contiguous segments, often called segment 'boundaries'. In this paper we focus on the computational modelling of *melody* segmentation, aiming to automatically detect the boundaries of melodic segments resembling music-theoretic *phrases*. That is, the boundaries of segments ranging roughly from 4–5 notes to 4–8 bars.

A factor often considered essential for human listeners to identify phrase boundaries in melodies is the (exact or approximate) repetition of fragments of

© Springer International Publishing Switzerland 2015
T. Collins et al. (Eds.): MCM 2015, LNAI 9110, pp. 73–84, 2015.
DOI: 10.1007/978-3-319-20603-5_7

the melody [1,3,5,6]. However, a known problem with automatic melody repetition identification is that the number of repetitions detected is generally much larger than the number of repetitions actually recognised by human listeners [8]. For segmentation modelling this issue is all the more acute, given that the number of repeated fragments relevant for boundary perception is likely to be even smaller [1,3,7]. Robust methods to select segmentation-determinative repetitions is thus crucial to the performance of repetition-based segmentation models. Repetition selection is most often modelled by enforcing constraints based on the frequency, length, and temporal overlap of/between detected repetitions [17]. In this paper we propose and quantify constraints based on the location of repetitions relative to (a) each other, (b) the whole melody, and (c) temporal gaps. To test our selection constraints, we incorporate them in a state-of-the-art repetition-based segmentation model initially proposed by Müller et al. in [12] (henceforth MUL). The original and constraint-extended versions of MUL are used to segment 400 (symbolically encoded) folk melodies. **Paper contribution:** Results show the constraint-extended version of MUL achieves a statistically significant 14 % average improvement over MUL's original version.

The rest of the document is organised as follows. In Sect. 2 we describe the MUL segmentation model. In Sect. 3 we introduce our location constraints and describe how they are integrated into MUL. In Sect. 4 we describe the experimental setting, present results, and discuss how location constraints affect the performance of MUL. Finally, in Sect. 5 we summarise conclusions and outline future work.

2   Description of the MUL Segmentation Model

The segmentation model used to test our constraints, MUL, is at the state-of-the-art of repetition-based melody segmentation [17]. MUL searches for the 'most representative' melody fragment and uses its repetitions to segment the melody. As shown in Fig. 1, MUL first computes a similarity matrix representation of the input melody, where repetitions can be visualised as diagonal or quasi diagonal stripes. Then, it uses an exhaustive stripe search technique to identify repetitions, and scores each repetition set according to the degree of similarity, frequency, and length of the repetitions contained in the set. Finally, it takes the highest scoring set of repetitions, and uses the start/end points of these repetitions as segment boundaries. Below we briefly describe the different processing stages of MUL (for a more detailed description we refer to [11]). In Sect. 2.1 we describe the similarity matrix construction, and in Sect. 2.2 we describe the repetition identification and selection stages.

2.1   Similarity Matrix Construction for Symbolic Data

In this paper MUL takes as input a symbolic representation of the melody, i.e. a sequence of temporally ordered symbols approximating note-like musical events. Each symbol in the sequence represents an attribute describing either a note's chromatic pitch or its quantized duration (or a combination of the two). Let then

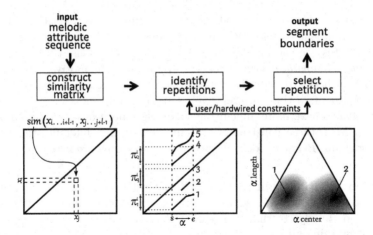

Fig. 1. Processing chain of the repetition-based segmentation model. Left: SM construction process. Middle: simplified SM depiction of a fragment α and five stripes, stripes $\{1, 3, 4\}$ constitute an 'accepted' set of repetitions $\mathcal{P}$. Right: scape plot representation of the space of fragments $\mathcal{A}^*$, shading depicts ϕ score (fitness), points 1 and 2 mark two fragments with high fitness.

$x = x_1 \ldots x_N$ be a sequence of melodic attribute symbols of length N, and let $x_{i\ldots j} = x_i \ldots x_j$ be a sub-sequence of x, with $i, j \in [1 : N]$. A similarity matrix SM of x corresponds to the matrix $\mathcal{S} = [s_{ij}]_{N \times N}$ of pairwise similarities between subsequences $s_{ij} = sim(x_{i\ldots i+l-1}, x_{j\ldots j+l-1})$, where l indicates the length of the subsequence and sim is a similarity measure. Figure 1 (left) depicts the construction process of an SM. In this paper we employ SMs that fulfil the normalisation properties $0 \leq \mathcal{S}(i,j) \leq 1$ for $i, j \in [1 : N]$, and $\mathcal{S}(i,i) = 1$ for $i \in [1 : N]$. In a SM repetitions are visualised as diagonal or quasi diagonal stripes. Figure 1 (middle) depicts a simplified SM, where five diagonal stripes mark potential repetitions of fragment α. The stripe structure of SMs computed from music data are often noisier than that shown in our simplified SM example. Thus, denoising and smoothing methods are commonly used to post-process SMs, aiming to enhance desired structural properties of the SM while suppressing unwanted ones [13].

The parameter settings to construct the SMs used in this paper (i.e. melodic representation, fragment length, similarity measure, denoising, and smoothing) are listed in Table 1, Sect. 4.3.

2.2 Constraint-Based Identification and Selection of Repetitions

In this section we first describe and motivate the constraints used within MUL to identify and select repetitions, and then describe how these constraints are employed to extract and score stripes from SMs.

Constraints to Identify and Select Repetitions: MUL uses constraints to define how similar two fragments need to be to be considered repetitions.

Also, it imposes constraints on attributes describing the repetitions detected for a given fragment to select which repetitions might be perceptually salient. These attributes are the number or *frequency* of repetitions, their *length*, and the amount of *temporal overlap* between repetitions. The motivation to impose constraints to these attributes is that human listeners seem to more easily notice repeated fragments if these are relatively long and frequent. Conversely, human listeners seem to have more difficulty recognising repetitions if these temporally overlap. In MUL these constraints are enforced during different stages of the processing chain. Below we describe when and how they are used.

Repetition Identification (Stripe Extraction): The goal is to identify and store repetitions for all fragments ranging in length from one event to all the events in the melody. To that end MUL defines the space of fragments $\mathcal{A}^*$ as a superset containing all sets $\mathcal{A} = \{\alpha_1, \alpha_2, \ldots \alpha_K\}$ of pairwise disjoint fragments $\alpha_h \cap \alpha_k = \varnothing$ for $h, k \in [1 : K]$ and $h \neq k$, where $\alpha = [b : e] \subseteq [1 : N]$ is a fragment of the melody. Repetitions of each melodic fragment are identified by extracting quasi diagonal stripes from $\mathcal{S}$ in the region encompassed by the fragment, e.g. Fig. 1 (middle SM) shows a fragment α and five stripes marking potential repetitions. If we take the tuple $(i_l, j_l) \in [1 : N]^2$, $l \in [1 : L]$ to denote a cell of $\mathcal{S}$, then a stripe of length L can be defined as any sequence $\pi = (i_l, j_l), \ldots, (i_L, j_L)$ forming a path within the region encompassed by fragment α. A path π has two projections $\pi^i = [i_l : i_L]$ and $\pi^j = [j_l : j_L]$. The constraints for a set of stripes $\mathcal{P} = \{\pi_1, \pi_2, \ldots \pi_Q\}$ to be a set of repetitions are:

1. stripe projections π^j must be of the same length as α (i.e. $j_1 = b$ and $j_L = e$),
2. stripes must be diagonal or quasi diagonal, for which user defined diagonal distortions are allowed (the default setting requires the slope of a stripe to lay within the bounds $1/2$ and 2), and
3. the set of stripe projections π^i must not temporally overlap.

In Fig. 1 (middle SM) we exemplify how MUL enforces these constraints. From the set of stripes $\{1, 2, 3, 4, 5\}$, the set of stripes complying with the criteria is $\{1, 3, 4\}$, since stripe 2 is unacceptably short, and stripe 5 is both unacceptably distorted and its π^i projection overlaps with that of stripe 4. Since a fragment can have more than one acceptable set of repetitions, MUL uses an optimisation procedure to search for the best possible set of repetitions. MUL defines the optimal set of repetitions $\mathcal{P}^o$ as that containing the most frequent and similar repetitions (identified using Eq. 1 below). The identification of repetitions and the search for the optimal set of repetitions is computed simultaneously, using a modification of the classic dynamic time warping algorithm.

Repetition Selection (Fitness Function): To select which fragment α to use for segmentation, MUL enforces constraints on the degree of similarity, length, and frequency of its associated set of repetitions $\mathcal{P}^o$. The main idea is to search for the 'most representative' fragment. MUL defines the most representative fragment as that which contains *the highest repeating and most similar set of repetitions*, which moreover *covers the largest portion of the melody*. To formalise this idea MUL employs two heuristic functions. The first is a *repetition* score function

$$\rho(\mathcal{P}) = \sum_{q=1}^{Q} \rho(\pi_q), \tag{1}$$

with $\rho(\pi) = \sum_{l=1}^{L} \mathcal{S}(i_l, j_l)$. The function $\rho(\mathcal{P})$ awards a high score to sets with highly similar and frequent repetitions. The second is a *coverage* score function

$$\kappa(\mathcal{P}) = \sum_{q=1}^{Q} |\pi_q|, \tag{2}$$

with $|\cdot|$ used to denote the length of π. The function $\kappa(\mathcal{P})$ awards a higher score to repetition sets that cover a large part of the melody. MUL uses normalised versions of $\rho(\cdot)$, $\kappa(\cdot)$. For brevity we omit a description of the normalisation procedures and refer to [11]. The normalised scoring functions (denoted by $\tilde{\rho}(\cdot)$, $\tilde{\kappa}(\cdot)$) are combined using a harmonic mean, i.e.

$$\phi(\alpha) = 2 \cdot \frac{\tilde{\rho}(\mathcal{P}^o) \cdot \tilde{\kappa}(\mathcal{P}^o)}{\tilde{\rho}(\mathcal{P}^o) + \tilde{\kappa}(\mathcal{P}^o)}, \tag{3}$$

MUL uses $\phi(\cdot)$ as a 'fitness' measure whose score represents a balance between having highly frequent/similar repetitions and covering large portions of the melody. The most representative fragment is that containing the repetition set of maximal fitness:

$$\alpha^m = \underset{\alpha}{\operatorname{argmax}} \, \phi(\alpha). \tag{4}$$

3   Location-Based Constraints for Repetition Selection

Just as MUL, most repetition-based models of melody segmentation use selection constraints based on repetition frequency, length, and temporal overlap [17]. Yet, empirical evidence [1,17,20] suggests that location constraints might also play an important role during repetition selection. Thus, we propose and quantify constraints based on the location of repetitions relative to (a) each other, (b) the whole melody, and (c) the location of temporal gaps. We formally define our location constraints below.

In respect to (a), we hypothesise that repetition sets in which instances are roughly evenly spaced (within the melody) are more salient than those that are not. We do so based on the observation that phrases tend to have a narrow distribution of possible phrase lengths [20]. Hence, if we assume that salient repetitions mark mainly the starting points of phrases [17], then the distribution of *inter-repetition-onset-intervals* (IROI) of salient repetitions should also be dominated by relative few and similar IROIs. We propose λ_1 (Eq. 5) as a scoring function that gives higher score repetition sets with low IROI dispersion.

$$\lambda_1(\mathcal{P}) = \begin{cases} \frac{1}{\sigma(\text{IROI})} & : \sigma(\text{IROI}) > 0 \\ 1 & : \sigma(\text{IROI}) = 0 \end{cases} \tag{5}$$

where $\text{IROI}(\pi_q) = \pi_{q+1} - \pi_q$, $\forall q = 1, \ldots, |\mathcal{P}| - 1$ and σ is the standard deviation.[1]

In respect to (b), we hypothesise that repetition sets in which the first instances occur earlier in the melody are more salient than those containing first instances appearing later in the melody. We do so based on the notion that melodic 'vocabulary' is mostly emergent, and so the earlier a 'vocabulary term' is introduced the more relevant [3,19]. To quantify this notion, we use λ_2 as a scoring function that prefers sets of repetitions with instances located both at the 'beginning' (I_b) and the 'rest' (I_r) of a melody.

$$\lambda_2(\mathcal{P}) = \sqrt{I_b \cdot I_r} \tag{6}$$

where $I_b = \frac{\mathcal{O} \cap \mathcal{B}}{|\mathcal{B}|}$ and $I_r = \frac{\mathcal{O} \cap \neg \mathcal{B}}{|\mathcal{O}|}$, $\mathcal{O}$ is the set of repetition onsets from $\mathcal{P}$, and $\mathcal{B}$ is the set of possible note locations at the beginning of the melody. We take $x_{1 \ldots \lfloor N/n \rfloor}$ to be the melody 'beginning', with n defined by the user (see settings in Table 1).[2]

In respect to (c), we hypothesise that repetition sets that better align to temporal gaps are more salient than those which do not. (Temporal gaps can be overly long note durations, musical rests, or a combination of the two.) The motivation for this hypothesis is based on the observation that temporal gaps often precede phrase starts [15,20], and repetitions often mark the starting points of phrases [4,7]. To quantify this notion, we use λ_3 as a scoring function that prefers sets of repetitions containing one or more instances starting right after temporal gaps.

$$\lambda_3(\mathcal{P}) = 2 \cdot \frac{T_p \cdot T_r}{T_p + T_r} \tag{7}$$

where $T_p = \frac{\mathcal{T} \cap \mathcal{O}}{|\mathcal{O}|}$ and $T_r = \frac{\mathcal{T} \cap \mathcal{O}}{|\mathcal{T}|}$, $\mathcal{O}$ is the set of repetition onsets from $\mathcal{P}$, and $\mathcal{T}$ is the set of temporal gap locations. To automatically obtain temporal gap locations, we use the temporal gap detection component of [2] (settings specified in Table 1). Each temporal gap location in $\mathcal{T}$ has been incremented on one note event to align with repetition onsets.

In our experiments we incorporate the arithmetic mean $\overline{\lambda}$ of the scores $\lambda_{1,2,3}$, in the fitness measure (Eq. 3) which results in

$$\phi(\alpha) = 3 \cdot \frac{\tilde{\rho}(\mathcal{P}^o) \cdot \tilde{\kappa}(\mathcal{P}^o) \cdot \overline{\lambda}(\mathcal{P}^o)}{\tilde{\rho}(\mathcal{P}^o) + \tilde{\kappa}(\mathcal{P}^o) + \overline{\lambda}(\mathcal{P}^o)} \tag{8}$$

To select a meaningful set of repetitions from the ϕ-space the same criterion used in Eq. 4 is employed, namely the most representative fragment is that containing the repetition set of maximal fitness.

---

[1] Since $\sigma \in \mathbb{R}_{\geq 0}$, normalisation of the λ_1 values is required.

[2] While theoretically $\lambda_2 \in [0,1]$, considering $\lim_{\mathcal{O} \to N} \lambda_2(\mathcal{O}) = 1$, in practice the values of λ_2 will never reach the maximum of the function's range, and so re-scaling is required.

4   Testing Variants of MUL

In this section we describe the test database and evaluation metrics, list experimental parameter settings, and present the results obtained in our experiments. For our experiments we use the implementation of MUL provided in the SM toolbox [10] for Matlab. We coded additional functions that compute SMs from symbolic data and implement the constraints described in Sect. 3.

4.1   Experimental Setting: Test Dataset

To test the ability of MUL and its extended versions to locate melodic phrase boundaries, we used a set of 200 instrumental folk songs randomly sampled from the Liederenbank collection[3] (LC) and 200 vocal folk songs randomly sampled from the German subset of the Essen Folk Song Collection[4] (EFSC). We chose to use the EFSC due to its benchmark status in the field of melodic segmentation. Additionally, we chose to use the LC to test generalisation to non-vocal melodies.[5,6]

The EFSC consists of ~6000 songs, mostly of German origin. The EFSC data was compiled and encoded from notated sources. The songs are available in EsAC and **kern formats. The origin of phrase boundary markings in the EFSC has not been explicitly documented (yet it is commonly assumed markings coincide with breath marks or phrase boundaries in the lyrics of the songs).

The instrumental (mainly fiddle) subset of the LC consists of ~2500 songs. The songs were compiled and encoded from notated sources. The songs are available in MIDI and **kern formats. Segment boundary markings for this subset comprise two levels: 'hard' and 'soft'. Hard (section) boundary markings correspond with structural marks found in the notated sources. Soft (phrase) boundary markings where annotated by two experts.[7] For our experiments we use the soft boundary markings.

---

[3] http://www.liederenbank.nl.

[4] http://www.esac-data.org.

[5] Vocal music has dominated previous evaluations of melodic segmentation (especially large-scale evaluations), which might give an incomplete picture of the overall performance and generalisation computational segmentation models.

[6] The samples are taken randomly from the EFSC and LC. However, following the corpus cleaning procedures of [18], we filtered out melodies which contained rests at annotated phrase markings, and also excluded melodies with just one phrase. The reason to exclude melodies with rests at annotated phrase markings is that, according to transcription research, sometimes musicologists transcribing the folk melodies would use rests at phrases as 'breath marks', regardless of whether performers would actually take breaths or not, making these rests an artefact of the transcription process (for a more detailed discussion on this topic see [18]).

[7] Instructions to annotate boundaries were related to performance practice (e.g. "where would you change the movement of the bow"). The annotators agreed on a single segmentation, so no inter-annotator-agreement analysis is possible.

Table 1. MUL parameter settings.

Parameters		Setting used for experimentation
SM construction		
Melodic fragment length	fl	fragment length of 4 notes
Similarity measures	sm	cosine similarity
Melody representation	mr	chromatic pitch interval, duration ratios
Matrix blending	smb	geometric mean, $w_{p,d} = 0.5$
Repetition identification/selection		
Allowed stripe distortion (step size)	sts	default=$\{(1,2), (2,1), (1,1)\}$
Minimum repetition length	mnl	minimum = 5 notes
Predict boundary	sb	starting points of selected repetitions
Setting for λ_3	$p\lambda_3$	temporal gap component of [2]; $k = 0.4$
Setting for λ_2	$p\lambda_2$	$n = 4$

4.2 Experimental Setting: Evaluation Measures

We use the same evaluation measures as used in previous comparative studies of melodic segmentation [14,17], i.e. the well known $F1 = \frac{2 \cdot p \cdot r}{p+r}$, with precision $p = \frac{tp}{tp+fp}$ and recall $r = \frac{tp}{tp+fn}$, with tp, fp, and fn corresponding to the number of true positives (hits), false positives (insertions), and false negatives (misses).

4.3 Experimental Setting: Parameters

In Table 1 we specify SMs construction parameters and repetition identification/selection parameters used for experimentation. The choice of parameters is the result of previous experimentation with MUL conducted in [17].

In [17] we showed that using either thresholding or smoothing methods is detrimental to the performance of MUL.[8] Hence, in our experiments we use clean, non post-processed SMs. Moreover, in [17] we also tested eight melody representation schemes and eight similarity measures, yet no combination of representation scheme and similarity more resulted in statistically significant improvements over other combinations. Hence, we opt for a commonly used representation scheme: chromatic pitch interval and inter-onset-interval-ratios, and measure similarity using the widely employed cosine similarity. We combine the pitch and duration representation using a geometric mean. That is, if we take $\mathcal{S}_p$ as an SM constructed using pitch information, $\mathcal{S}_d$ as an SM constructed using duration information, the geometric mean is computed as $(\mathcal{S}_p w_p \circ \mathcal{S}_d w_p)^{\circ \frac{1}{2}}$, with $\circ$ denoting the Hadamard or element-wise product.

---

[8] We tested both standard thresholding and the thresholding method provided in the SM toolbox (with the default parameters). We also tested Gaussian smoothing with window sizes $\in \{2, 3, 6\}$ notes.

Table 2. Performance of MUL variants and baselines (abbreviations are defined in the text). From left to right: mean recall $\overline{R}$, precision $\overline{P}$, and $\overline{F1}$ with standard deviation in parenthesis. Highest performances are marked in bold. * indicates performances that *are* significantly different ($\alpha = 0.05$) to the *highest* performance. ○ indicates performances that *are* significantly different ($\alpha = 0.05$) to the ORG performance.

Database	Vocal (200 mels.)			Instrumental (200 mels.)		
MUL variants & baselines	$\overline{R}$	$\overline{P}$	$\overline{F1}$	$\overline{R}$	$\overline{P}$	$\overline{F1}$
ORG	0.36	0.34*	0.33* (0.28)	0.32	0.25*	0.26* (0.20)
ORG$\overline{\lambda}_{12}$	0.43	0.49	0.42 (0.37)	0.27*	0.37*	0.28* (0.29)
ORG$\overline{\lambda}_{13}$	0.44	°0.54	0.45 (0.35)	0.33	°0.51	0.37 (0.30)
ORG$\overline{\lambda}_{23}$	0.43	0.38	0.38 (0.26)	°**0.43**	0.33	°0.33 (0.20)
FIT$\overline{\lambda}_{123}$	0.39	°**0.59**	0.43 (0.34)	0.27*	°**0.57**	0.34 (0.27)
FITλ_1	0.29	0.40	0.31 (0.37)	0.16	0.31	0.20 (0.26)
FITλ_2	0.37	0.25	0.29 (0.20)	0.42	0.18	0.25 (0.15)
FITλ_3	0.38	0.44	0.35 (0.24)	0.27	0.52	0.31 (0.22)
FITλ_2	0.37	0.25	0.29 (0.20)	0.42	0.18	0.25 (0.15)
FITλ_3	0.38	0.44	0.35 (0.24)	0.27	0.52	0.31 (0.22)
RND10 %	0.17	0.25	0.20 (0.20)	0.17	0.19	0.17 (0.15)
ALWAYS	1.00	0.10	0.17 (0.04)	1.00	0.02	0.04 (0.02)
NEVER	0.00	0.00	0.00 (0.00)	0.00	0.00	0.00 (0.00)

4.4 Results, Baselines, and Significance Testing

In Table 2 we present mean recall $\overline{R}$, precision $\overline{P}$, and $\overline{F1}$ results obtained by the different variants of MUL. Phrase boundaries are considered as predicted correctly (a *tp*) if the prediction identifies either the last event of an annotated phrase or the first event of the following phrase.

The variants of MUL in Table 2 are abbreviated as follows: ORG corresponds to the original version of MUL, and hence computes the fitness ϕ using Eq. 3; ORG$\overline{\lambda}_{123}$ computes ϕ using Eq. 8; ORG$\overline{\lambda}_{ij}$ also computes ϕ using Eq. 8, but this time the mean $\overline{\lambda}$ is computed over the pairs $ij \in \{12, 13, 23\}$; FIT$\overline{\lambda}_{123}$ uses the mean $\overline{\lambda}$ of scores $\lambda_{1,2,3}$ instead of ϕ as a fitness function; finally, FITλ_i, for $i \in \{1, 2, 3\}$, uses λ_i instead of ϕ as a fitness function.

To define a lower bound of performance we tested three naïve baselines: RND10 %, which predicts a segment boundary at random in 10 % of the melody (10 % approximates the mean number of estimated boundaries produced by the tested variants of MUL); ALWAYS, which predicts a segment boundary at every melodic event position; NEVER, which does not make predictions (for completeness). We also tested the statistical significance of the paired $F1$, P, and R differences between the compared configurations of MUL and the baselines. For the statistical testing we used a non-parametric Friedman test ($\alpha = 0.05$).

Furthermore, to determine which pairs of measurements significantly differ, we conducted a post-hoc Tukey HSD test.

4.5 Discussion

In this section we analyse the results shown in Table 2. We first discuss aspects related to the general performance of MUL. Then we discuss more specific aspects of performance: the possible benefits of using our proposed location constraints, and the relative importance of each location constraint scoring function $\lambda_{1,2,3}$.

General Performance Observations: First, the performances obtained for vocal melodies are in general higher than those obtained for instrumental melodies. However, the $F1$ performance differences between each MUL variant for vocal and instrumental melodies are not statistically significant. This suggests that MUL generalises to these two sets. Second, for both vocal and instrumental melodies $\overline{P}$ is tends to be higher than $\overline{R}$ ($\sim$9 % higher for vocal and $\sim$11 % for instrumental melodies). This can be explained by recalling that MUL models only repetition-based segmentation cues, while the annotated boundaries might have been perceived taking into account other cues. Third, all pairwise $F1$ performance differences between MUL variants and baselines showed to be significant at the 5 % level.

Benefits of Location Constraints: For both vocal and instrumental melodies $\text{ORG}\overline{\lambda}_{123}$ obtains the highest performance. The $\overline{F1}$ improvements of $\text{ORG}\overline{\lambda}_{123}$ over ORG are of 14 % in the vocal set and of 13 % in the instrumental set. For both sets their $F1$ performance differences are statistically significant. These significant improvements support our hypothesis, suggesting that location constraints are an important addition when attempting to discern which repetitions human listeners might recognise and use for segmentation. Furthermore, the fact that the differences in $F1$ performances between $\text{FIT}\overline{\lambda}_{123}$ and ORG (for both sets) are *not* significant stresses the level of importance of location constraints. To be more precise, while for both (vocal/instrumental) sets the $\overline{R}$ of $\text{FIT}\overline{\lambda}_{123}$ is comparable to that of ORG (R differences are not significant), the $\overline{P}$ of $\text{FIT}\overline{\lambda}_{123}$ shows large and statistically significant improvements over ORG, suggesting that the human annotators of the melodic datasets might be recognising repetitions by using location constraints in a greater degree than constraints on repetition frequency or length.

Role of Location Constraints 1, 2, and 3: In both vocal and instrumental sets the $\overline{F1}$ performances of each variant $\text{FIT}\overline{\lambda}_{1,2,3}$ is similar, with the best one for both sets being $\text{FIT}\overline{\lambda}_3$. For both sets the difference between the $F1$ performances of $\text{FIT}\overline{\lambda}_3$ and $\text{FIT}\overline{\lambda}_1$ is significant, and the one between $\text{FIT}\overline{\lambda}_3$ and $\text{FIT}\overline{\lambda}_2$ is not. Moreover, when $\lambda_{1,2,3}$ are used in combination in $\text{FIT}\overline{\lambda}_{123}$, the $F1$ performances of $\text{FIT}\overline{\lambda}_{123}$ are not significantly different to those of $\text{FIT}\lambda_2$ and $\text{FIT}\lambda_3$. This suggests that the impact of repetitions aligned to temporal gaps λ_3 and repetitions with instances at the beginning of the melody λ_2 is higher than that of having evenly distributed repetitions λ_1. That said, it is only when all location constraints are used ($\text{ORG}\overline{\lambda}_{123}$) that a significant performance increase over ORG is obtained. This suggests that, even though in isolation $\lambda_{2,3}$ seem to have higher importance than λ_1, when associated to other constraints, such as repetition frequency and length, all location constraints become essential.

5 Conclusions

In this paper we have proposed a set of location constraints for repetition-based modelling of melody segmentation. Our proposed constraints aim to enhance repetition selection of repetition-based segmenters. To test our constraints, we quantified and incorporated them in a state-of-the-art repetition-based segmentation model [9,11]. The original and constraint-extended versions of the model are used to segment 400 (symbolically encoded) folk melodies. Results show the constraint-extended version of the model achieves a statistically significant 14 % average improvement over the model's original version.

In future work the role of metrical structure has to be taken into consideration. As shown in [1], even exact repetitions of melodic material might not be recognised by humans if these are not congruent with the metric structure of the melody. We also plan to extend our analysis to audio data, given that the constraints proposed in this paper are independent of the representation scheme (although the robustness of temporal gap detection in automatically extracted onset information would need to be assessed). Lastly, we plan to use the extended repetition-based model in a multiple-cue segmentation model like the one proposed in [16].

Acknowledgements. We thank the anonymous reviewers for the useful comments on earlier drafts of this document. Marcelo Rodríguez López and Anja Volk are supported by the Netherlands Organization for Scientific Research (NWO-VIDI grant 276-35-001).

References

1. Ahlbäck, S.: Melodic similarity as a determinant of melody structure. Musicae Sci. **11**(1), 235–280 (2007)
2. Cambouropoulos, E.: The local boundary detection model (LBDM) and its application in the study of expressive timing. In: Proceedings of the International Computer Music Conference (ICMC 2001), pp. 232–235 (2001)
3. Cambouropoulos, E.: Musical parallelism and melodic segmentation. Music Percept. **23**(3), 249–268 (2006)
4. Huron, D.: Sweet Anticipation: Music and the Psychology of Expectation. MIT press, Cambridge (2006)
5. Lartillot, O.: Reflections towards a generative theory of musical parallelism. Musicae Sci. Discuss. Forum **5**, 195–229 (2010)
6. Lerdahl, F., Jackendoff, R.: A Generative Theory of Tonal Music. MIT press, Cambridge (1983)

7. Margulis, E.H.: Musical repetition detection across multiple exposures. Music Percept. Interdisc. J. **29**(4), 377–385 (2012)
8. Meredith, D., Lemström, K., Wiggins, G.A.: Algorithms for discovering repeated patterns in multidimensional representations of polyphonic music. J. New Music Res. **31**(4), 321–345 (2002)
9. Müller, M., Grosche, P., Jiang, N.: A segment-based fitness measure for capturing repetitive structures of music recordings. In: ISMIR, pp. 615–620 (2011)
10. Müller, M., Jiang, N., Grohganz, H.: SM toolbox: matlab implementations for computing and enhancing similarity matrices. In: Audio Engineering Society Conference: 53rd International Conference: Semantic Audio. Audio Engineering Society (2014)
11. Müller, M., Jiang, N., Grosche, P.: A robust fitness measure for capturing repetitions in music recordings with applications to audio thumbnailing. IEEE Trans. Audio Speech Lang. Process. **21**(3), 531–543 (2013)
12. Müller, M., Grosche, P.: Automated segmentation of folk song field recordings. In: Proceedings of the ITG Conference on Speech Communication, Braunschweig, Germany (2012)
13. Müller, M., Kurth, F.: Enhancing similarity matrices for music audio analysis. In: IEEE International Conference on Acoustics, Speech and Signal Processing (ICASSP), vol. 5 (2006)
14. Pearce, M., Müllensiefen, D., Wiggins, G.: The role of expectation and probabilistic learning in auditory boundary perception: a model comparison. Perception **39**(10), 1365 (2010)
15. Rodríguez-López, M., Volk, A.: Symbolic segmentation: a corpus-based analysis of melodic phrases. In: Aramaki, M., Derrien, O., Kronland-Martinet, R., Ystad, S. (eds.) CMMR 2013. LNCS, vol. 8905, pp. 548–557. Springer, Heidelberg (2014)
16. Rodríguez-López, M., Bountouridis, D., Volk, A.: Multi-strategy segmentation of melodies. In: Proceedings of the 15th Conference of the International Society for Music Information Retrieval (ISMIR), pp. 207–212 (2014)
17. Rodríguez-López, M., Volk, A., de Haas, W.: Comparing repetition-based melody segmentation models. In: Proceedings of the 9th Conference on Interdisciplinary Musicology (CIM), pp. 143–148 (2014)
18. Shanahan, D., Huron, D.: Interval size and phrase position: a comparison between german and chinese folksongs (2011)
19. Takasu, A., Yanase, T., Kanazawa, T., Adachi, J.: Music structure analysis and its application to theme phrase extraction. In: Abiteboul, S., Vercoustre, A.-M. (eds.) ECDL 1999. LNCS, vol. 1696, pp. 92–105. Springer, Heidelberg (1999)
20. Temperley, D.: The Cognition of Basic Musical Structures. MIT press, Cambridge (2004)

Modeling Musical Structure
with Parametric Grammars

Mathieu Giraud[1] and Sławek Staworko[1,2,3]([⊠])

[1] Algomus, CRIStAL (UMR CNRS 9189, Université de Lille), Lille, France
mathieu.giraud@lifl.fr, staworko@gmail.com
[2] LINKS, Inria Lille and CRIStAL (UMR CNRS 9189, Université de Lille),
Lille, France
[3] Diachron Project, LFCS, University of Edinburgh, Edinburgh, Scotland

Abstract. Finding high-level structure in scores is one of the main
challenges in music information retrieval. Searching for a formalization
enabling variety through fixed musical concepts, we use parametric grammars, an extension of context-free grammars with predicates that take
parameters. Parameters are here small patterns of music that will be
used with different roles in the piece. We investigate their potential use in
defining and discovering the structure of a musical piece, taking example
on Bach inventions. A measure of conformance of a score with a given
parametric grammar based on the classical notion of edit distance is
investigated. Initial analysis of computational properties of the proposed
formalism is carried out.

1 Introduction

Finding high-level structure in scores is one of the fundamental research challenges in music information retrieval. Listeners are capable of discerning structure in music through the identification of common parts and their relative
organization. Capturing musical structure with formal grammars is an old idea,
taking roots in linguistics [1,12,18–20]. A grammar consists of a collection of
productions, transforming *non-terminal* symbols into other symbols, and eventually producing *terminal* symbols that can be the actual notes or other elements
of the musical surface. Grammars can be used as a music analysis tool, to find
the right grammar modeling a piece, as well as a composition tool, to generate
pieces following a grammar. Typically, for a grammar to be used as a generating
tool, the productions are additionally labeled with probabilities [3].

The Shenkerian analysis [15] and the Lerdahl and Jackendoff Generative Theory of Tonal Music (GTTM) [9] share similar ideas with formal grammars. Some
studies tried to put these approaches into computational models [5,10]. Other
authors use different kinds of formal languages and high-level descriptions to
encode music [4,11]. Many of these formal approaches propose a derivation tree
over musical surface that can be understood as parse tree of a grammar. In this
view a parse tree identifies the structure of a musical piece while a grammar models the structure of a set of related pieces that share the same structural footprint.

© Springer International Publishing Switzerland 2015
T. Collins et al. (Eds.): MCM 2015, LNAI 9110, pp. 85–96, 2015.
DOI: 10.1007/978-3-319-20603-5_8

Indeed, there exists works that attempt to automatically infer a context-free grammar from a piece [17].

All such approaches may suffer from a lack of generality: The precise signification of a non-terminal is often very dependent on the piece. But it should be possible to develop more generic tools, where the concepts of *chorus, verse, theme, variation,* and *development* exist independently of the underlying music data. Several studies proposed techniques to infer or discover patterns used in different roles, even with some variations [2,8]. Further research should be carried to propose formalisms able to encode notions on musical structure on potentially different musical material, and to be able to assert the compliance of these formalisms to the music. This paper proposes two steps in these directions:

- We use *parametric grammars* as an extension of context-free grammars where non-terminals have *parameters* that take values in the set of terminals. Parameters are here small patterns of music that will be used with different roles in the piece. Theses patterns are used both in measure-level or phrase-level production rules (texture arising from the local organization of the patterns) as well as in high-level organization of the piece into parts (each part involving similar or different phrase-level organizations). The formalization with parametric grammars enables us to model some musical concepts as the notion of *development* which can take several short patterns as parameters, even if the actual order of patterns is not specified. We show how to model Bach inventions with such grammars.
- We propose an algorithm to compute the optimal distance between a piece and a parametric grammar while building a derivation tree of the piece. Checking compliance between a musical piece and the model is the first step to more elaborated tasks, for instance, find the right grammar, or learning grammars from a set of examples. We show, however, that the fundamental problem of constructing the optimal alignment is intractable. To alleviate the negative impact of this result, we propose a number of practical restrictions (bounded height, width and errors) that renders computation feasible.

We prove the adequacy of our approach to model the music structure and find the optimal derivation on Bach inventions within a music analysis framework that attempts to capture only *certain* aspects of the music. We do not model everything. In fact, we do not work with single notes of the musical pieces but instead work on a level of abstraction and represent the pieces with a small set of repeated patterns. As such the generative aspect of this use of parametric grammar may be limited. The analysis perspective allows us to focus on the high-level structure. Moreover, as we try to have generic production rules in the grammar, the matching between the derivation of our grammars and the actual music is far from being exact. Nevertheless, we believe that this fuzziness reflects some aspect of the music complexity, as a fragment from any music material may have several different – and sometimes contrasting – roles.

The following section presents the idea of the modeling on inventions by J.S. Bach. Section 3 contains a formal definition of parametric grammars and proposes a method for constructing an optimal alignment between a musical piece and a parse tree that captures the structure in a musical piece. Section 4 comes back to the musical examples, presenting results of computing the parse trees that represent the music structure.

Fig. 1. Patterns used to model musical surface of Bach invention #01 in C major, taken from the first four measures of the sopran voice. All patterns have a duration of a quarter. The main patterns are a/A: four sixteenths in a upwards (a) or downwards (A) movement; b/B: four sixteenths (two successive thirds) in a upwards (b) or downwards (B) movement; c: two eights, large intervals; e: two eights, small intervals.

```
abce|abce|ABAB|ABAB|c?AB|BBzz||--ab|--ab|--AB|--AB|eece|cees|  ...
--ab|--ab|eece|cees|abce|ec?z||abce|abce|ABec|ABec|ABAB|ABAB|  ...

...  |abAA|BbAz||ABss|abss|ABss|abss|abab|abcz|AB??|----||
...  |c?AB|BBcz||--AB|ssab|ssAB|ssab|eece|ceAB|ceaz|----||
```

Fig. 2. Reduction of the whole invention using patterns of length of a quarter.

2   Modeling Bach Inventions with Parametric Grammars

2.1   Low-Level Paradigmatic Analysis

Most Bach inventions can be decomposed into motivic patterns, which is very convenient to test models on higher structural levels. We thus choose here to reduce the score with a paradigmatic analysis, based on our analysis and inspired by [13,16]. Figure 1 shows the decomposition of the first four measures of both voices of Bach invention #01 using several patterns of length of a quarter note. The four main patterns (a, b, c, e) are taken from the first measure in the soprano voice. Of course, there are some arbitrary choices in this analysis – patterns could have been be longer, and many of these patterns are related. For example, pattern B is the mirror of pattern b, and pattern e can be seen as a condensed pattern B. Ultimately, most of these patterns are derived from the base patterns a and b. Figure 2 shows the decomposition of all the piece: About 80 % of quarters can been seen as occurrence of either a, b, c, or e, sometimes with slight variations, including mirroring (A and B).

2.2 High-Level Structural Analysis

We propose to roughly model the invention #01 with the following grammar G_1,

$$G_1 \begin{cases} S_0() & \rightarrow P(x,y,z,w) + P(x,y,z,w) + P(z,w,x,y) \\ P(x,y,z,w) & \rightarrow T(x,y) + D(z,w) + I(w) \\ T(x,y) & \rightarrow (x/\_ + y/\_ + \_/x + \_/y) * 2 \\ & \mid (\_/x + \_/y + x/\_ + y/\_) * 2 \\ D(x,y) & \rightarrow (x/\_ + y/\_) * 4 \mid (\_/x + \_/y) * 4 \\ I(x) & \rightarrow (x/\_) * 3 \mid (\_/x) * 3 \end{cases}$$

The formalism – parametric grammars – is described in the next section. The grammar ultimately generates a stream of terminals such as "$x/\_$" (pattern given by parameter x at the sopran voice, and any pattern at the alt voice). Here, we explain informally how this grammar allows to capture the structure of the invention #01.

The piece S_0 has three parts. Each part P is split in three sub-parts with an increased perceived pulse. In the *thematic* sub-part $T(x,y)$, two patterns x and y are used during two consecutive quarters, then on again two consecutive quarters but on the other voice. This pattern is repeated twice, giving a feeling of a large repeat of period one measure. For example, the first two measures (soprano: abce|abce / alt: --ab|--ab) are modeled with $T(a,b)$. In the *development* sub-part $D(x,y)$, there is a voice where the patterns x and y are repetitively used during two measures (such as ABAB|ABAB / eece|cees with $D(A,B)$). Here, the individual *halves* of each measure become more pronounced. Finally, in the *intensification* sub-part $I(x)$, there is a voice where a pattern x is played three times *at every quarter note* (such as BBB / eec with $I(B)$). This intensification is concluded by a cadential element before the start of the following part.

What are the advantages of using parametric grammar over using the standard context-free grammars? The P_1 part could be roughly defined by explicit rules of the following standard context-free grammar, without parameters:

$$G_{non-parametric} \begin{cases} S_0 & \rightarrow P_1 + P_2 + P_3 \\ P_1 & \rightarrow T_1 + D_1 + I_1 \\ T_1 & \rightarrow (a/\_ + b/\_ + \_/a + \_/b) * 2 \\ D_1 & \rightarrow (a/\_ + b/\_) * 4 \\ I_1 & \rightarrow (B/\_) * 3 \\ \dots \end{cases}$$

Similar modeling can be done for the others parts. The complete non-parametric grammar resulting from this modeling is, however, less concise, does not allow to capture the structural similarities in elements of the same function (part, theme, development, intensification), and fails to identify the connections among structural elements of the piece that are established by using the same of pieces of musical material.

In the parametric grammar, the rule deriving TDI sub-parts from $P(x,y,z,w)$ links the different sub-parts of a part. The fact that the same or similar musical material is used throughout the part contributes the unity of this part. For instance, the pattern x for the D sub-part is reused in the I sub-part. Similarly, at the top level S_0, the different parts use the same material, but the last part, $P(z, w, x, y)$, uses the music material in a different order.

The parametric grammar G_1 can be made more flexible and closer to actual pieces by relaxing some production rules as in the following grammar G_2:

$$
G_2 \begin{cases}
S_0() & \to P(x, y, z, w) + P(x, y, z, w) + P(z, w, x, y) \\
P(x, y, \{z, w\}) \to T(x, y) + D(z, w) + I(w) & \leq 2 \\
P(x, y, \{z, w\}) \to T(x, y) + I(w) & \leq 2 \\
P(x, y, \{z, w\}) \to T(x, y) + D(z, w) & \leq 2 \\
T(x, y) & \to (x/\_ + y/\_ + \_/x + \_/y) * [1; 2] \\
& \mid (\_/x + \_/y + x/\_ + y/\_) * [1; 2] \\
D(x, y) & \to (x/\_ + y/\_) * [3; 4] \mid (\_/x + \_/y) * [3; 4] \\
I(x) & \to (x/\_) * [3; 4] \mid (\_/x) * [3; 4]
\end{cases}
$$

Now a P part can be composed from only TI or TD sub-parts instead of all three TDI sub-parts. Moreover, the number of repeats in the individual T, D and I sub-parts is variable. To limit the combinatorial explosion, a limit has been set to 2 for each P part. This bounds the "alignment distance" to the actual musical content, also limiting the number of candidate P "fragments" inside the score. Distance and fragments are formally defined in the next section.

Note also that the w pattern playing a specific role in the I rule may be any of the two patterns of the D rule. This choice is here modeled by the set $\{z, w\}$ in the rules for P.

Table 1 details the structure of four Bach inventions that can be seen as productions of this grammar. As flexibility is inherent to the proposed grammar model, it can define a large number of different musical pieces and many, if not most, are unlikely to satisfy any reasonable aesthetic requirements of *good music*. Using parametric grammars for generative purposes would therefore require adding constraints on the flow of the piece and then generating a piece that satisfies them. This is, however, beyond the scope of this paper whose main focus is to provide an analytic framework capable of exploring certain high-level aspects of the musical structure.

Table 1. Reference analysis derivation for some Bach inventions. These inventions were manually modeled as successive parts (P) composed of thematic (T), and possibly development (D) and intensification (I) sub-parts. Some structures also include special transition (W, R) and coda (C) parts that will not be discussed here.

	Length	Complete piece	part P_1	part P_2	part P_3	part P_4
#01 C major	88	$S_0 \to P_1 P_2 P_3$	TDI	$TTDI$	TD	
#03 D major	65	$S_0 \to P_1 P_2 P_3 W P_4 C$	TI	TI	TI	TI
#04 D minor	54	$S_0 \to P_1 P_2 P_3$	TDD	$TDDD$	T	
#13 A minor	104	$S_0 \to P_1 P_2 R P_3 P_4$	TDI	TDI	TR	TI

3 Parametric Grammars

In this section, we formally define parametric grammars and we present how to align scores to parse trees of parametric grammars. A parametric grammar is essentially an extension of context-free grammar whose non-terminals take *parameters*. They can be viewed as a specialized attribute grammars [7]. They have the same expressiveness as context-free grammars but can be significantly more concise [6].

3.1 Definitions

Let Σ be a finite set of symbols and $k > 0$ the number of output voices. A *(k-voice) output atom* is an vector of k symbols i.e., an element of Σ^k, and we use $\bar{a}, \bar{b}, \ldots$ to range over output atoms. For an output atom $\bar{a} \in \Sigma^k$ by a_i we denote the symbol at the i-th voice of a i.e., $\bar{a} = (a_1, \ldots, a_k)$. A *($k$-voice) string* is a sequence of atoms i.e., an element of $(\Sigma^k)^*$, and we use $w, v, \ldots$ to range over strings. By $|w|$ we denote the length of the string w and by w_i we denote the i-th atom of w for $i \in \{1, \ldots, |w|\}$ i.e., $w = (w_1, \ldots, w_{|w|})$.

A *grammar signature* is a tuple $\mathcal{S} = (\Sigma, k, X, V, arity)$, where Σ is a finite set of symbols, $k > 0$ is the number of output voices, X is a finite set of parameters, and V is a finite set of non-terminals together with the function $arity : V \to \mathbb{N}$ that assigns to every transition symbol the number of its parameters. We assume a fixed grammar signature $\mathcal{S}$ and define a number of concepts over $\mathcal{S}$. An *output term* is a vector of k elements from $\Sigma \cup X$. A *intermediate term* is $N(\tau_1, \ldots, \tau_n)$, where $N \in V$ is a non-terminal of arity $n = arity(N)$ and $\tau_i \in \Sigma \cup X$ for $i \in \{1, \ldots, n\}$. A *term* is either an output term or a intermediate term. A term is *ground* if it does not use any parameter. Note that a ground output term is an output atom i.e., an element of Σ^k, and similarly, a sequence of ground output atoms is a string i.e., an element of $(\Sigma^k)^*$. A *substitution* is a function θ that maps parameters in X to symbols in Σ (this function may be partial). The result of applying a substitution θ to t, in symbols $\theta(t)$, is obtained by replacing every parameter x by the symbol $\theta(x)$ that θ assigns to x. Applying substitution is extended to sequences of terms in the canonical fashion: we apply the substitution to every element of the sequence. For example, on the grammar G_2, the output atoms are $\{a, b, c, e, A, B, \ldots\}$, the intermediate terms are $P(\ldots)$, $T(\ldots)$, $D(\ldots)$ and $I(\ldots)$, and the parameters are $\{x, y, z, w\}$. The substitution mapping $T(x, y)$ to $T(a, b)$ is $\{\theta(x) = a, \theta(y) = b\}$.

A *parametric grammar* is a tuple $G = (\mathcal{S}, S_0, P)$, where $\mathcal{S} = (\Sigma, k, X, V, arity)$ is its signature, $S_0 \in V$ is a distinguished *starting non-terminal*, and P is a set of productions of the form $t \to s$, where t is a term and s a sequence of terms. A *derivation tree* of G is a tree T such that:

1. the leaves of T are labeled by ground output terms;
2. the non-leaf nodes are labeled by ground intermediate terms;
3. for every non-leaf node there exists a production $t \to s$ and a valuation θ such that, the node is labeled by $\theta(t)$ and the consecutive labels of its children give the sequence $\theta(s)$.

The *foliage* of a derivation tree T, denoted $yield(T)$, is the sequence obtained by taking the leaf labels in the standard left-to-right traversal of the tree. Note that $yield(T) \in (\Sigma^k)^*$ since the leaves of a derivation tree can be labeled by ground output terms only. A *parse tree* of $w \in (\Sigma^k)^*$ (w.r.t. G) is any derivation tree T whose root node is labeled by an intermediate term $S_0(a_1, \ldots, a_k)$ for some $a_i \in \Sigma$ and $yield(T) = w$. The *language* of G, denoted $L(G)$, is the set of all strings $w \in \Sigma^*$ that have a parse tree (w.r.t. G).

The grammar G_2 defined in the previous section further uses elements of syntactic sugar, allowing to represent a parametric grammar in a compact way:

1. disjunction e.g., $t \rightarrow s_1 \mid s_2$ equivalent to two rules $t \rightarrow s_1$ and $t \rightarrow s_2$,
2. numerical repetition e.g., $t \rightarrow s * [1; 2]$ equivalent to $t \rightarrow s|s+s$ ($+$ is concatenation operator, omitted in the formal definition). Many repetitions in music have between 2 and 4 occurrences.
3. Grouped set of parameters on the left-hand side e.g., $N(x_1, \{x_2, x_3\}) \rightarrow s$ equivalent to $N(x_1, x_2, x_3) \rightarrow s$ and $N(x_1, x_3, x_2) \rightarrow s$, as well on the right-hand side e.g., $t \rightarrow N(x_1, \{x_2, x_3\})$ equivalent to $t \rightarrow N(x_1, x_2, x_3)$ and $t \rightarrow N(x_1, x_3, x_2)$. Indeed, the same basic music material is often used in differents parts *with different roles*. For example, in the grammar G_2, the secondary patterns $\{z, w\}$ play different roles in the three P parts.

3.2 Aligning Scores to Parse Trees

While a parametric grammar allows to define strings, hence scores, that exhibit a very specific structure defined by the grammar, real-life musical pieces rarely adhere to this structure. Consequently, we propose a method of aligning a given string w, that represents the musical surface, to a parse tree T of another string $v \in L(G)$, that exhibits the structure defined by G. In particular, we do not assume that w is recognized by G, which may be too strict, but instead we introduce a measure of distance between w and T that we aim to minimize. This measure captures two types of operations performed on the parse tree T and the string v:

1. Basic *string editing operations*, which include inserting and deleting an atom in v as well as renaming a symbol of a single voice. For example, the cost of renaming one symbol can be equal to 1 and the cost of inserting and deleting an atom can be equal to k, the number of voices. The cost can also be linked to the actual musical content of the pattern.
2. *Move operations* that introduce gaps or overlaps between the outputs (foliage) of two sibling nodes in the parse tree T. While gaps (overlaps) in T can be captured with insertion (deletion resp.) operations in v, their cost can be different. Setting this cost to something smaller than the length of the gap will favour the move of these output blocks.

We now fix a k-voice parametric grammar G and a string $w \in (\Sigma^k)^*$ that needs not belong to $L(G)$. Let $w = (w_1, \ldots, w_n)$ and define the set of *positions* in w as $Pos(w) = \{1, \ldots, n+1\}$ with $n+1$ denoting a virtual end-of-string position.

A *fragment* of w is pair $F = (s, e) \in Pos(w)^2$ of positions of w such that $s \le e$ and $start(F) = s$ is called its *start* of F and $end(F) = e$ its *end*. The fragment F represents a substring $(w_{start(F)}, w_{start(F)+1}, \ldots, w_{end(F)-1})$ and we point out that the end position of F is not included in the substring w_F but intuitively it is the position of w that immediately follows the last position of w_F. A fragment is *empty* if its start and end are the same. We define a number of relations on pairs of fragments F and F' of w: (1) F' is *included* in F if $start(F) \le start(F')$ and $end(F') \le end(F)$; (2) F and F' *overlap* if $start(F) \le start(F') < end(F)$ or $start(F') \le start(F) < end(F')$; (3) F' *follows* F if $start(F) \le start(F')$. Note that if F' follows F, the fragments may overlap.

Now, let T be a derivation tree of some string v w.r.t. the grammar G and let N_T be the set of nodes of T. An *alignment* of w to T is an assignment A of a fragment of w to every node of T that satisfies the following two conditions:

1. The fragment of the root node spans the whole string w i.e., $A(root_T) = (1, n + 1)$, where $root_T$ is the root node of T.
2. The fragment of any non-root node is included in the fragment of its parent i.e., if n' is a child of n, then $A(n')$ is included in $A(n)$.
3. The fragment of any inner node follows the fragment of any of its preceding siblings i.e., if n has children $n_1, \ldots, n_m$, then $A(n_j)$ follows (and possibly overlaps with) $A(n_i)$ for any $i \in \{1, \ldots, j-1\}$.

We denote by $start_A(n) = start(A(n))$ and $end_A(n) = end(A(n))$ the start and the end of the fragment of the node n in the alignment A. We define the measure of distance of an alignment A of the string w to a parse tree T recursively on the structure of T. That is, we define a function $cost_A$ that assigns the alignment cost to every node of the parse tree T of some string $v \in L(G)$. We start in the leaf nodes of the parse tree, where we attempt to identify a position within the assigned fragment of w, that is closest to the output atom of the leaf node, and any possible editing operations that need to be performed on the string v. For any leaf note n, the cost $cost_A(n)$ has to take into account whether the fragment $A(n)$ is empty (deletion of material) or not (identity or substitution of material). The latter case may involve distance comparing atoms, as for example the standard Hamming distance equal to the number of renaming operations necessary to obtain the atom $\bar{a}$ from the atom $\bar{b}$.

For an inner node n with m children $n_1, \ldots, n_m$, the definition of the cost is more involved. First of all, the cost of n must include the cost of all its children. Additionally, it has to incorporate the possible overlaps and gaps between fragments of any pair of two consecutive nodes. Note that the length of overlap/gap for n_i and n_{i+1} is equal to $|end_A(n_i) - start_A(n_{i+1})|$. Also, the cost needs to incorporate the possible left margin between the fragment of the first child n_1 and the fragment of its parent n as well as the right margin between the fragment of the last child n_m and the fragment of its parent. Altogether, we obtain the following formula:

$$cost_A(n) = (start_A(n_1) - start_A(n)) + (end_A(n) - end_A(n_m))$$
$$+ \sum_{i=1}^{m-1} |end_A(n_i) - start_A(n_{i+1})| + \sum_{i=1}^{m} cost_A(n_i),$$

possibly with weights on the different cost contributions. Now, the *alignment distance* between a grammar G and a string w is the minimum cost of an alignment of w to a parse tree of G (we assume that G has at least one parse tree):

$$dist(G, w) = \min\{cost_A(root_T) \mid A \text{ is an alignment of } w \text{ to a parse tree } T \text{ of } G\}.$$

Note that an alignment with the minimal cost needs not be unique just as there is more than one way of transforming the string ab to ba with the standard editing operations of deleting and inserting a character.

3.3 Computational Challenge

For a given word w and a given parametric grammar G, we want to construct an *optimal alignment* (which includes a parse tree) that minimizes the alignment distance of w to G. The intractability of this problem follows from the high complexity of a much simpler problem of *membership*: given G and w, check whether $w \in L(G)$:

Theorem 1. *Membership for parametric grammars is NP-complete.*

This can be proven with a reduction from a variant of SAT [14]. It is easy to see that $w \in L(G)$ if and only if $dist(G, w) = 0$. Therefore, even measuring the alignment distance alone is intractable and the task becomes more complex when the construction of an optimal alignment is required. Observe that a given parametric grammar can be converted to an equivalent context-free grammar by grounding the nonterminals i.e., substituting the parameters with all possible values. This procedure may yield a context-free grammar of size exponential in the number of occurrences of parameters. Consequently, the number of occurrences of parameters is one source of complexity and bounding it by a constant renders the membership problem tractable.

3.4 Constructing Optimal Alignment

We now outline an algorithm that constructs optimal alignment for a given input string w and a given parametric grammar G. The basic data structure we employ is a *link* which represents a node of a parse tree of G aligned to a fragment of the input w with a given cost. Consequently, a link shall constitute of an start and end position, a cost value, a ground term, and if the term is intermediate, also a set of pointers to other links which capture the structure of the parse tree.

During the computation we maintain a set of links which represent partially constructed parse trees each with an optimal alignment to a fragment of the input tree. Initially, this set contains only links with output terms that correspond to aligning the leaves nodes to the output atoms while deleting a number of adjacent atoms and links with output terms inserted at a specific positions of the input string. Then, iteratively, we saturate the set of links with links with intermediate terms that are obtained from applying production rules of G together with any possible move operations. A collection of pointers to the links that triggered using a production rule is stored in the newly created link, and its start, end, and cost is calculated appropriately from the starts, ends, and costs of those links. This process continues until no further link can be added and at this point we search for a link with the smallest cost with start 0 and end $|w|$ that is labeled with the start nonterminal. The cost of this link is the cost an optimal alignment that can be constructed by following the pointers to child links.

4   Computing Alignment Distances on Bach Inventions

We computed the distance computation on a number of inventions using the grammars described in Sect. 2. When constructing an optimal alignment, an important computational factor is its distance from the input string (i.e., the cost at the root node of the alignment tree). Essentially, the cost of finding an alignment is exponential in its cost. To render the computation feasible we allow the grammar to additionally specify:

– bounds on the length of the fragment derived from a given rule,
– and bounds on the overall cost of aligning any fragment to a given rule (such as ≤ 2 in G_2).

In our experiments we employ a slightly modified cost function that for moving operations does not penalize gaps between elements but only overlaps.

Figure 3 details the derivation tree found for Bach invention #01 by taking a simplified version of the grammar G_2. The invention #01 is decomposed into three parts (P), each one including full TDI sub-parts, even if the grammar

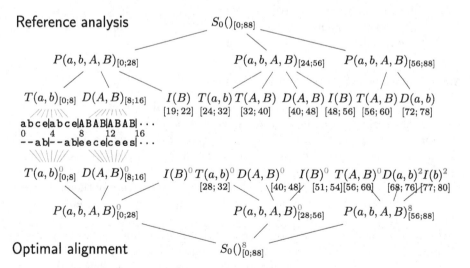

Fig. 3. Decomposition of Invention #01, in C major, with a simplified version of grammar G_2. Each node n is displayed together with, in subscript, the $[start_A(n); end_A(n)]$ values and, in superscript, the alignment distance $cost_A(n)$.

allows to skip some of these parts. The computed derivation succeeds thus in finding this 3-part structure, with relevant patterns, even if the computed derivation is not always identical to the reference analysis. By further adjustments in the grammar, it is possible to make the computed derivation even closer to the reference analysis. However, that is not our goal, but rather we show that a single generic grammar can be used to model several music pieces.

Finding a unique grammar that can parse several real pieces is quite difficult: further research need to be conducted to find constraints that are musically relevant while allowing more flexibility in the grammar. However, we show that the same grammar applied on the first three parts of the invention #03 (Fig. 4) predicts almost correctly bounds to these parts and corresponding sub-parts.

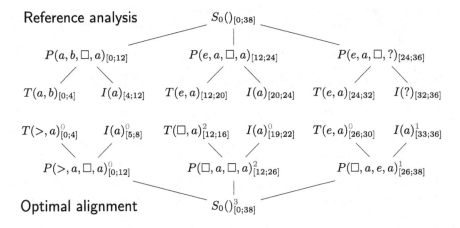

Fig. 4. Decomposition of an extract of Invention #03, in D major.

5   Conclusions and Future Work

The proposed formalism of parametric grammars and derivations obtained with optimal alignments offer an attractive way of modeling and identifying the music structure. Our results indicate that our technique is adequate for describing some elements of high-level features of the score.

It should be noted that we started with a paradigmatic analysis giving a first abstract representation of the musical surface. Such an intermediate representation, providing low-level semantics to the music, is here more appropriate that working on a raw stream of notes. Naturally, this kind of patterns could be inferred directly from the musical surface (notes, pitch), or the grammar could produce individual notes. This, however, would add a layer of complexity and its impact needs to be studied further. Also, a number of assumptions and simplifications we have made in our work comes from our intent to use the parametric grammars as a descriptive rather generative model. Possible applications of parametric grammars for music generation need to be explored further.

References

1. Chemillier, M.: Grammaires, automates et musique. In: Briot, J.-P., Pachet, F. (eds.) Informatique Musicale, pp. 195–230. Hermès, Paris (2004)
2. Conklin, D.: Distinctive patterns in the first movement of Brahms' string quartet in C minor. J. Math. Music 4(2), 85–92 (2010)
3. Cope, D.: Virtual Music: Computer Synthesis of Musical Style. MIT Press, Cambridge (2004)
4. Deutsch, D., Feroe, J.: The internal representation of pitch sequences in tonal music. Psychol. Rev. 88(6), 503–522 (1981)
5. Hamanaka, M., Hirata, K., Tojo, S.: Implementing "a generating theory of tonal music". J. New Music Res. 35(4), 249–277 (2006)
6. Hopcroft, J.E., Motwani, R., Ullman, J.D.: Introduction to Automata Theory, Languages, and Computation, 2nd edn. Addison Wesley, Boston (2001)
7. Knuth, D.E.: Semantics of context-free languages. Math. Syst. Theory 2(2), 127–145 (1968)
8. Lartillot, O.: Taxonomic categorisation of motivic patterns. Music. Sci. 13 (1 suppl), 25–46 (2009)
9. Lerdahl, F., Jackendoff, R.S.: A Generative Theory of Tonal Music. MIT Press, Cambridge (1983)
10. Marsden, A.: Schenkerian analysis by computer. J. New Music Res. 39(3), 269–289 (2010)
11. Meredith, D.: A geometric language for representing structure in polyphonic music. In: International Society for Music Information Retrieval Conference (ISMIR 2012) (2012)
12. Mesnage, M., Riotte, A.: Formalisme et modèles musicaux (2006)
13. Neumeyer, D.: The two versions of J.S. Bach's A-minor invention, BWV 784. Indiana Theory Rev. 4(2), 69–99 (1981)
14. Schaefer, T.J.: The complexity of satisfiability problems. In: ACM Symposium on Theory of Computing (STOC), pp. 216–226 (1978)
15. Schenker, H.: Der freie Satz. Universal Edition, Wien (1935)
16. Shafer, J.: The two-part and three-part inventions of Bach: a mathematical analysis. Honors project, East Texas Baptist University, March 2010
17. Sidorov, K., Jones, A., Marshall, D.: Music analysis as a smallest grammar problem. In: ISMIR 2014 (2014)
18. Steedman, M.J.: A generative grammar for jazz chord sequences. Music Percept. 2(1), 52–77 (1984)
19. Sundberg, J., Lindblom, B.: Generative theories in language and music descriptions. Cognition 4, 99–122 (1976)
20. Winograd, T.: Linguistics and the computer analysis of tonal harmony. J. Music Theory 12, 2–49 (1948)

Perfect Balance: A Novel Principle for the Construction of Musical Scales and Meters

Andrew J. Milne[1]([✉]), David Bulger[2], Steffen A. Herff[1],
and William A. Sethares[3]

[1] MARCS Institute, University of Western Sydney, Penrith, NSW 2751, Australia
{a.milne,s.herff}@uws.edu.au
[2] Macquarie University, Sydney, NSW 2109, Australia
david.bulger@mq.edu.au
[3] University of Wisconsin-Madison, Madison, WI 53706, USA
sethares@wisc.edu

Abstract. We identify a class of periodic patterns in musical scales or meters that are *perfectly balanced*. Such patterns have elements that are distributed around the periodic circle such that their 'centre of gravity' is precisely at the circle's centre. *Perfect balance* is implied by the well established concept of *perfect evenness* (e.g., equal step scales or isochronous meters). However, we identify a less trivial class of perfectly balanced patterns that have no repetitions within the period. Such patterns can be distinctly uneven. We explore some heuristics for generating and parameterizing these patterns. We also introduce a theorem that any *perfectly balanced* pattern in a discrete universe can be expressed as a combination of regular polygons. We hope this framework may be useful for understanding our perception and production of aesthetically interesting and novel (microtonal) scales and meters, and help to disambiguate between balance and evenness; two properties that are easily confused.

Keywords: Music · Scales · Meters · Balance · Evenness · Microtonal · Discrete Fourier transform

1 Introduction

A *perfectly balanced* pattern is a set of points on a circle whose mean position, or centre of gravity, is the centre of the circle (see Fig. 1). A *perfectly even* pattern is one in which the elements are equally spaced around the periodic circle (see Fig. 1(a)). More generally, the *balance* or *evenness* of a pattern is a measure of how closely it conforms, respectively, to perfect evenness or to perfect balance (formal definitions are provided in Sect. 2). Evenness has been identified as an important principle for the construction and analysis of scales and meters [1–3]. However, much research involving evenness has proceeded seemingly unaware that, in many common examples (e.g., well-formed scales [4]), it is strongly

© Springer International Publishing Switzerland 2015
T. Collins et al. (Eds.): MCM 2015, LNAI 9110, pp. 97–108, 2015.
DOI: 10.1007/978-3-319-20603-5_9

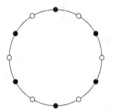
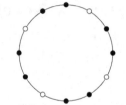

(a) Perfectly balanced, perfectly even, sub-periodic.

(b) Perfectly balanced, uneven, sub-periodic.

(c) Perfectly balanced, uneven, irreducibly periodic (no sub-periods).

Fig. 1. Three perfectly balanced periodic patterns exhibiting different classes of evenness and sub-periodicity. The small circles represent a universe of available pitch classes or metrical times (in these examples, there are twelve, which might correspond to twelve chromatic pitch classes or twelve metrical pulses). The filled circles are the notes or beats making the pattern under consideration.

associated with balance – indeed, perfect evenness implies perfect balance. For instance, it may be that the musical utility of well-formed patterns derives from them having high balance as well as high evenness. In order to tease apart these two properties, we will demonstrate a number of interesting patterns that are perfectly balanced but also distinctly uneven and *irreducibly periodic*. In this paper, we do not empirically test the recognizability, utility, likeability, and so forth, of balance; rather, we lay down some of the mathematical and conceptual framework around which future empirical work may be conducted.

Figure 1 presents some simple patterns to elucidate the above-mentioned properties; they are all perfectly balanced, but they exhibit different classes of evenness and reducibility of the period. The pattern in Fig. 1(a) is perfectly even (it might represent a whole tone scale or a $\frac{6}{4}$ meter).

The pattern in Fig. 1(b) is different because it is uneven (it can represent a diminished scale or a triplet shuffle). However, both (a) and (b) have rotational symmetries; for example, if (a) is compared with a version that has been rotated by 60°, the locations of all the filled circles will perfectly align; the same follows if (b) is rotated by 90°. This is because (a) has a *fundamental sub-period* subtending 60°, while (b) has a fundamental sub-period subtending 90°. A *fundamental sub-period* is *fundamental* in that it is the smallest-sized period of repetition in the pattern, and all other periods are multiples of it; it is a *sub-period* because it subtends an angle smaller than the full circle. A circular pattern with sub-periods is described as *reducibly periodic*, a circular pattern with no sub-periods is described as *irreducibly periodic*. Importantly, although both (a) and (b) are perfectly balanced over the whole circle, neither is perfectly balanced over its fundamental sub-period (this is explained in greater detail in Sect. 3).

The pattern in Fig. 1(c) is particularly interesting because it is perfectly balanced, uneven, and it has no sub-periods. This means the pattern is perfectly balanced over its fundamental period and, hence, over all its possible periods. We describe such a pattern as having *irreducibly periodic perfect balance*, and this is the class of patterns this paper focuses on. Such patterns may form useful templates for novel microtonal scales and meters. Furthermore, the clear separation of evenness and balance may allow the impact of both properties, with regards to perception and action, to be independently measured.

As previously mentioned, evenness and balance are closely intertwined. The next section aims to demonstrate some connections and differences between them, while the section after investigates balance itself.

2 Evenness and Balance

A way to demonstrate the relationships between evenness and balance is to express a periodic pattern as a complex vector and take its discrete Fourier transform. Vector $x \in [0,1)^K$ has K real-numbered pitch or time values between 0 and 1, ordered by size so $x_0 < x_1 < \cdots < x_{K-1}$ (the period has a size of 1). For instance, for the diatonic scale in a standard 12-TET tuning, the vector $x = \left(\frac{0}{12}, \frac{2}{12}, \frac{4}{12}, \frac{5}{12}, \frac{7}{12}, \frac{9}{12}, \frac{11}{12}\right)$. The elements of this vector are mapped to the unit circle in the complex plane with $z[k] = e^{2\pi i x[k]} \in \mathbb{C}$, so $z \in \mathbb{C}^K$. Each complex element $z[k]$ of z has unit magnitude, and its angle represents its time location or pitch as a proportion of the period (whose angle is 2π). We term this vector the *scale vector*.

We will also use an alternative vector representation of a periodic pattern that is suitable for patterns whose scale vector comprises only rational values. This *indicator vector* is denoted $a \in \{0,1\}^N$ and is given by $a[n] = [n/N \in x]$ (for $n = 0, \ldots, N-1$), where N is the cardinality of the chromatic universe (which must be some multiple of $1/\gcd(x_0, x_1, \ldots, x_{N-1})$) and the square (Iverson) brackets denote an indicator function that is unity when the enclosed relation is true, otherwise zero. Hence the previous 12-TET diatonic scale is $a = (1,0,1,0,1,1,0,1,0,1,0,1)$.

The tth coefficient of the discrete Fourier transform of the scale vector is given by

$$\mathcal{F}z[t] = \frac{1}{K} \sum_{k=0}^{K-1} z[k]\, e^{-2\pi i t k/K}. \tag{1}$$

We will use the zeroth and first coefficients to characterize balance and evenness.

2.1 Evenness – the First Coefficient

As first shown by Amiot [5], the magnitude of the first coefficient gives the *evenness* of the pattern:

$$evenness = |\mathcal{F}z[1]| \in [0,1]\,, \text{ where}$$

$$\mathcal{F}z[1] = \frac{1}{K} \sum_{k=0}^{K-1} z[k]\, e^{-2\pi i k/K}. \tag{2}$$

In statistical terms, evenness is equivalent to unity minus the circular variance [6] of the circular displacements between each successive term of z and each successive kth-out-of-K equal division of the period – if the displacements are all equal, their circular variance is zero and the pattern is *perfectly even*.

2.2   Balance – the Zeroth Coefficient

Unity minus the magnitude of the zeroth coefficient gives the *balance* of the pattern:

$$balance = 1 - |\mathcal{F}\boldsymbol{z}[0]| \in [0, 1], \text{ where}$$

$$\mathcal{F}\boldsymbol{z}[0] = \frac{1}{K} \sum_{k=0}^{K-1} \boldsymbol{z}[k]. \tag{3}$$

In statistical terms, balance is equivalent to the circular variance of the pattern itself. When a pattern's balance is 0 (i.e., it is maximally unbalanced), the K elements all have the same pitch or occur at the same time, so they are maximally clustered; when the balance is 1 (a condition we term *perfect balance*), they have the maximal possible circular variance. Importantly, as we will prove below, perfect balance does not imply evenness; hence these are two distinct properties.

An equivalent definition, for rational-valued patterns in an N-fold universe, can be calculated from the indicator vector, this time using the first coefficient:

$$balance = 1 - \frac{N|\mathcal{F}\boldsymbol{a}[1]|}{K} \in [0, 1], \text{ where}$$

$$\mathcal{F}\boldsymbol{a}[1] = \frac{1}{N} \sum_{n=0}^{N-1} \boldsymbol{a}[n] \, \mathrm{e}^{-2\pi i n/N}. \tag{4}$$

2.3   Relationships Between Evenness and Balance

Theorem 1. *Perfect evenness implies perfect balance.*

Proof. Under Parseval's theorem, $\sum_{t=0}^{K-1} |\mathcal{F}\boldsymbol{z}[t]|^2 = \frac{1}{K} \sum_{k=0}^{K-1} |\boldsymbol{z}[k]|^2$. By definition, all $|\boldsymbol{z}[k]| = 1$, hence $\sum_{t=0}^{K-1} |\mathcal{F}\boldsymbol{z}[t]|^2 = 1$. When $|\mathcal{F}\boldsymbol{z}[1]| = 1$ (perfect evenness), all other coefficients of $\mathcal{F}\boldsymbol{z}$ must, therefore, be zero. $\square$

Theorem 2. *Maximal imbalance implies maximal unevenness.*

Proof. The proof follows the same line of argument as that for Theorem 1 but using the zeroth coefficient of $\mathcal{F}\boldsymbol{z}$ instead of the first. $\square$

Theorem 3. *Perfect balance does not imply perfect evenness.*

Proof. This can be simply proven by example (as shown in Sects. 1 and 3). $\square$

Theorem 4. *In a perfectly even N-fold universe, the complement of a perfectly balanced pattern is also perfectly balanced.*

Proof. The proof is trivial. $\square$

Remark 1. This theorem parallels how, in an N-fold chromatic universe, the complement of a maximally even scale of K pitches is the maximally even scale of $N - K$ pitches [7, Proposition 3.2]. For example, the complement of the diatonic scale in the 12-fold chromatic scale is the pentatonic – more prosaically, the piano's black notes fill in all the gaps between the white notes.

Having established the above relationships between evenness and balance, we now turn our attention to the principle of balance itself.

3   Balance

Here is a physical analogy of *balance*. Imagine a vertically oriented bicycle wheel that can rotate freely about a horizontal axle. The wheel has N equally spaced slots around its circumference. We also have K weights all of the same mass and, into each slot, a single weight may be placed. Each slot represents a periodic pitch or time, and each weight represents an event at that pitch or time. In totality, therefore, they may be thought of as representing a scale or a meter (as described earlier). When the K weights are placed in the wheel's slots, after any perturbation, the wheel will always rotate into a stable position so that its 'heaviest' part is pointing vertically down. Phrased more mathematically, it will rotate (under the action of gravity) until the sum of the K vectors – each pointing from the wheel's centre to a weight – is pointing vertically downwards.

However, there is a class of *perfectly balanced* patterns where the wheel has no preferred orientation in that, provided it is not spun when released, the wheel will remain in whatever rotational position it is left at. This arises from a pattern whose sum of vectors is nil – as shown in (3). An alternative visualization is to think of a horizontal disk resting on a vertical pole at its centre. As alluded to in Sect. 1, the disk will balance only if the centre of gravity is at the disk's centre.

As shown in (4), the balance of any K-element pattern in an N-fold universe can be calculated from the indicator vector $\boldsymbol{a}$. Using this method, Lewin [8] describes a scale where this coefficient is zero as having the 'exceptional' property. This is the property we call perfect balance. Building on Lewin's insights, Quinn [9] clearly describes the meaning of this and the other coefficients as representing different types of 'balance'. However, often the distinction between evenness and balance has not been adequately explored. For example, Callender [10] describes this coefficient as 'a measure of how unevenly a set divides the octave' which is correct (as shown in Theorem 2) but does not mention the more interesting property that, when this value is minimized (i.e., balance is maximized), evenness is not implied (as shown in Theorem 3). As we discuss later, Amiot [11] has recently conducted a search for perfectly balanced patterns in a 30-fold universe.

A simple and graphical way to obtain perfect balance is to place the weights at the vertices of regular polygons – e.g., a digon, equilateral triangle, square, regular hexagon, and so forth, as shown in Fig. 2. Clearly, the greater the number of divisors of N, the greater the number of different regular polygons available.

But these are rather trivial patterns in that they are perfectly even and each actually comprises K smaller identical patterns of length N/K – in other words, their fundamental periods are $1/K$ of the circle (put differently, they have rotational symmetry of order K). How might we create more interesting, less even, irreducibly periodic, but still perfectly balanced structures?

We could take a copy of our polygon, rotate it by a distance less than that separating its vertices, then add it to the original. For instance, we can take an equilateral triangle and rotate it by one chromatic step and add it to the original to give an augmented (hexatonic) scale as illustrated in Fig. 3(a). This appears to create a perfectly balanced and interestingly uneven pattern; however, the

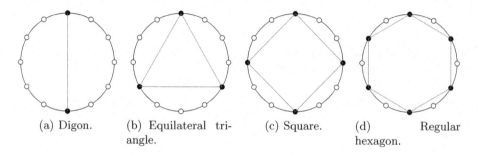

(a) Digon.   (b) Equilateral tri-   (c) Square.   (d)   Regular
              angle.   hexagon.

Fig. 2. Perfectly balanced regular polygons in a twelve-fold period.

resulting scale still has sub-periods (of length N/n, where n is the number of vertices in the repeated polygon). Indeed, the pattern in Fig. 3(a) consists of a fundamental sub-period that repeats three times within the circle. And if we stretch out this smaller pattern so it takes up a full circumference, as shown in Fig. 3(b), we can see how it is actually unbalanced (the sum of vectors is non-zero) over its fundamental period.

Similarly – as shown in Fig. 4 – we can take the $(N-K)$-element complement of any of the K-vertex regular polygons in Fig. 2 (Theorem 4). But these patterns also have sub-periods, also of lengths N/K, as the complement of any of the K-vertex regular polygons is a simple combination of different rotations of the original polygon.

The impact of such sub-periods may differ depending on whether the context is scalic or metrical. In a scalic context, the octave is an interval over which periodicity is often perceived (pitches an octave apart are typically heard as being, in some sense, equivalent). This means that the smaller sub-periods within the octave may be perceptually subsumed by the periodicity of the larger octave. For example, even though the augmented scale in Fig. 3(a) has repetition every quarter-octave, the most dominant perceived period of repetition may still be heard at the octave. In a metrical context, however, there is no specific duration

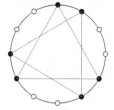

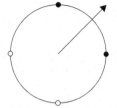

(a) Two displaced equilateral   (b) The augmented scale over its
triangles make the augmented   fundamental (third-octave) pe-
scale.   riod.

Fig. 3. A reducible pattern, which is perfectly balanced over a 12-fold period, but not over its fundamental 4-fold sub-period (as shown by the resultant vector in (b)).

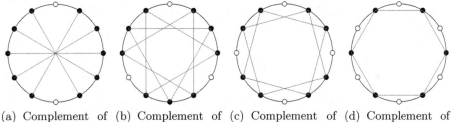

(a) Complement of digon.   (b) Complement of equilateral triangle.   (c) Complement of square.   (d) Complement of regular hexagon.

Fig. 4. Perfectly balanced, but sub-periodic, complements of regular polygons in a twelve-fold period. Because they are sub-periodic, they are 'modes of limited transposition' corresponding to Messiaen's: (a) seventh mode, (b) third mode, (c) second mode (diminished scale or triplet shuffle), (d) first mode (whole tone scale or $\frac{6}{4}$ meter) [12].

that is perceptually privileged, hence sub-periods may be more easily perceived as perceptually dominant. This might suggest that irreducible periods are more obviously related to a metrical rather than a scalic context. However, these possible different impacts of sub-periodicity do not imply that perfect balance – as a general principle – is not equally applicable to meters and scales. In both, irreducibly periodic patterns may be more useful due to their greater complexity and less obvious construction. Furthermore, irreducibly periodic scales have musically useful properties not found in reducibly periodic scales; for example, they have N distinct transpositions, and every different scale degree is surrounded by a different sequence of intervals (Balzano's property of *uniqueness* [13]).

So, is there a way to create an uneven and *irreducibly periodic* pattern that is also perfectly balanced? In the following subsection we will describe a simple heuristic method. In the subsection after that, we will describe an extension that enables us to find a different class of perfectly balanced structures.

3.1 Heuristics for Irreducibly Periodic Perfect Balance

Coprime Disjoint Regular Polygons. Add regular polygons (each expressed as an indicator vector) such that no two vertices have the same location (if their vertices did coincide, the resulting magnitude at that location would be greater than 1, which is not a 'legal' element of the indicator vector a defined in Sect. 2). This ensures balance, but such patterns may contain sub-periods as in the augmented scale shown in the previous subsection. To avoid sub-periods, we must additionally ensure that the numbers of vertices of the polygons used is coprime (their greatest common divisor is 1). For example, in a twelve-fold period (e.g., an equally tempered chromatic scale or a twelve-pulse meter), there are only two such patterns – as illustrated in Fig. 5. The first is created by adding a digon and a triangle; the second by adding two digons and a triangle. Note that, in a twelve-fold period, a third disjoint digon cannot be added because the three digons would take the form of a regular hexagon, which is not coprime with the triangle (the resulting pattern would have a sub-period).

(a) Five-element perfectly balanced pattern comprising a
digon and a triangle.

(b) Seven-element perfectly balanced pattern comprising two
digons and a triangle.

Fig. 5. The only two irreducibly periodic perfectly balanced patterns available in a
twelve-fold period. Note how uneven these patterns are. The seven-element pattern is
equivalent to a scale variously denoted the double harmonic, Arabic, or Byzantine. In
the North Indian tradition this scale is the Bhairav that, and in the Carnatic tradition
it is the scale used in the Mayamalavagowla raga.

The resulting patterns are perfectly balanced, whilst also being uneven and
irreducibly periodic. They might be thought of as *displaced polyrhythms* – take a
standard polyrhythm containing two isochronous beats of different and coprime
interonset intervals (e.g., 3 against 2), but displace one of the beats with respect
to the other so they never coincide.

These heuristics imply that, for two regular polygons, the period must comprise
$N = jk\ell$ equally tempered chromatic pitches or isochronous pulses, where j, k, ℓ
are integers all greater than 1 and $\gcd(k, \ell) = 1$ (a $k\ell$-fold universe is the smallest
that can embed two polygons with coprime k and ℓ vertices, but there must be at
least twice as many so one of the polygons can be rotated to make it disjoint to the
other). The smallest possible N are, therefore, $2 \times 2 \times 3 = 12, 2 \times 3 \times 3 = 18,$
$2 \times 2 \times 5 = 20, 2 \times 3 \times 4 = 24, 2 \times 2 \times 7 = 28, 2 \times 3 \times 5 = 30, 3 \times 3 \times 4 = 36,$
$2 \times 4 \times 5 = 40$, and so forth.

Similar to Fig. 4, the complement of such a pattern is irreducibly periodic
and perfectly balanced as well, as it is a combination of polygons in which at
least one polygon is coprime to at least one other. This can be seen in Fig. 5,
where the 5- and 7-element patterns are complementary (if one is rotated 180°).

Interestingly, perfectly balanced patterns do not have to be derived from the
addition of disjoint regular polygons; this is merely one method to ensure perfect
balance.

Searching Across Dihedral Groups of Order K. As Amiot has demonstrated, it is feasible to conduct a brute-force search for perfectly balanced patterns of size K in a cardinality of N [11]. To increase speed, Amiot factored out
the dihedral group; that is, his search did not separately consider rotationally or
reflectionally equivalent patterns. The search shows that patterns that are not
the sum of disjoint regular polygons do indeed exist. When $N = 30$ and $K = 7$,
there is one such pattern out of a total of 17 perfectly balanced patterns (it
would seem, therefore, that such patterns are comparatively rare in a discrete
universe of relatively low cardinality). Amiot's scale is illustrated in Fig. 6, and
has elements at $\left(\frac{0}{30}, \frac{6}{30}, \frac{7}{30}, \frac{13}{30}, \frac{17}{30}, \frac{23}{30}, \frac{24}{30}\right)$.

Fig. 6. Amiot's scale, which is not composed of disjoint polygons.

Integer Combinations of Intersecting Regular Polygons. However, contrary to first appearances, Amiot's scale is actually composed of regular polygons. But this time it is a linear combination of ten vertex-sharing (non-disjoint or intersecting) regular polygons where five have a weight of −1 and five have a weight of 1. So long as the sum of weights, at each location, and across all the polygons is either zero or unity, the resulting pattern is 'legal' (all its elements have a magnitude of 1 and therefore lie on the unit circle) and will be perfectly balanced.

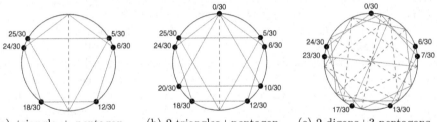

(a) triangle + pentagon − digon make a 6-element pattern in a 30-fold period.

(b) 2 triangles+pentagon− digon make a 9-element pattern in a 30-fold period.

(c) 2 digons+3 pentagons− 3 digons − 2 triangles make Amiot's scale.

Fig. 7. Perfectly balanced integer combinations of intersecting regular polygons in a thirty-fold period. When the vertex of one positive-weighted polygon coincides with the vertex of one negative-weighted polygon they cancel out to zero.

We will illustrate with some simple examples. First, let us start with a digon with a weight of −1. We can cancel out both its vertices by adding coprime unit-weighted polygons that share its vertices – as shown in Fig. 7(a), where we add an equilateral triangle and a regular pentagon. Indeed, we can add another intersecting polygon to one of the digon's vertices which gives, at that location, a combined weight of unity – as shown in Fig. 7(b). In Fig. 7(c), we show precisely how Amiot's scale can be derived from ten intersecting polygons with positive and negative unity weights.

In forthcoming work, we will show that any perfectly balanced subset of an equally tempered (or isochronous) universe can be constructed in the same way; that is, as an integer-weighted sum of regular polygons.

Theorem 5. *Let $N \in \mathbb{N}$. Any perfectly balanced vector $\boldsymbol{a} \in \{0,1\}^N$ can be expressed as an integer combination of regular polygons; that is,*

$$\boldsymbol{a} = \sum_{m=1}^{M} j_m \boldsymbol{p}_m \tag{5}$$

for some $M \in \mathbb{N}$, integers j_m, and N-fold regular n-gons $\boldsymbol{p}_m$ with $n > 1$.

This theorem shows that the method described in the next section, which provides a simple parameterization to connect a variety of perfectly balanced scales across a continuum, can generate any possible perfectly balanced scale or rhythm embedded in any N-fold universe.

Smoothly Rotating Polygons. To explain the proposed method for navigating over useful continua of perfectly balanced scales, it may be helpful to first consider an analogous approach for evenness. The approach is to maximize evenness under the constraint of a given j and k, where j is the number of large steps all of size ℓ and k is the number of small steps all of size s such that $s < l$, and j and k are coprime. The maximally even pattern of these step sizes is irreducibly periodic and, for a given j and k, is invariant over all ℓ and s (the word with alphabet l and s is a conjugate of a Christoffel word encoding an integer path of j/k). The choice of j and k essentially constrains the space into a one-dimensional form that can be parameterized by ℓ/s, the ratio of the large and small step-sizes. Such scales are typically called *well-formed* and are discussed in depth in [14].

An analogous constraint can be applied to perfectly balanced scales. We can specify a small number k of regular polygons (or perfectly balanced integer combinations of intersecting polygons like the examples in Fig. 7) such that their numbers of vertices are coprime, and then simply smoothly rotate the polygons independently between intersections. This results in a bounded $(k - 1)$-dimensional continuum that can be smoothly navigated.

As demonstrated in [15], we can take a single well-formed scale, characterized by (j, k), and search for ℓ/s values that give numerous good approximations of privileged structures (e.g., just intonation intervals, which have low integer frequency ratios). An analogous process can be applied to the perfectly balanced scales as parameterized by the relative phases of their constituent polygons. This may, therefore, be a useful method for determining a novel class of musically interesting microtonal scales. We intend to identify such scales in future work.

Optimizing Against the DFT. Another intriguing possibility is to use optimization to find perfectly balanced patterns in the continuum. This requires randomly initializing a K-element pattern (the phase values in $\boldsymbol{z}$), and optimizing it against a loss function defined as $|\mathcal{F}\boldsymbol{z}[0]|$, so as to converge to a perfectly balanced pattern. Early experiments have shown that such patterns take a wide variety of forms and describe an interesting manifold. We are currently investigating the use of loss functions incorporating additional factors, such as evenness and symmetry, so as to impose more regularity on this distribution. This technique provides a natural match for an æsthetic that embraces unpredictability.

4   Conclusion

We have shown that perfect evenness implies perfect balance, but that perfect balance does not necessarily imply perfect evenness. Creative work and research that has targeted evenness may, therefore, have inadvertently targeted balance too. By disentangling these two properties we hope to have opened up a new method for analysing, constructing, and understanding scales and meters.

We have demonstrated an analytical method, using the discrete Fourier transformation, as well as geometrically driven approaches, using integer combinations of disjoint or intersecting regular polygons, to construct perfectly balanced rhythms and scales. The methods suggested in this article are a first attempt to give both musicians and researchers the opportunity to create and manipulate balance within music.

In addition to the points already mentioned, future work could investigate differently weighting each scale degree or time event. Using this method, the principle of perfect balance could be applied to any conceivable pattern. For example, we might weight the events by their probability of occurring in a stochastic process (or prevalence in a composition), by their loudness, or by any conceivable musical parameter.

It might also be of considerable interest to investigate human perception and production of perfectly balanced but uneven rhythms and scales in order to further elucidate the impact of balance.

Acknowledgements. The first author would like to thank Emmanuel Amiot for invigorating conversations about evenness and balance, and also for opening his eyes to the possibility of perfectly balanced patterns not derived from disjoint regular polygons.

References

1. Clough, J., Douthett, J.: Maximally even sets. J. Music Theory **35**, 93–173 (1991)
2. Johnson, T.A.: Foundations of Diatonic Theory: A Mathematically Based Approach to Music Fundamentals. Scarecrow Press, USA (2008)
3. London, J.: Hearing in Time: Psychological Aspects of Musical Meter. Oxford University Press, Oxford (2004)
4. Carey, N., Clampitt, D.: Aspects of well-formed scales. Music Theory Spectrum **11**, 187–206 (1989)
5. Amiot, E.: Discrete Fourier transform and Bach's good temperament. Music Theory Online **15**(2) (2009)
6. Fisher, N.I.: Statistical Analysis of Circular Data. Cambridge University Press, Cambridge (1993)
7. Amiot, E.: David Lewin and maximally even sets. J. Math. Music **1**, 157–172 (2007)
8. Lewin, D.: Re: intervallic relations between two collections of notes. J. Music Theory **3**, 298–301 (1959)
9. Quinn, I.: A Unified Theory of Chord Quality in Equal Temperaments. Ph.D. Dissertation, University of Rochester (2004)
10. Callender, C.: Continuous harmonic spaces. J. Music Theory **51**, 277–332 (2007)

11. Amiot, E.: Sommes nullés de racines de l'unité. Bulletin de l'Union des Professeurs de Spéciales **230**, 30–34 (2010)
12. Messiaen, O.: Technique De Mon Langage Musical. Leduc, Paris (1944)
13. Balzano, G.J.: The pitch set as a level of description for studying musical perception. In: Clynes, M. (ed.) Music, Mind, and Brain. Plenum Press, New York (1982)
14. Milne, A.J., Carlé, M., Sethares, W.A., Noll, T., Holland, S.: Scratching the scale labyrinth. In: Agon, C., Andreatta, M., Assayag, G., Amiot, E., Bresson, J., Mandereau, J. (eds.) MCM 2011. LNCS, vol. 6726, pp. 180–195. Springer, Heidelberg (2011)
15. Milne, A.J., Sethares, W.A., Laney, R., Sharp, D.B.: Modelling the similarity of pitch collections with expectation tensors. J. Math. Music **5**, 1–20 (2011)

Characteristics of Polyphonic Music Style and Markov Model of Pitch-Class Intervals

Eita Nakamura[✉] and Shinji Takaki

National Institute of Informatics, Tokyo 101-8430, Japan
eita.nakamura@gmail.com, takaki@nii.ac.jp

Abstract. For the purpose of quantitatively characterising polyphonic music styles, we study computational analysis of some traditionally recognised harmonic and melodic features and their statistics. While a direct computational analysis is not easy due to the need for chord and key analysis, a method for statistical analysis is developed based on relations between these features and successions of pitch-class (pc) intervals extracted from polyphonic music data. With these relations, we can explain some patterns seen in the model parameters obtained from classical pieces and reduce a significant number of model parameters (110 to five) without heavy deterioration of accuracies of discriminating composers in and around the common practice period, showing the significance of the features. The method can be applied for polyphonic music style analyses for both typed score data and performed MIDI data, and can possibly improve the state-of-the-art music style classification algorithms.

Keywords: Polyphonic music analysis · Pitch class interval · Statistical music model · Music style recognition · Composer discrimination

1 Introduction

Harmonic and melodic features of polyphonic music have long been recognised to characterise music styles in and around the common practical period [1–3]. A quantitative and computational method of analysing these features would yield applications such as music style/genre recognition. However, a direct analysis is not easy because it requires chord and key recognition techniques, which are still topics of developing research [4,5]. Music style/genre classification has recently been gathering attentions in music information processing (e.g. [6–11]), but there is still much room for researches in incorporating/relating and traditional knowledge in music theory and musicology to computational models. How to extract effective features from a generic polyphonic music data including performed MIDI data with temporal fluctuations of notes is also an open problem. In this study, we relate four traditionally recognised features of polyphonic music to computationally extractable elements of generic polyphonic MIDI data and develop a method for statistical analysis based on these elements.

© Springer International Publishing Switzerland 2015
T. Collins et al. (Eds.): MCM 2015, LNAI 9110, pp. 109–114, 2015.
DOI: 10.1007/978-3-319-20603-5_10

2   Polyphonic Features and Successions of PC Intervals

We list four commonly studied features of polyphonic tonal music regarding harmony and melody [1–3], which will be called polyphonic features:

F1 Dissonant Chords and Motions: Use of dissonant chords and motions is generally more severely constrained in older music.

F2 Non-diatonic Motions: These include successive semi-tone-wise motions and a succession of major third, etc. and characterise music styles.

F3 Modulations: The type and frequency of modulations characterise composers and periods.

F4 Non-harmonic Notes: Their usage and frequency characterise composers and periods.

In order to study these features efficiently for generic music data, we only consider intervals of the pitch classes (pcs) and disregard other elements including durations as the subject of analysis. We assume that the data is represented as a sequence of integral pitches (with the identification of enharmonic equivalents) ordered according to their onset times. If there are several notes with simultaneous onset times, we prescribe that they can be ordered in any way. Any data, either typed scores or recorded performances, given in MIDI format can be used. The sequence of pc intervals is obtained by applying the modulo operation of divisor 12 and then taking intervals. Because data points with a zero pc interval express little about the polyphonic features, they are dropped and we have a reduced sequence of pc intervals denoted by $\boldsymbol{x} = (x_n)_{n=1}^{N}$ ($x_n = 1, \cdots, 11$).

A dissonant interval in a chord can be expressed by a pc interval $x = 1, 2, 6, 10, 11$ within the chord. Since only tritone is a definite dissonant interval in a melodic motion with the identification of enharmonic equivalents, $x = 6$ is the only direct indication of a dissonant motion. Based on these facts, the distribution of pc intervals is used to characterise music styles in Ref. [6].

Extending this basic result, more abundant information on the polyphonic features can be extracted from the successions of pc intervals (hereafter PCI successions). For example, a chord containing G, B, and F could be represented by a succession (F,G,B), and correspondingly, $(2, 4)$ in the sequence of pc intervals. The tritone is implicit as a pc interval but appears indirectly as a composite interval of $2 + 4 = 6$. A similar case appears in an indirect melodic motion involving a tritone and in false relations involving a tritone. These cases be generalised to a succession of two pc intervals (x_n, x_{n+1}) with $x_n + x_{n+1} = 1, 2, 6, 10, 11$.

A diatonic motion of pitches can be defined as a sequence of pitches which can be embedded in a diatonic (or major) scale, and a non-diatonic motion is defined conversely. Any pc interval can result from a diatonic motion: A pc interval 1 can correspond to m2[1] (we express this as $1 \rightarrow$ m2), and similarly, $2 \rightarrow$ M2, $3 \rightarrow$ m3, $4 \rightarrow$ M3, $5 \rightarrow$ P4, $6 \rightarrow$ a4, d5, $7 \rightarrow$ P5, $8 \rightarrow$ m6, $9 \rightarrow$ M6, $10 \rightarrow$ m7, $11 \rightarrow$ M7. By contrast, certain successions appear only in non-diatonic

---

[1] We use abbreviations for diatonic intervals such as m2 is minor second, M2 is major second, P4 is perfect fourth, a4 is augmented fourth, d5 is diminished fifth, etc.

Table 1. Classes of successions of two pc intervals and their relation with the polyphonic features.

Label	Name	Member
C1	Succession to tritone/ semitone/whole tone	$\{(x,y)\|y=6,1,11,2,10\}$
C2	Indirect octave	$\{(x,y)\|x+y=0 \bmod 12\}$
C3	Indirect tritone/ semitone/whole tone	$\{(x,y)\|$ $x+y=6,1,11,2,10 \bmod 12\}$
C4	Non-diatonic succession	Given in the text
C5	Major/minor triad	Given in the text

Polyphonic feature	Related class(es)
F1	C1, C3
F2	C4
F3	C4
F4	C2, C5

motions. With some calculation, we can find all such "non-diatonic successions" (x_n, x_{n+1}) as (1,1), (1,3), (1,8), (1,10), (2,11), (3,1), (3,8), (4,4), (4,9), (4,11), (8,1), (8,3), (8,8), (9,4), (9,11), (10,1), (11,2), (11,4), (11,9), (11,11). Although it is not always true, a modulation often involves a non-diatonic motion (within a voice or across voices), which induces a non-diatonic PCI succession if it occurs in a small range.

Non-harmonic notes cannot be expressed simply in the sequence of pc intervals without preliminary chord analysis. Nevertheless we can find related PCI successions by paying attention to the opposite notion of "harmonic notes". The condition of a harmonic note in a strict sense is that the note and the other notes sounding at the same time are contained in a major or minor triad. A major/minor triad can be expressed with a pair of pc intervals. For example, a C major chord can appear as (E,G,C) and its permutations in the sequence of pcs, which is expressed as PCI successions (3,5), (4,3), (5,4), (7,9), (8,7), or (9,8). Similarly a minor triad is expressed as (3,4), (4,5), (5,3), (7,8), (8,9), or (9,7). Another class of PCI successions related to harmonic notes is indirect octave, which is represented as (x_n, x_{n+1}) with $x_n + x_{n+1} = 0 \bmod 12$. Many simultaneous notes in octaves imply that they are harmonic notes. They are related to the number of voices or the chord density, which characterise polyphonic textures.

Table 1 summarises the discussed classes of successions of two pc intervals and their relation to the polyphonic features. The same argument can be applied to successions of three or more pc intervals. Longer successions provide more information, but it is harder to obtain statistically meaningful results when dealing with numerical data. We here concentrate on successions of two pc intervals.

3   Markov Model of PC Intervals

The polyphonic features characterise music styles in terms of their frequencies, and statistical models can be used to describe their quantitative nature. A simple statistical model that can describe relations between successions of two data points in a sequence is the first-order Markov model. The (stationary) Markov model of pc intervals is described with an initial probability and transition probabilities, which are given by $P^{\mathrm{ini}}(x) = P(x_1 = x)$ and $P(x|y) = P(x_{n+1} = x|x_n = y)$. Due to the ergodicity of a Markov model,

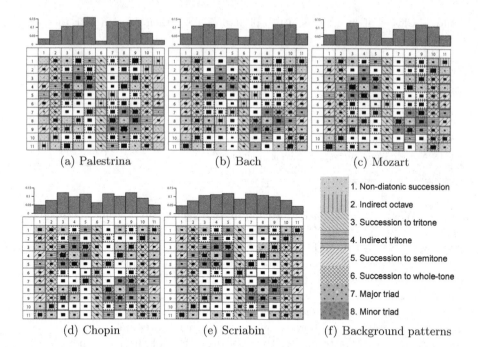

Fig. 1. Transition probabilities obtained from pieces of Palestrina, Bach, Mozart, Chopin, and Scriabin (a)–(e). Each black square at the centre of the (x,y)-th cell shows the transition probability $P(y|x)$ in proportional to the value. For each row in each table, the three highest (resp. lowest) values are indicated with blue dashed (resp. red bold) square frames. The list (f) explains the background patterns and colours of the cells. The distribution of pc intervals $P(x)$ is also shown above each table (Color figure online).

the initial probability has little effect for a long sequence, and we here mainly consider the transition probabilities, which have $11 \cdot 10 = 110$ independent parameters. The distribution of pc intervals $P(x) = P(x_n = x)$ can be derived from the transition probabilities by the equilibrium equation: $P(x) = \sum_y P(x|y)P(y)$. The relative frequencies of a succession (x, y) is described with $P(y|x)$.

Figure 1 illustrates the values of transition probabilities obtained from MIDI data of pieces by five composers, Palestrina, J. S. Bach, Mozart, Chopin, and Scriabin, whose works are usually associated with the period of the Renaissance, the Late Baroque, the Classical, the Early Romantic, and the Late Romantic/Early 20th century. The number of pieces and the data size are shown in Table 2. The background patterns of the cells indicate the classes of the corresponding PCI successions as summarised in Fig. 1(f). When a succession belongs to more than one classes, the upper most class in the list is indicated.

We see that some patterns in the transition probabilities accord with the background patterns of the cells. For example, non-diatonic successions, successions to tritone, and indirect tritones generally have small probabilities. Probability values corresponding to these PCI successions are generally larger for composers of later

Table 2. Results of discriminating pieces by five composers with the Markov model with the constrained (resp. full) parametrisation. Each value indicates the rate (%) of pieces recognised as the corresponding composer.

Composer	Data size	Palestrina	Bach	Mozart	Chopin	Scriabin
Palestrina	175 pcs (1.95 MB)	94.9 (99.4)	2.9 (0.6)	1.7 (0)	0.6 (0)	0 (0)
Bach	108 pcs (0.93 MB)	2.8 (0)	77.8 (80.6)	13.0 (13.9)	1.9 (5.6)	4.6 (0)
Mozart	77 pcs (2.40 MB)	2.6 (1.3)	5.2 (3.9)	71.4 (84.4)	15.6 (10.4)	5.2 (0)
Chopin	90 pcs (1.51 MB)	3.3 (1.1)	10.0 (1.1)	15.6 (14.4)	44.4 (67.8)	26.7 (15.6)
Scriabin	102 pcs (0.79 MB)	9.8 (5.9)	15.7 (6.9)	3.9 (2.0)	14.7 (19.6)	55.9 (65.7)

periods, which is a consequence of the time evolution in the use of dissonances. Similarly, other patterns of transition probabilities can be associated with the classes discussed in the previous section, and their tendencies for each composer reflect the quantitative nature of the polyphonic features in different music styles. We omit further details of the analysis for the lack of space.

4   Constrained Parametrisation and Composer Discrimination

To quantitatively examine how much the polyphonic features provide information to characterise different music styles, we compare results of composer discrimination with the Markov model and a reduced model with constrained parameters that are related to the classes of PCI successions. In the constrained model, we introduce five parameters p(non-diatonic), p(indirect-octave), p(tritone), p(second), and p(triad), which parametrises transition probabilities of class 1, 2, $\{3,4\}$, $\{5,6\}$, and $\{7,8\}$ in Fig. 1(f). The rest probabilities $P(x|y)$ are assumed to be uniform for each y and determined by the normalisation of probabilities $\sum_x P(x|y) = 1$. An algorithm to discriminate composers can be developed from these models with the maximum likelihood estimation.

Results of composer discrimination are shown in Table 2. To avoid statistical artefacts by overfitting, the piece-wise leave-one-out method was used. The composer-wise averaged accuracy was 68.9 % (resp. 79.6 %), and the mean reciprocal rank (MRR), which is the averaged reciprocal rank of the correct composer, was 1/1.20 (resp. 1/1.11) for the constrained (resp. full) parametrisation. Compared to the reduction of parameters (110 to five), there was a small decrease of the accuracy for Palestrina and Bach, and a rather large (but not very large) decrease for the other composers. For Chopin and Scriabin, we see that a large proportion of misclassified pieces are associated with adjacent composers in the table for both models.

The rather high accuracies of the full Markov model indicate that the model parameters well capture characteristics of the composers, and the not-heavy deterioration of accuracies with the reduced model indicates that a significant part of the characteristics is associated with the polyphonic features. The overall tendency that misclassified pieces were more frequently classified to a composer that is near in the lived period implies that the features capture not only particular styles of the composers but also a generic style of the composed period to some extent, which confirms the general intuition about the evolution of music styles.

5   Discussion

It is interesting to apply the present analysis for music style classification problems. The current state-of-the-art classification algorithms naturally employ many features related to pitch and rhythm [8–10], and the use of the polyphonic features and the pcs intervals would improve the accuracy, computational efficiency, and generality. It would be possible to construct an effective classification algorithm applicable for general polyphonic MIDI data including performance recordings, for which information on voice and rhythm cannot be extracted directly. To our knowledge such an algorithm has not been proposed so far.

Acknowledgement. This work is supported in part by Grant-in-Aid for Scientific Research from Japan Society for the Promotion of Science, No. 25880029 (E.N.).

References

1. Jeppesen, K.: The Style of Palestrina and the Dissonance, 2nd edn. Dover Publications, New York (2005). Originally published by Oxford Univ. Press in 1946
2. Kostka, S., Payne, D., Almén, B.: Tonal Harmony, 7th edn. McGraw-Hill, New York (2004)
3. Tymoczko, D.: A Geometry of Music. Oxford University Press, New York (2011)
4. Hu, D.: Probabilistic topic models for automatic harmonic analysis of music. Ph. D. Assertion, UC SanDiego (2012)
5. Handelman, E., Sigler, A.: Key induction and key mapping using pitch-class set assertions. In: Yust, J., Wild, J., Burgoyne, J.A. (eds.) MCM 2013. LNCS, vol. 7937, pp. 115–127. Springer, Heidelberg (2013)
6. Honingh, A., Bod, R.: Clustering and classification of music by interval categories. In: Agon, C., Andreatta, M., Assayag, G., Amiot, E., Bresson, J., Mandereau, J. (eds.) MCM 2011. LNCS, vol. 6726, pp. 346–349. Springer, Heidelberg (2011)
7. Wolkowicz, J., Kulka, Z.: N-gram based approach to composer recognition. Arch. Acoust. **33**(1), 43–55 (2008)
8. Hillewaere, R., Manderick, B., Conklin, D.: String quartet classification with monophonic models. In: Proceedings of the ISMIR, pp. 537–542 (2010)
9. Hasegawa, T., Nishimoto, T., Ono, N., Sagayama, S.: Proposal of musical features for composer-characteristics recognition and their feasibility evaluation. J. Inf. Process. Soc. Jpn. **53**(3), 1204–1215 (2012). (in Japanese)
10. Conklin, C.: Multiple viewpoint systems for music classification. J. New Music Res. **42**(1), 19–26 (2013)
11. MIREX (Music Information Retrieval Evaluation eXchange) homepage: http://www.music-ir.org/mirex/wiki/MIREX_HOME

A Corpus-Sensitive Algorithm for Automated Tonal Analysis

Christopher Wm. White[✉]

The University of Massachusetts Amherst, Amherst, USA
cwmwhite@music.umass.edu

Abstract. A corpus-sensitive algorithm for tonal analysis is described. The algorithm learns a tonal vocabulary and syntax by grouping together chords that share scale degrees and occur in the same contexts and then compiling a transition matrix between these chord groups. When trained on a common-practice corpus, the resulting vocabulary of chord groups approximates traditional diatonic Roman numerals. These parameters are then used to determine the key and vocabulary items used in an unanalyzed piece of music. Such a corpus-based method highlights the properties of common-practice music on which traditional analysis is based, while offering the opportunity for analytical and pedagogical methods more sensitive to the characteristics of individual repertoires.

Keywords: Computation · Corpus analysis · Cognitive modeling · Tonality · Harmony · Analysis · Alphabet · Vocabulary · Syntax

1 Introduction

Figure 1 shows the first phrase from Mozart's Theme and Variations, K. 284, iii. A human would analyze the two chords within the dotted boxes as vi in D major and ii^6 in A major, respectively. However, while this task may be straightforward for human musicians, the computational task is anything but. First, the computer must recognize that a B-minor triad underlies the two circled < 1, 2, 4, 6, 11 > pc-sets, even though the ordering and metric position of the triad's pcs is not consistent between the two events. Second, the computer must recognize the ideal keys in which the two B-minor triads function. This task is approximated in the rows below the example: we might imagine the program comparing various tonal interpretations, settling on the most probable D major and A major analyses.

This paper will introduce a corpus-based approach to algorithmic key finding by creating a model that learns a chord vocabulary from a common-practice corpus and then uses this vocabulary to analyze musical surfaces. Our approach will be similar to Ponsford et al. [1] and Hedges et al. [2], Cope [3, 4], Rohrmeier and Cross [5], Quinn [6], and Quinn and Mavromatis [7] in its processing of a style-specific corpus, and to Krumhansl [8], Temperley [9] and Chew [10] by modeling tonal orientations using surface information; however, the unsupervised learning of "classroom-style" Roman numerals from score-based data uniquely situates this work to contribute to music

© Springer International Publishing Switzerland 2015
T. Collins et al. (Eds.): MCM 2015, LNAI 9110, pp. 115–121, 2015.
DOI: 10.1007/978-3-319-20603-5_11

Fig. 1. Mozart's K. 284, iii, mm. 1–8

theoretical discussions of chords and chord types, along with pedagogical discussions of how students learn chord syntax when being exposed to a new corpus.

2 Materials and Method

This study relies on the Yale Classical Archive Corpus, or YCAC. This corpus, as described at www.ycac.yale.edu and White and Quinn [11], consists of "salami slices" of MIDI files (vertical slices each time a pitch is added or subtracted from the texture); the corpus is analyzed using a windowed key-profile analysis, such that each stretch of music is either identified with a key and mode or labeled as ambiguous. The files were divided into lists of unordered scale-degree sets (thereby ignoring the bass note and voicing), and divisions were made at key changes. To simplify the dataset, if a set of cardinality <3 was a subset of either adjacent set, the subset was deleted, as were repeated sets. (In the YCAC, a slice's most frequent transition is almost always to itself: this fact was incorporated into the reduction function, rather than the training set.) Inversionally nonequivalent prime forms of each slice were also compiled (i.e. [037] ≠ [047]). This process was implemented in the Python language using the music21 software package (Cuthbert and Ariza [12]).

3 The Reduction Formalism

The algorithm transforms a *raw vocabulary* of chords drawn from a musical surface into a *reduced vocabulary* using the original chords' contexts (i.e. their transition probabilities) and their scale-degree content, "reducing" a musical surface based on contextual probabilities and some notion of edit distance. The formalization of (1), adapted from White [13], shows that the series of observations o in the raw vocabulary O can be related to a reduced vocabulary S in which s and o at timepoint i share at least one set member.

The two parameters of this equation are the *contextual probability*, or $P\left(s_i \middle| k\left(o_i\right)\right)$ and the *set proximity*, or $\pi(s_i, o_i)$. The relationship between the two categories maximizes the product of the two parameters throughout the sequence of n events.

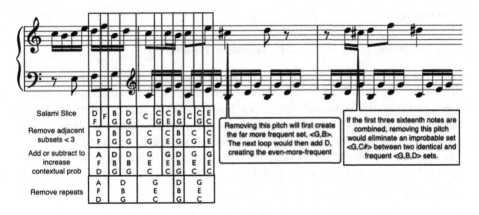

Fig. 2. An example of the reduction process in Mozart K. 279

$$O = (o_1, o_2 \ldots o_n) \; ; S = (s_1, s_2 \ldots s_n)$$
$$s_i \cap o_i \geq 1 \tag{1}$$
$$S = \arg\max \prod_{i=1}^{n} P\left(s_i | k\left(o_i\right)\right) \pi\left(s_i, o_i\right)$$

In what follows, we take a simple approach to these general relationships, using n–gram probabilities with $n = 1$ (i.e., first-order Markov probabilities) drawn from the transitions in the raw vocabulary. The set proximity was determined by the number of common tones shared by the two chords s_i and o_i. In the current study, at each stage of reduction, we take the simplifying step of only considering chords related by the addition or subtraction of one scale degree.

4   The Reduction Process and Reduced Vocabulary

In order to maximize the two parameters over the whole series of observations within the corpus, an iterative dynamic program was used. Figure 2 shows an example of a reduction, and python-based pseudo-code can be found in this study's online supplement at chriswmwhite.com/research. To accommodate our earlier assumptions, repeats are removed at each iteration. Furthermore, to simplify the task (as well as remove encoding errors) improbable chords were removed, here defined as chords not within the top 85 % of the zero-eth order probability mass. Once these steps have been taken, the above reduction formalism is applied. To maximize the probability of the entire sequence, the Viterbi algorithm is used (Jurafsky and Martin [14]). This process them produces a *reduced vocabulary*, along with the transitions between those vocabulary members. The properties of this reduced vocabulary are discussed in White [13] and in this study's online supplement.

5   Analysis

5.1   Conforming to the Reduced Prime-Form Vocabulary

In early versions of this algorithm, the myriad ways that an undifferentiated string of salami slices could be organized into scale-degree sets proved both computationally expensive and produced inconsistent results. It was therefore determined to divide the analytical task into two sub-routines: first, an unanalyzed musical surface is parsed to conform to the prime forms of the reduced vocabulary; second, the surface is conformed to the reduced vocabulary's scale-degree sets in order to determine the passage's key. Figure 3 illustrates the steps involved in the first sub-process.

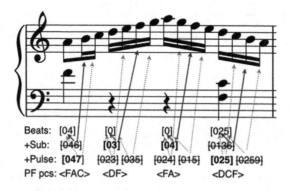

Fig. 3. Conforming Mozart K. 545, i, m. 5 to the prime-form vocabulary

This subroutine divides the music into metric levels and, favoring the stronger metric levels, finds combinations of pitches that produce the most probable prime forms, here reckoned as zero-eth order (unigram) probability-of-occurrence within the reduced vocabulary. Performing a meter-finding analysis on the music surrounding Fig. 4 indicates the quickest division of the music to be the sixteenth note, followed by the eighth note, and then the quarter note. (In principle, any meter-finding algorithm may be used; however, since we are committed to an analysis that begins with no prior knowledge or templates, this study uses Zikanov [15]'s autocorrelation approach, an algorithm that finds recurrent periodicities in note onset patterns.) The rows below Fig. 3 list the prime forms produced at each metric level with the arrows indicating the combination used to produce these sets: after finding the beat's prime forms, we add the pitches of the following subdivision and pulses. Once all possible combinations are considered, the pcs of the most probable prime forms are collected, as shown in the example's last row. Again, the process is outlined in python-based pseudo-code in this study's online supplement.

5.2   Conforming to the Reduced Scale-Degree Set Vocabulary

The process then conforms the surface to each of the twelve possible scale-degree levels. A window of music is subjected to the reduction procedures described in Sect. 4. If the reduction process causes the window to fall below some preset number of chords x, the window is extended. (This ensures sufficient context – a long enough series of transitions – for successful key finding.) As in the reduction process, the Viterbi algorithm chooses the sequence of chords that returns the maximum probability, now using scale-degree sets from the reduced vocabulary; in turn, the transposition level whose sequence returns the highest probability is chosen as the analysis. Figure 1 can now be understood as an example of this process. At four different scale-degree transpositions, the program returns relatively higher and lower overall probabilities, ultimately choosing that with the highest results. The current program tracks the overall probabilities in bits: the first window chooses D major (39.115 bits) over G major (52.38 bits) and A major (119.394 bits); in the second passage, A major is now selected (13.558 bits) over D major (207.95 bits).

6 Discussion

6.1 Overlaps with Traditional Analysis

Perhaps the most striking characteristic of this work is its overlap with traditional notions of Roman numeral analysis. As seen in the chosen analyses of Figs. 1 and 2, the reduced vocabulary distills most non-traditional structures present in the raw vocabulary into triads and seventh chords. The reduced prime forms and syntax are also recognizable analogues to traditional textbook tonal harmony, especially in their utility to produce intuitive tonal analyses. By tracing a pathway between surface statistics and analytical intuition, the reduction process seems to illustrate properties of common-practice music (as represented by the YCAC) that invite analysts to adopt and are highlighted by Roman-numeral analysis. However, these connections are bolstered only by intuition and not experimentation. Further work would quantify the success of this algorithm by either having its output graded by a series of human experts or comparing its analyses to the Roman numerals of hand-tagged corpora.

Notably this process speaks to pedagogical considerations: teaching music theory at an undergraduate level often involves introducing students to common-practice music who have had little prior experience with that repertoire. These students' task – learning the norms and vocabulary of new repertoire – is exactly the computational task described here. While not advocating for undergraduate music theory to be taught in an algorithmic manner, a program that relies on the properties of the corpus to produce a textbook-like chord vocabulary can teach us about the process through which someone might learn this vocabulary by being exposed to common-practice music.

6.2 Cultural Implications

As described in Huron [16], Byros [17], and White [18], different time periods, social groupings, and geographies produce corpora with different tonal properties. It not only seems possible, but expected that these corpora might differ to the point that they exhibit different underlying syntaxes. When we import an analytical method tailored to one repertoire onto another, we are at best using incorrect technology and at worst imposing a problematic canon/periphery power dynamic into our analyses, a dynamic mitigated by our data-driven modeling of chord vocabularies and syntaxes.

7   Conclusions and Future Goals

An algorithm was presented that learns a tonal vocabulary and syntax by grouping together chords that share scale degrees and occur in the same contexts and compiling a transition matrix between these chord groups. This training was then used by an analysis algorithm to determine the key and chords of an unanalyzed piece of music. Due to its overlap with traditional Roman-numeral analysis, it was suggested that this method illustrates properties of common-practice music highlighted by traditional classroom analysis. Pedagogical and cultural implications were also discussed.

References

1. Ponsford, D., Wiggins, G., Mellish, C.: Statistical learning of harmonic movement. J. New Music Res. **28**(2), 150–177 (1999)
2. Hedges, T., Roy, P., Pachet, F.: Predicting the composer and style of jazz chord progressions. J. New Music Res. **43**(3), 276–290 (2014)
3. Cope, D.: Experiments in music intelligence. In: Proceedings of the International Computer Music Conference, pp. 170–173. Computer Music Association, San Francisco (1987)
4. Cope, D.: Computer Models of Musical Creativity. MIT Press, Cambridge (2005)
5. Rohrmeier, M., Cross, I.: Statistical properties of tonal harmony in Bach's Chorales. In: Proceedings of the International Conference on Music Perception and Cognition, ICMPC, pp. 619–627. Sapporo (2008)
6. Quinn, I.: Are pitch-class profiles really key for key? Zeitschrift der Gesellschaft der Musiktheorie **7**, 151–163 (2010)
7. Quinnn, I., Mavromatis, P.: Voice leading and harmonic function in two chorale corpora. In: Agon, C., Andreatta, M., Assayag, G., Amiot, E., Bresson, J., Mandereau, J. (eds.) MCM 2011. LNCS, vol. 6726. Springer, Heidelberg (2011)
8. Krumhansl, C.L.: The Cognitive Foundations of Musical Pitch. Oxford University Press, Oxford (1990)
9. Temperley, D.: Music and Probability. The MIT Press, Cambridge (2007)
10. Chew, E.: Mathematical and Computational Modeling of Tonality: Theory and Applications. Springer, New York (2014)
11. White, C., Quinn, I.: The Yale-classical archives corpus. Poster presented at: International Conference for Music Perception and Cognition. Seoul, South Korea (2014)
12. Cuthbert, M.S., Ariza, C.: Music21: a toolkit for computer-aided musicology and symbolic music data. In: Proceedings of the International Symposium on Music Information Retrieval, ISMIR, Utrecht, pp. 637–42 (2010)
13. White, C.: An alphabet reduction algorithm for chordal N-grams. In: Yust, J., Wild, J. (eds.) MCM 2013. LNCS, vol. 7937. Springer, Heidelberg (2013)
14. Jurafksy, D., Martin, J.H.: Speech and Language Processing: An Introduction to Natural Language Processing, Computational Linguistics, and Speech Recognition. Prentice Hall, Upper Saddle River (2000)
15. Zikanov, K.: Metric Properties of Mensural Music: An Autocorrelation Approach. National Meeting of the American Musicological Society, Milwaukee (2014)
16. Huron, D.: Sweet Anticipation: Music and the Psychology of Expectation. The MIT Press, Cambridge (2006)

17. Byros, V.: Foundations of tonality as situated cognition, 1730–1830. Ph.D dissertation. Yale University (2009)
18. White, C.: Changing styles, changing corpora, changing tonal models. Music Percept. **31**(2), 244–253 (2014)

Finding Optimal Triadic Transformational Spaces with Dijkstra's Shortest Path Algorithm

Ryan Groves[✉]

McGill University, Montreal, Canada
ryan.groves@mail.mcgill.ca

Abstract. This paper presents a computational approach to a particular theory in the work of Julian Hook—Uniform Triadic Transformations (UTTs). A UTT defines a function for transforming one chord into another, and is useful for explaining triadic transitions that circumvent traditional harmonic theory. By combining two UTTs and extrapolating, it is possible to create a two-dimensional chord graph. Meanwhile, graph theory has long been studied in the field of Computer Science. This work describes a software tool which can compute the shortest path between two points in a two-dimensional transformational chord space. Utilizing computational techniques, it is then possible to find the optimal chord space for a given musical piece. The musical work of Michael Nyman is analyzed computationally, and the implications of a weighted chord graph are explored.

Keywords: Neo-Riemannian theory · Shortest path algorithm · Transformational spaces · Computational musicology

1 Introduction

Innovation has always been the impetus for further innovation; the creation of new concepts and new ideas immediately demands new methods of analysis, and new tools for that analysis. Contemporary Art Music is no exception. Indeed, innovative compositions of the late 20th century have thwarted the methods of analysis on which many theorists rely. In the late 1980s, a new sub-discipline of music theory arose out of the need to describe triadic and intervallic motions irrespective of a common key tonic. This sub-discipline has been labelled as transformational theory [7], and has inspired a handful of different theories that are concerned with triadic and intervallic relationships outside of the more common diatonic framework—those that relate pitch-based units directly by transforming one into the other.

The somewhat narrower field of neo-Riemannian theory is similarly focused on the transformation of musical objects, specifically triads. Neo-Riemannian theory encapsulates the work of Richard Cohn [2], and more recently Julian

R. Groves—Many thanks to Professor Robert Hasegawa for much encouragement and guidance during the process of this research.

T. Collins et al. (Eds.): MCM 2015, LNAI 9110, pp. 122–127, 2015.
DOI: 10.1007/978-3-319-20603-5_12

Hook [6]. All of these theorists are concerned with the relationships between musical objects independent of a centralized tonic reference. Together, the theories have been expanded to analyze composers in different eras, for example Milton Babbitt, Béla Bartók and György Ligeti [1,5,8].

For the purposes of this research, the definition of an optimal chord space is the transformational space which most efficiently represents the transitions of the triadic sequence of an entire piece. Since there are many possible harmonic spaces, this paper seeks to find the optimal space for a given piece. It is a logical extension, then, to employ one of the most effective tools in computational analysis to analyse the possible harmonic spaces of new music: the computer. Specifically, this essay will outline the application of UTTs to the first three pieces of Michael Nyman's *Six Celan Songs*, and a computational approach to discovering the most appropriate space.

1.1   Uniform Triadic Transformations

In Lewin's Generalized Musical Intervals and Transformations [7], he defines sets of pitches (labelled "spaces"), as well as corresponding groups of transformations. Based on Lewin's foundational work, Julian Hook similarly uses transformational spaces to define movement between triads [6]. Hook developed the concept of Uniform Triadic Transformations (UTT), which is based on Riemann's fundamental idea of major and minor triads' being mirror images of one another. The format for defining a UTT is <+/−, m, n>. The plus / minus sign defines whether or not the resulting triad will be the same (+) or the opposite (−) type of triad, and the m and n variables define the movement of the triad's root for the major and minor subsets, respectively [3].

1.2   Song Harmonies

Nyman's Six Celan songs are very triadic in their harmonic structure, but also don't follow standard tonal syntax. Thus, they are a good starting point for the utilization of digital UTT spaces. Most of the harmonic content in Nyman's songs have relatively regular durations, which is ideal since the UTT space does not consider durations of the triads. One drawback of the UTT space is that only major or minor triads can be considered. Therefore, if certain parts of the piece contain more complicated harmonies than simple triads, they must be simplified to the closest approximation of a major or minor triad. This is an unfortunate drawback, however most of the content in the pieces are simple triads, so the majority of the harmonies will be represented correctly.

The reason this piece is so relevant is that there are chord movements which cannot be ascribed to traditional forms of functional harmony. For example, the opening progression is Dm→ Fm→Am. The harmony rests on Am for more than 1 measure, which suggests that the tonic is Am. However, this movement could only be described as a iv→vi→i movement in harmonic minor, which is not a very common nor functional progression. We select the first three songs from Nyman's Six Celan songs as a proof of concept for the new optimal chord space method.

2   Finding the Optimal Harmonic Space of a Song

Hook describes certain groups of UTTs that have a unique transformation for every triadic transition in the group. These are called *simply transitive* groups, and Hook explains that "Simply transitive groups also bring a measure of clarity to situations involving apparently redundant transformations" [6]. Since our approach is statistical—that is, we are looking for the most repeated transformations of a particular piece—a simply transitive group would be a good starting point. We consider the simply transitive group K(1,1) as defined by Hook, shown in Fig. 1. Here, "T" and "E" represent the integers 10 and 11.

This subgroup was also the focus of analysis in [3]. In that research, Cook assembles his own toroidal space by creating a 2-dimensional graph for which each axis represents the repeated application of a particular UTT in K(1,1). Cook measures distance on that 2-dimensional space as the total number of transformations required to traverse the minimum path from one triad to another on that triadic graph. The approach presented here will be much the same.

Mode Preserving:				Mode Reversing:		
<+, 0, 0>	<+, 4, 4>	<+, 8, 8>		<−, 0, 1>	<−, 4, 5>	<−, 8, 9>
<+, 1, 1>	<+, 5, 5>	<+, 9, 9>		<−, 1, 2>	<−, 5, 6>	<−, 9, T>
<+, 2, 2>	<+, 6, 6>	<+, T, T>		<−, 2, 3>	<−, 6, 7>	<−, T, E>
<+, 3, 3>	<+, 7, 7>	<+, E, E>		<−, 3, 4>	<−, 7, 8>	<−, E, 0>

Fig. 1. The UTTs which comprise the K(1,1) simply transitive subgroup.

To select a pair of UTTs from the K(1,1) we look to the shortest path algorithm. Conceived in 1956 by Edsger Dijkstra, the shortest path algorithm will find the shortest path between two vertices in a graph, as long as a path exists between the two vertices [4]. Intuitively, the pair of UTTs that best represent a sequence of triads should have the shortest cumulative distance between each pair of triads in the sequence. This is a purely computational challenge. We have a list of UTTs provided. For every UTT, we must do the following procedure[1]:

1. Loop through every UTT that is not identical to the current UTT.
2. Create the 2-dimensional UTT space that represents all the triads that can possibly be visited given the two current transformations.
   a. Ensure that the created space contains every triad in the set of all major and minor triads.
3. Step through the triads in the given piece, and compute the shortest distance between each pair of triads, sequentially.
4. Add up the total number of transformations that were required for every triadic transition in the piece, based on its path in the 2-D space.

---

[1] The code for the generation of UTT spaces can be found at Ryan Groves' github landing page at http://github.com/bigpianist/UTTSpaces.

Every possible space created from pairs in the K(1,1) subgroup was computed for each piece, and the distances between each triadic transition computed. The most efficient spaces for the three analyzed Nyman songs are:

Song title	Optimal UTT pair	Dist
Chanson einer Dame im Schatten	(<−, 4, 5>, 1, A), (<+, 2, 2>, 1, B)	246
Es war Erde in ihnen	(<−, 1, 2>, 1, A), (<−, 2, 3>, 1, B)	172
Psalm	(<−, 1, 2>, 1, A), (<+, 2, 2>, 1, B)	107

3   Application and Investigation

The efficient spaces discovered for each song can then be used to explore the transformational motion going on in each song. Take the first song, *Chanson einer Dame im Schatten*, for example. First, it is useful to also visualize the space.

In Fig. 2, the opening phrase of *Chanson einer Dame im Schatten* is shown with its transformations in the UTT space of [(<−, 4, 5>, 1, A), (<+, 2, 2>, 1, B)].

It is difficult to tell how effective this new space is at first glance. However, with deeper inspection, one can start to understand the effectiveness of representing the harmony in such a way. Take, for example, the ending of most phrases. Often, each chord phrase ends in two successive chords, often on the last two beats of the last bar of the phrase. The opening phrase defines this clearly with its F→Am movement. Given the quickness of the transition between these two chords, it seems that an accurate harmonic space would likely find these chords to be close in harmonic distance. Grouping each of these

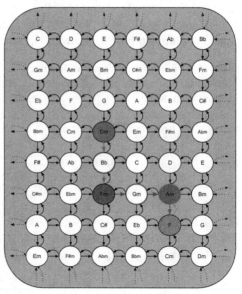

Fig. 2. UTT space of [(<−, 4, 5>, 1, A), (<+, 2, 2>, 1, B)], with the transformations of the opening chord sequence [Dm, Fm, Am, F, Am] shown in alternating colors.

quick transitions with their corresponding transformation(s) show that this is, indeed, the case (Table 1).

Two things are interesting about the transformations that the phrase-ending chord transitions create. For one, they are mostly a single transformation. The transitions with a single transformation also represent single transformations in *both* axes of the UTT space (i.e., there exists both a single 'A' transformation, as well as a single 'B' transformation), which implies that both UTTs are effective. Secondly, these transformations occur from different starting points, showing that it is not just the same chord sequence occurring repeatedly in the song.

The UTT space for *Es war Erde in ihnen* is necessarily an approximation because of its large quantity of triad simplifications (roughly 20 % were augmented or diminished, and had to be converted to major and minor triads, respectively). Thus, it is more useful to consider the space of *Psalm*, and to investigate the commonalities between the space of *Psalm* and *Chanson einer Dame im Schatten*. One of the most striking characteristics of the transformations of the chord transitions in *Psalm* is the combination of three identical transformations. For example, for a single chord transition in the song, the combination of three 'B' ($<+, 2, 2>$) transformations in sequence, or the inversely equivalent transformation of three 'B^{-1}' transformations in sequence occurs a total of 8 times in the song. Similarly, the total number of times the combination of three 'A' transformations ('A|A|A') or three 'A^{-1}' transformations ('A^{-1}|A^{-1}|A^{-1}') occurs is also 8. This is very convincing evidence that the UTT space is appropriate, since similar movements in both axes are equally likely. Furthermore, simple movements of 'A' and 'A^{-1}' occur a total sum of 13 times. The exception is the simple movement of a single 'B' or a single 'B^{-1}' transformation, both of which are completely missing from the song.

Table 1. Table of quick, phrase-ending chord transitions with their corresponding transformations.

Start chord	End chord	Transformation
F	Am	A
Am	C	$B^{-1}\|A$
Eb	Gm	A
Gm	Eb	A^{-1}
Eb	F	B
Dm	F	$B^{-1}\|A$
Am	Em	A^{-1}
Bb	Dm	A

Still, there is a distinct motion of three consecutive, identical transformations in both the 'A' and 'B' axes for the UTT space $[(<-, 1, 2>, 1, A), (<+, 2, 2>, 1, B)]$ in *Psalm*. Let us consider the UTT created when applying the UTT of $<-, 1, 2>$ three times consecutively. From a major chord, the root movement will be 1, then 2 (since the mode was reversed), then 1 again, for a total of 4. Starting on a minor chord, the root movement will be 2, then 1, then 2, for a total of 5, with the mode changing three times (which is equivalent to one reversal). The resulting UTT is exactly that from *Chanson einer Dame im Schatten*, $<-, 4, 5>$!

4   Conclusion

These new methods for transforming triads and intervals provide a new perspective on the possible harmonic spaces in which songs may have been composed. Furthermore, computational models of harmonic spaces provide immense power when considering multiple spaces. Commonly, the approach for utilizing these new transformational spaces has been a top-down approach, where the musicologist would identify what she thought were the most important harmonic movements, and shape the harmonic space around those. The computational method, on the other hand, provides a bottom-up approach, where every transition is considered when deciding on an optimal space. However, one must not rely entirely on this new tool set; still, there is a need for the application of musical intuition. As was evident in the analysis of the first three of Nyman's *Six Celan Songs*, the shortest distance algorithm found the most efficient harmonic space for each song. In the end, it was still up to the analyst to investigate the possible consequences of the identified spaces.

There is no doubt that the computer can serve music theorists for representing complex and abstract concepts. The current state of computer technologies affords a wide range of tools for the application of a computational approach to even the most cutting-edge of musical theories. It is with these tools that one can gain a new perspective on the music, and lead music analysts to new theories and innovative insights.

References

1. Callender, C.: Interactions of the lamento motif and jazz harmonies in György Ligeti's arc en ciel. Intégral **21**, 41–77 (2007)
2. Cohn, R.: Introduction to Neo-Riemannian theory: a survey and historical perspective. J. Music Theor. **42**(2), 167–180 (1998)
3. Cook, S.A.: Moving through triadic space: an examination of Bryars's seemingly haphazard chord progressions, March 2009. http://www.mtosmt.org/issues/mto.09.15.1/mto.09.15.1.cook.html
4. Dijkstra, E.W.: A note on two problems in connexion with graphs. Numer. Math. **1**, 269–271 (1959)
5. Gollin, E.: Some unusual transformations in Bartóks minor seconds, major sevenths. Intégral **12**(2), 25–51 (1998)
6. Hook, J.: Uniform triadic transformations. J. Music Theor. **46**(1/2), 57–126 (2002)
7. Lewin, D.: Generalized Musical Intervals and Transformations. Yale University Press, New Haven (1987)
8. Lewin, D.: Generalized interval systems for Babbitt's lists, and for Schoenberg's string trio. Music Theor. Spectr. **17**(1), 81–118 (1995)

A Probabilistic Approach to Determining Bass Voice Leading in Melodic Harmonisation

Dimos Makris[1]([✉]), Maximos Kaliakatsos-Papakostas[2],
and Emilios Cambouropoulos[2]

[1] Department of Informatics, Ionian University, Corfu, Greece
c12makr@ionio.gr
[2] School of Music Studies, Aristotle University of Thessaloniki, Thessaloniki, Greece
{maxk,emilios}@mus.auth.gr

Abstract. Melodic harmonisation deals with the assignment of harmony (chords) over a given melody. Probabilistic approaches to melodic harmonisation utilise statistical information derived from a training dataset to harmonise a melody. This paper proposes a probabilistic approach for the automatic generation of voice leading for the bass note on a set of given chords from different musical idioms; the chord sequences are assumed to be generated by another system. The proposed bass voice leading (BVL) probabilistic model is part of ongoing work, it is based on the hidden Markov model (HMM) and it determines the bass voice contour by observing the contour of the melodic line. The experimental results demonstrate that the proposed BVL method indeed efficiently captures (in a statistical sense) the characteristic BVL features of the examined musical idioms.

Keywords: Voice leading · Hidden Markov model · Bass voice · Conceptual blending

1 Introduction

Melodic harmonisation systems assign harmonic material to a given melody. Harmony is expressed as a sequence of chords, but the overall essence of harmony is not concerned solely with the selection of chords; an important part of harmony has to do with the relative placement of the notes that comprise successive chords, a problem known as *voice leading*. Voice leading places focus on the horizontal relation of notes between successive chords, roughly considering chord successions as a composition of several mutually dependent voices. Thereby, each note of each chord is considered to belong to a separate melodic stream called a *voice*, while the composition of all voices produces the chord sequence.

Regarding melodic harmonisation systems, there are certain sets of "rules" that need to be taken under consideration when evaluating voice leading. However, these "rules" are defined by musical systems, called *idioms*, with many differences. The work presented in this paper is a part of an ongoing research

© Springer International Publishing Switzerland 2015
T. Collins et al. (Eds.): MCM 2015, LNAI 9110, pp. 128–134, 2015.
DOI: 10.1007/978-3-319-20603-5_13

within the context of the COINVENT project [10], which examines the development of a computationally feasible model for conceptual blending. Therefore, the inclusion of many diverse musical idioms in this approach is required for achieving bold results that blend characteristics from different layers of harmony across idioms.

The aspect of harmony that this paper addresses is voice leading of the bass voice, which is an important element of harmony. Experimental evaluation of methodologies that utilise statistical machine learning techniques demonstrated that an efficient way to harmonise a melody is to add the bass line first [11]. To the best of our knowledge, no study exists that focuses only on generating voice leading contour of the bass line independently of the actual chord notes (i.e. the actual chord notes that belong to the bass line are determined at a later study).

2   Probabilistic Bass Voice Leading

The proposed methodology aims to derive information from the melody voice in order to calculate the most probable movement for the bass voice, hereby referred to as the *bass voice leading* (BVL). This approach is intended to be harnessed to a larger modular probabilistic framework where the selection of chords (in GCT form [2]) is performed on an other probabilistic module [6]. Therefore, the development of the discussed BVL system is targeted towards providing indicative guidelines to the overall system about possible bass motion rather than defining specific notes for the bass voice.

The level of refinement for representing the bass and melody voice movement for the BVL system is also a matter of examination in the current paper. It is, however, a central hypothesis that both the bass and the melody voice steps are represented by abstract notions that describe pitch direction (up, down, steady, in steps or leaps etc.). Several scenarios are examined in Sect. 3 about the level of refinement required to have optimal results. Table 1 exhibits the utilised refinement scales in semitone differences for the bass and melody voice movement. For example, by considering a refinement level 2 for describing the melody voice, the following set of seven descriptors for contour change are considered: $mel_2 = \{st\_v, s\_up, s\_down, sl\_up, sl\_down, bl\_up, bl\_down,\}$ while an example of refinement level 0 for the bass voice has the following set of descriptors: $bass_0 = \{st\_v, up, down\}$. On the left side of the above equations, the subscript of the melody and the bass voice indicators denotes the level of refinement that is considered. Under this representation scheme, a given chord sequence in MIDI pitch numbers, such as: $[67, 63, 60, 48]$, $[67, 62, 65, 47]$, $[63, 60, 65, 48]$, $[65, 60, 60, 56]$ gives bass and melody (soprano) voice leading: $[-1, 0]$, $[+1, -4]$, $[+8, +2]$, which eventually becomes: [down, st_v], [up, bl_down], [up, sl_up].

The main assumption for developing the presented BVL methodology is that bass voice is not only a melody itself, but it also depends on the piece's melody. Therefore, the selection of the next bass voice note is dependent both on its previous note(s), as well as on the current interval between the current and the previous notes of the melody. This assumption, based on the fact that a

Table 1. The pitch direction refinement scales considered for the development of the proposed BVL system, according to the considered level of refinement.

Refinement description	Short name	Semitone difference range	Refinement level
Steady voice	st_v	0	0, 1, 2
Up	up	above 0	0
Down	down	below 0	0
Step up	s_up	between 1 and 2	1, 2
Step down	s_down	between −2 and −1	1, 2
Leap up	l_up	above 2	1
Leap down	l_down	below −2	1
Small leap up	sl_up	between 3 and 5	2
Small leap down	sl_down	between −3 and −5	2
Big leap up	bl_up	above 5	2
Big leap down	bl_down	below −5	2

probabilistic framework is required for the harmonisation system, motivates the utilisation of the *hidden Markov model* (HMM) methodology. According to the HMM methodology, a sequence of observed elements is given and a sequence of (hidden) states is produced as output. The training process of an HMM incorporates the extraction of statistics about the probabilities that a certain state (bass direction descriptor) follows an other state, given the current observation element (melody direction descriptor). These statistics are extracted from a training dataset, while the state sequence that is generated by an HMM system, is produced according to the maximum probability described by the training data statistics – considering a given sequence of observation elements.

3   Experimental Results

Aim of the experimental process is to evaluate whether the presented approach composes bass voice leading sequences that capture the intended statistical features regarding BVL from different music idioms. Additionally, it is examined whether there is an optimal level of detail for grouping successive bass note differences in semitones (according to Table 1), regarding BVL generation. To this end, a collection of five datasets has been utilised for training and testing the capabilities of the proposed BVL-HMM, namely: (1) a set of Bach Chorales, (2) several chorales from the 19th and 20th centuries, (3) polyphonic songs from Epirus, (4) a set of medieval pieces and (5) a set of modal chorales. These pieces are included in a dataset composed by music pieces (over 400) from many diverse music idioms (seven idioms with sub-categories). The Bach Chorales have been extensively employed in automatic probabilistic melodic harmonisation [1,3,8,9], while the polyphonic songs of Epirus [5,7] constitute a dataset that has hardly been studied. Several refinement level scenarios have been examined for the melody and the bass voices that are demonstrated in Table 2.

Table 2. The examined scenarios concerning bass and melody voice refinement levels. According to Table 1, each refinement level is described by a number of states (bass voice steps) and observations (melody voice steps).

Scenario	Bass refinement	Melody refinement	States × Observations
1	1	1	5 × 5
2	1	2	5 × 7
3	0	2	3 × 7
4	0	1	3 × 5
5	0	0	3 × 3

Each idiom's dataset is divided in two subsets, a *training* and a *testing* subset, with a proportion of 90 % to 10 % of the entire idiom's dataset. The training subset is utilised to train a BVL-HMM according to the selected refinement scenario. A model trained with the sequences (bass movement transitions and melody movement observations) of a specific idiom, X, will hereby be symbolised as M_X while the testing pieces denoted as D_X. The evaluation of whether a model M_X predicts a subset D_X better than a subset D_Y is achieved through the cross-entropy measure. The measure of cross-entropy is utilised to provide an entropy value for a sequence from a dataset, $\{S_i, i \in \{1, 2, \ldots, n\}\} \in D_X$, according to the context of each sequence element, S_i, denoted as C_i, as evaluated by a model M_Y. The value of cross-entropy under this formalisation is given by $-\frac{1}{n}\sum_1^n \log P_{M_Y}(S_i, C_{i,M_Y})$, where $P_{M_Y}(S_i, C_{i,M_Y})$ is the probability value assigned for the respective sequence element and its context from the discussed model.

The magnitude of the cross entropy value for a sequence S taken from a testing set D_X does not reveal much about how well a model M_Y predicts this sequence – or how good is this model for generating sequences that are similar to S. However, by comparing the cross-entropy values of a sequence X as predicted by two models, D_X and D_Y, we can assume which model predicts S better: the model that produces the *smaller* cross entropy value [4]. Smaller cross entropy values indicate that the elements of the sequence S "move on a path" with greater probability values. The effectiveness of the proposed model is indicated by the fact that most of the minimum values per row are on the main diagonal of the matrices, i.e. where model M_X predicts D_X better than any other D_Y.

Results indicated that scenarios 3 and 4 constitute more accurate refinement combinations for the melody and bass voices. Table 3 exhibits the cross-entropy values produced by the BVL-HMM under the refinement scenario 3, which is among the best refinement scenarios, where the systems are trained on each available training datasets for each test set's sequences. The presented values are averages across 100 repetitions of the experimental process, with different random divisions in training and testing subsets (preserving a ratio of 90 %-10 % respectively for all repetitions).

Table 3. Mean values of cross-entropies for all pairs of datasets, according to the refinement scenario 3.

	M_{Bach}	$M_{19th-20th}$	M_{Epirus}	$M_{Medieval}$	M_{Modal}
D_{Bach}	**2.4779**	2.5881	31.0763	16.0368	5.3056
$D_{19th-20th}$	13.8988	**5.0687**	70.1652	31.6096	15.9747
D_{Epirus}	3.3127	3.1592	**2.8067**	2.9990	3.0378
$D_{Medieval}$	3.0988	3.0619	3.1845	**2.7684**	2.8539
D_{Modal}	3.0037	2.9028	3.3761	2.9611	**2.7629**

An example application of the proposed BVL system is exhibited in Fig. 1, where GCT chords were produced by the cHMM [6] system. The chordal content of the harmonisation is functionally correct and compatible with Bach's style. The proposed bass line exhibits only two stylistic inconsistencies, namely the two $^6_4$ chords in the first bar. The overall voice leading is correct, except for the parallel octaves (first two chords) - note that the inner voices have been added by a very simple nearest position technique and that no other voice leading rules are accounted for. The presented musical example, among other examples, strongly suggests that further (statistical) information about the voicing layout of chords is required for generating harmonic results that capture an idioms style.

Fig. 1. Bach chorale melodic phrase automatically harmonised, with BVL generated by the proposed system (roman numeral harmonic analysis done manually).

4   Conclusions

This paper presented a methodology for determining the bass voice leading (BVL) given a melody voice. Voice leading concerns the horizontal relations between notes of the harmonising chords. The proposed bass voice leading (BVL) probabilistic model utilises a hidden Markov model (HMM) to determine the most probable movement for the bass voice (hidden states), by observing the soprano movement (set of observations). Many variations regarding the representation of bass and soprano voice movement have been examined, discussing different levels of representation refinement expressed as different combinations for the number of visible and hidden states. Five diverse music idioms were trained creating the relevant BVLs, while parts of these idioms were used for testing every system separately. The results indicated low values in term of cross entropy for each trained BVL system with the corresponding testing dataset and high values for examples from different music idioms. Thereby, it is assumed that the proposed methodology is efficient, since some characteristics of voice leading are captured for each idiom.

For future work, a thorougher musicological examination of the pieces included in the dataset will be pursued, since great difference were observed for the voice leading of pieces included in some idioms (e.g. $M_{19th\text{-}20th}$ set). Additionally, our aim is the development of the overall harmonisation probabilistic system that employs additional voicing layout statistical information, while chord selection (based on a separate HMM module) will be also biased by the adequacy of each chord to fulfil the voice leading scenario provided by the voice leading probabilistic module – part of which is presented in this work.

Acknowledgements. This work is founded by the COINVENT project. The project COINVENT acknowledges the financial support of the Future and Emerging Technologies (FET) programme within the Seventh Framework Programme for Research of the European Commission, under FET-Open grant number: 611553.

References

1. Allan, M., Williams, C.K.I.: Harmonising chorales by probabilistic inference. In: Advances in Neural Information Processing Systems 17, pp. 25–32. MIT Press (2004)
2. Cambouropoulos, E., Kaliakatsos-Papakostas, M., Tsougras, C.: An idiom-independent representation of chords for computational music analysis and generation. In: Proceeding of the Joint 11th Sound and Music Computing Conference (SMC) and 40th International Computer Music Conference (ICMC), ICMC-SMC 2014 (2014)
3. Jordan, M.I., Ghahramani, Z., Saul, L.K.: Hidden markov decision trees. In: Mozer, M., Jordan, M.I., Petsche, T. (eds.) NIPS, pp. 501–507. MIT Press, Cambridge (1996)
4. Jurafsky, D., Martin, J.H.: Speech and Language Processing. Prentice Hall, New Jersey (2000)
5. Kaliakatsos-Papakostas, M., Katsiavalos, A., Tsougras, C., Cambouropoulos, E.: Harmony in the polyphonic songs of Epirus: representation, statistical analysis and generation. In: 4th International Workshop on Folk Music Analysis (FMA) 2014, June 2011

6. Kaliakatsos-Papakostas, M., Cambouropoulos, E.: Probabilistic harmonisation with fixed intermediate chord constraints. In: Proceeding of the Joint 11th Sound and Music Computing Conference (SMC) and 40th International Computer Music Conference (ICMC), ICMC-SMC 2014 (2014)
7. Liolis, K.: To Epirótiko Polyphonikó Tragoúdi (Epirus Polyphonic Song). Ioannina (2006)
8. Manzara, L.C., Witten, I.H., James, M.: On the entropy of music: an experiment with bach chorale melodies. Leonardo Music J. **2**(1), 81–88 (1992)
9. Paiement, J.-F., Eck, D., Bengio, S.: Probabilistic melodic harmonization. In: Lamontagne, L., Marchand, M. (eds.) Canadian AI 2006. LNCS (LNAI), vol. 4013, pp. 218–229. Springer, Heidelberg (2006)
10. Schorlemmer, M., Smaill, A., Kühnberger, K.U., Kutz, O., Colton, S., Cambouropoulos, E., Pease, A.: Coinvent: towards a computational concept invention theory. In: 5th International Conference on Computational Creativity (ICCC) 2014, June 2014
11. Whorley, R.P., Wiggins, G.A., Rhodes, C., Pearce, M.T.: Multiple viewpoint systems: time complexity and the construction of domains for complex musical viewpoints in the harmonization problem. J. New Music Res. **42**(3), 237–266 (2013)

Performance

Hypergestures in Complex Time: Creative Performance Between Symbolic and Physical Reality

Maria Mannone and Guerino Mazzola[(✉)]

School of Music, University of Minnesota, Minneapolis, USA
{manno012,mazzola}@umn.edu

Abstract. Musical performance and composition imply hypergestural transformation from symbolic to physical reality and vice versa. But most scores require movements at infinite physical speed that can only be performed approximately by trained musicians. To formally solve this divide between symbolic notation and physical realization, we introduce complex time ($\mathbb{C}$-time) in music. In this way, infinite physical speed is "absorbed" by a finite imaginary speed. Gestures thus comprise thought (in imaginary time) and physical realization (in real time) as a world-sheet motion in space-time, corresponding to ideas from physical string theory. Transformation from imaginary to real time gives us a measure of artistic effort to pass from potentiality of thought to physical realization of artwork. Introducing $\mathbb{C}$-time we define a musical kinematics, calculate Euler-Lagrange equations, and, for the case of the elementary gesture of a pianist's finger, solve corresponding Poisson equations that describe world-sheets which connect symbolic and physical reality.

Keywords: Complex time · Performance theory · Euler-Lagrange equation · String theory · Hypergestures · World-sheets of space-time

1 Introduction to the Problem

In his famous question: "If I am at s and wish to get to t, what characteristic gesture should I perform to get there?", David Lewin in [7] was probably thinking of notes, pitch classes, and the like. But the question remains virulent if we generalize the objects s and t to, for example, harmonies, tonalities, and thereby thematizing gesturally driven tonal modulations. This has been accomplished in recent research [13]. Or to contrapuntal interval movements, which has been accomplished in [1].

In this paper, we focus on a third generalization, one which is quite akin to Lewin's wording ("to perform") and also to his metaphor in [7] of a dancer in such a gestural performance, namely the question *"If I am considering a score s and wish to create one of its performances $p = \wp(s)$, what characteristic gesture should I perform to get there?"* The symbol $\wp$ is the standard notation for a performance transformation from symbolic reality to physical reality ("symbolic"

© Springer International Publishing Switzerland 2015
T. Collins et al. (Eds.): MCM 2015, LNAI 9110, pp. 137–148, 2015.
DOI: 10.1007/978-3-319-20603-5_14

being understood as synonymous to "mental", as opposed to psychological and physical, but see [8, Ch. 33.2.2]).

We should add that our present objective is the general relation between symbolic reality (as typically represented in a Western score) and physical reality of a musical performance (of such a score). We are viewing such a relation in both directions: the performance of a given symbolic object as well as the creation of such an object as a result of a creative performance in improvisation-driven composition, such as Beethoven's creation of Sonata Op.109 (see [6]).

1.1 Performance Stemmata and Gesture Theory

The gesturalization of performance theory has been initiated in a paper [12] where the stemmatic unfolding of a performance as described in [8, Chap. 38] was gesturalized in the refinement arrows (driven by performance operators) that lead from a mother performance to one of its daughter performances. However, the stemma in that investigation still consists of a tree whose vertices are essentially performance transformations $\wp$, not gestures, but diffeomorphisms between symbolic and physical spaces, i.e., classical Fregean functions. Our present objective is to gesturalize these transformations, which, as rightly observed by Gilles Châtelet in [3], reduce the functional movement metaphor to the separated input and output entities. For the time being, we do not complete the gesturalization of stemmata insofar as the gesturalization of refinement arrows between *gesturalized* $\wp$ transformations is not realized yet. This is due to the still open question of how to conceptualize performance operators between gesturalized $\wp$ transformations.

In classical performance theory, the double ontology of symbolic versus physical reality is respected, and the diffeomorphism $\wp$ is dealt with by the formalism of performance vector fields $\mathbf{\Sigma}$ in the symbolic reality, giving rise to so-called *performance scores*. It is the central challenge of the present project to introduce an enriched ontology that comprises symbolic and physical perspectives, and where a gesturalization of $\wp$ transformations can successfully be conceived. In [10, p. 128], a first sketchy idea about gesturalization of $\wp$ transformations could be conceived. Figure 1 shows a commutative diagram which represents a lifting of the $\wp$ level to a gestural performance level where a symbolic gesture, sitting above the score object is deformed into a physical gesture that sits over the performed score object.

1.2 Musical Thought and Action in Analysis, Composition, and Performance

Music psychologist Helga de la Motte-Haber in [4, p. 232] states "Musikalisches Denken ist grundsätzlich Denken in Musik." Musical thinking is fundamentally a thinking *within* music. This general statement is meant to be valid for analysis, composition, and performance. Its radical position makes it impossible to understand music without making music, which implies that music has an irreducible essence of ineffability. The question of how music is thought is an ontological

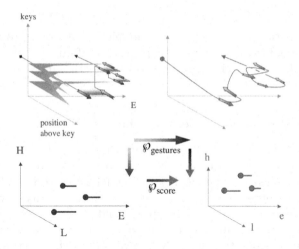

Fig. 1. A gestural performance lifting of a score performance. The axes mean: E symbolic onset, L symbolic loudness, H symbolic pitch, e physical onset, l physical loudness, h physical pitch; on top left for symbolic gestures: pitch is replaced by "keys", loudness by "position above key"; on top right the corresponding physical gesture space is presented. The arrows on the top level represent gestural movements. The four arrows in the middle represent maps between the different spaces of notes and gestures.

one, and de la Motte-Haber implicitly claims that there is no ontology where thinking and making music can coexist without enforcing ineffability. We agree insofar that music necessitates a specific ontology, and this is also the insight of Paul Valéry in [16, I]: "La musique mathématiquement discontinue peut donner les sensations les plus continues." Music in its symbolically discrete notation can generate completely continuous sensations. An ontology that comprises these discrepancies must be of a singular type.

1.3   The Paradoxical Time Structure of Classical Score Notation

Valéry's astonishment is not only abstract reflection, but supported by concrete conflicts that become evident when trying to perform a Western score *as written*. In fact, most scores cannot be played as written. Let us see two simple examples. In the first, take a piano score, where a given note q is notated twice, the second time after the first notation, but without rest, two consecutive quavers on middle C, say. To play these two notes means to push the middle C key for the eighth note duration (given whatever tempo), then lift that finger and push it down a second time on the same key, keeping it there for another eighth note duration. This repetition requires a physical movement (lifting and pushing down) of the finger in zero seconds since no rest is written. This is a physical movement with infinite velocity, an impossibility. The second example is the consecutive notation of a third interval, middle C and E, with quaver duration, to be played with third and fifth finger of the right hand, and then followed without rest by the third interval D and F of same duration, to be played with first and third finger. This necessitates the third finger to move infinitely fast from C to F, a physical absurdity. Although we have chosen the pianist's performance, the above divide between symbolic and physical gesture subsists for other instruments, including the human voice.

1.4   Artistic Presence and Complex Time in Physics

De la Motte-Haber's verdict is also critical with respect to the time dimension of musical thoughts for the artist's performative presence. In [10,11], this phenomenon has been discussed in detail. Let us summarize those insights as follows: The performative presence of a musician is a complex entity, involving three dimensions of body interfaces, gesture spaces, and structural flow. That such an involved activity happens in the physical presence of zero seconds (or even in a very short non-vanishing time interval) contradicts the huge amount of artistic consciousness when performing as a thinking agent.

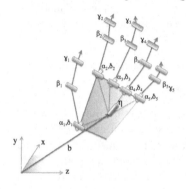

Fig. 2. The spatial coordinates of a hand.

In that discussion it was proposed to introduce a second, imaginary time dimension, orthogonal to real physical time in the sense of complex numbers, where the imaginary axis $i\mathbb{R}$ is orthogonal to the real axis $\mathbb{R}$, enabling an extended time ontology at each physical moment. This idea is known from cosmology and has been popularized by Stephen Hawking and others [5]. In physics imaginary time is related to real time by the Wick rotation, the multiplication by the imaginary unit $i = \sqrt{-1}$. To the authors' knowledge, no relation has been made to this date of physical imaginary time to imaginary time of human consciousness as proposed. However, Roger Penrose claims that human consciousness should be integrated in a more complete physical theory [15]. The Cartesian divide $RC = res\ cogitans/RE = res\ extensa$ could be solved by complex time, setting $RC = i\mathbb{R} \oplus \mathbb{R}^3, RE = \mathbb{R} \oplus \mathbb{R}^3$ with spatial intersection $RC \cap RE = \mathbb{R}^3$.

2   Introducing Complex Musical Time for an Extended Musical Ontology

In what follows we shall discuss a pianist's gesture as a prototype for gesturalization of performance transformations. The hand's dynamic has two aspects, a spatial and a temporal one. Refer to Fig. 2 for the spatial aspect. As shown in Fig. 2, a hand's position is defined by four angles per finger, a vector $b \in \mathbb{R}^3$ for the position of the carpus center, two real numbers for the orientation of the carpus plane, and one real number for the rotational position of the carpus plane around the normal vector through the center of the carpus plane, a total of 26 real numbers. Denote this position by $s \in \mathbb{R}^{26}$. We may also assume that for every position, a small open neighborhood is also a set of possible (however small) variations, which means that the spatial information of the hand's dynamic is a connected open set $M \subset \mathbb{R}^{26}$. For the definition of an extended ontology comprising symbolic and physical perspectives, the spatial information should not differ, therefore we shall use M for both realities.

2.1   Solving the Divide Between Symbolic and Physical Reality

The paradoxical situations mentioned above stem from the relation between space and time, infinite velocities emerging if no time (zero duration) is allowed to perform certain movements in the symbolic reality. Moreover, the symbolic movement is a possibly non-differentiable curve that violates the finiteness of forces (or, equivalently: masses and accelerations) that are available in any realistic physical context.

The immediate consequence of such conflicts is that one should consider two time qualities: a physical time and a "symbolic" time. We shall comply with this requirement in a mathematically consistent way, i.e., embedding both time qualities in one space, namely the Gaussian complex numbers $\mathbb{C}$. This means that we now consider complex time $t = t_r + it_i$, having a *real time* component t_r and an *imaginary time* component t_i. This means that the hand's spatio-temporal position is now defined by a vector $st = (s, t) \in M_{\mathbb{C}}$, where $M_{\mathbb{C}} = M \oplus \mathbb{C}$.

With this setup, one may now discuss the gestural dynamics of a hand that moves from symbolic reality to physical reality as a hypergesture $h \in \uparrow \overrightarrow{@} \Delta \overrightarrow{@} M_{\mathbb{C}}$, where Δ is a skeleton digraph of the hand's articulated movement, and where $\uparrow$ is the skeleton digraph with two vertices and one connecting arrow[1]; we shall see examples later in this paper. Here, the gesture $h(0) \in \Delta \overrightarrow{@} M_{\mathbb{C}}$ is the initial symbolic gesture whereas the gesture $h(1) \in \Delta \overrightarrow{@} M_{\mathbb{C}}$ denotes the final physical gesture. Relating to time, the initial gesture may move in imaginary time, whereas the final one should move in real time.

In what follows we shall first focus on the situation where $\Delta = \uparrow$, which is basic in view of the fact that (1) any Δ is the colimit of a diagram of $\uparrow$s and (2) the Escher Theorem (see [9, Proposition 2.4]) that allows to permute the skeletal factors in hypergesture spaces. This means that our hypergesture is $h \in \uparrow \overrightarrow{@} \uparrow \overrightarrow{@} M_{\mathbb{C}}$, or, equivalently, it is a singular chain $st : I^2 \to M_{\mathbb{C}} : (x, y) \mapsto st(x, y) = (s(x, y), t(x, y))$ that is defined on the unit square $I^2 = [0, 1]^2 \subset \mathbb{R}^2$. Parameter x stands for the hypergesture argument of the left vertical arrow, y stands for the gesture parameter for every fixed x. This situation is analogous to the situation in physical string theory, where the singular chain st would denote a world-sheet connecting two strings $st|_{x=0}$ and $st|_{x=1}$ (see also [17]). Observe that now, time is no longer a genuine parameter, but stands side by side with the spatial parameters, the hypergesture parameters x, y being the connecting

---

[1] The symbol $\uparrow \overrightarrow{@} \Delta \overrightarrow{@} M_{\mathbb{C}}$ denotes the space of hypergestures, i.e., of gestures with skeleton $\uparrow$ and body in the space of gestures $\Delta \overrightarrow{@} M_{\mathbb{C}}$ with skeleton Δ and body in $M_{\mathbb{C}}$.

background arguments. This is a typical situation in musical gesture theory, namely that gestural parameters are not time parameters in general, i.e., time, much as space, can appear as a function of gestural parameters.

2.2   Redefining Elementary Concepts of Kinematics for Complex Time

To understand the dynamics of the world-sheet hypergesture st, it is straightforward to recall the Lagrangian formalism and its role in the Hamiltonian variational principle that yields a number of fundamental laws in physics (see [17]). To define the usual Lagrangian formula $L = V - U$, namely the difference between kinetic energy V and potential energy U, one first needs to define velocity $\frac{ds}{dt}$.

In view of the fact that the complex number $t(x, y)$ is not a primordial parameter, the immediate formal definition of velocity $\frac{ds}{dt} = \partial_x s \frac{dx}{dt} + \partial_y s \frac{dy}{dt}$ is critical. Since $\frac{dx}{dt}, \frac{dy}{dt}$ are not defined, we should replace them by $\frac{1}{\partial_x t}, \frac{1}{\partial_y t}$, whence the definition with quotients of partial derivatives

$$\frac{ds}{dt} = \frac{\partial_x s}{\partial_x t} + \frac{\partial_y s}{\partial_y t}.$$

In this definition, the actual gestural velocity at a given hypergesture parameter x is the right summand $\frac{\partial_y s}{\partial_y t}$. Writing this vector as $\frac{\partial_y s}{\partial_y t_r + i \partial_y t_i}$, one recognizes that it has two components: the real velocity $\frac{\partial_y s}{\partial_y t_r}$ and the imaginary one $\frac{\partial_y s}{\partial_y t_i}$. The advantage of complex time now becomes evident: it may happen that the real time derivative $\partial_y t_r$ vanishes and the real velocity would go to infinity, while the imaginary derivative $\partial_y t_i$ doesn't vanish, and the velocity $\frac{\partial_y s}{\partial_y t_r + i \partial_y t_i} = \frac{\partial_y s}{i \partial_y t_i}$ is still a finite vector. A Wick rotation from imaginary to real velocity could be understood as a continuous movement on the world-sheet that rotates imaginary velocity into real velocity, or, vice versa: transforming finite real velocity (to connect two spatial positions, say) into infinite real velocity, but increasing imaginary velocity as compensation for the real singularity, a situation that is well-known in physics for the Dirac function, see Fig. 3.

Fig. 3.   The Dirac-function shape of real velocity with constant distance trajectory in a time converging to zero.

2.3   Analytic Time Functions

The nature of the time function $t(x, y)$ turns out to be critical in this complex setup. We shall see in Sect. 2.5 that a general Euler-Lagrange function is complicated, and it may also happen without stronger assumptions that our complex velocity could diverge to infinity. We therefore make the following assumption:

The time function t is analytic with non-vanishing derivative t'.

We view the map $t : I \to \mathbb{C}$ as a map that is defined on an open neighborhood $O \subset \mathbb{C}$ of I^2 which is seen as a subset of the space $\mathbb{R}^2 = \mathbb{C}$. Recall that analyticity is equivalent to the Cauchy-Riemann differential equation $\partial_x t = -i\partial_y t$, and the derivative is given by $t'(x,y) = \partial_x t(x,y)$. With this assumption, keeping real-valued space coordinates, and taking the usual sesquilinear scalar product $\langle , \rangle$ on $\mathbb{C}^n$, the square of the velocity is

$$\langle \frac{ds}{dt}, \frac{ds}{dt} \rangle = \frac{1}{|t'|^4}(\langle \partial_x s \overline{\partial_x t} + \partial_y s \overline{\partial_y t}, \partial_x s \overline{\partial_x t} + \partial_y s \overline{\partial_y t} \rangle)$$

$$= \frac{1}{|t'|^2}(|\partial_x s|^2 + |\partial_y s|^2) + \frac{1}{|t'|^4}(\langle \partial_x s, \partial_y s \rangle (\overline{\partial_x t} \partial_y t + \overline{\partial_y t} \partial_x t)).$$

But since $\partial_x t = -i\partial_y t$, $\overline{\partial_x t}\partial_y t + \overline{\partial_y t}\partial_x t = 0$, i.e., $\langle \frac{ds}{dt}, \frac{ds}{dt} \rangle = \frac{1}{|t'|^2}(|\partial_x s|^2 + |\partial_y s|^2)$.

2.4   The Lagrange Function and Its Action for Musical Performance

With the above definition of velocity and its norm, we may introduce the Lagrange density function for a potential $U(s(x,y))$ and a density μ. For the time being these two quantities need to be understood and interpreted in musical terms, this will be discussed in the final Sect. 4. Lagrangian density is defined by the standard formula

$$\mathcal{L}(x,y) = \frac{\mu}{2|t'|^2}(|\partial_x s|^2 + |\partial_y s|^2) - U(x,y).$$

The Lagrangian action is defined by

$$S = \int_y \int_x \mathcal{L}(x,y),$$

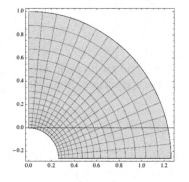

Fig. 4. A time map for the function $t(x,y)$ described in the text.

and we have to calculate its variation from a variation $M_\delta(x,y) = M(x,y) + \delta M = (s(x,y), t(x,y)) + (\delta s, \delta t)$, to obtain the Euler-Lagrange equations. Calculating with the standard methods using partial integration, one gets these Euler-Lagrange equations

$$-\nabla U = \mu \left(\frac{s''}{|t'|^2} - 2\frac{\langle t', t'' \rangle}{|t'|^4}s' + \frac{\ddot{s}}{|\dot{t}|^2} - 2\frac{\langle \dot{t}, \ddot{t} \rangle}{|\dot{t}|^4}\dot{s} \right)$$

with the usual notation symbols in physics: $\dot{s} = \partial_y s$, $\ddot{s} = \partial_y^2 s$, $s' = \partial_x s$, $s'' = \partial_x^2 s$, etc. This complex formula can be simplified with the following assumption, namely that $t' \perp t''$ and $\dot{t} \perp \ddot{t}$. This is equivalent to the condition that $|t'|^2 = |\dot{t}|^2$ is only a function of y, i.e., the time velocity modulus does not change with the hypergestural parameter x.

In fact, $t' \perp t''$ is equivalent to $\dot{t} \perp \ddot{t}$ since $\dot{t} = it'$ because t is complex analytical. The orthogonality $t' \perp t''$ is a special case of the condition that for an analytical function g, we have $g \perp g'$, and this means that g'/g is imaginary. But this is evidently equivalent to the differential equation $0 = \partial_x(|g|^2)$, i.e., $|g|^2$ is only a function of y. Applying this result to $g = t'$, our claim follows. An example of such a time function is

$$t(x, y) = -0.2624i + \frac{1}{3.8104} e^{\frac{1}{2}\pi i(1-x) + \frac{\pi y}{2}},$$

$$3.8104 \approx e^{\pi/2} - 1, \quad 0.2624i \approx e^{\frac{1}{2}\pi i}/(e^{\pi/2} - 1),$$

mapping the unit imaginary time interval at $x = 0$ to the unit real time interval at $x = 1$, see Fig. 4. In the following section, we will use the time function in the range between 0 and 6.

2.5   The Euler-Lagrange Equations and Their Solution

The Euler-Lagrange equations with the above orthogonality condition are as follows:

$$\frac{-|t'|^2}{\mu}\nabla U = \Delta s,$$

where $|t'|^2/\mu$ can be seen as a factor of temporal versus material density. This is a Poisson equation, and using Green functions, one proves that

$$s = P\left(\nabla U \frac{|t'|^2}{\mu}\right) + S(\partial s),$$

where $\partial s = \sum(-1)^i s_i$ is the homological boundary of the singular chain s, and P, S are linear functions. This means that s is determined by its values on the boundary, including the symbolic initial gesture ($x = 0$), the physical final gesture ($x = 1$), and the curves for $y = 0, 1$, the x-parametrized transition of symbolic to physical gesture from initial and final gestural positions.

We have calculated the solution of the world-sheet for a simple movement of one finger: push the key, lift the finger, keep it up, go down again on that key, keep it on the key. The symbolic gesture is the linear edgy curve to the left in Fig. 5, the physical gesture is the smooth curve to the right. The potential is taken as a test case function $\nabla U(s(x, y)) = (x + 4y - 12)e^{-\pi y/6}$, and $\mu = 1$. In this case we have considered equal time intervals for each part of the splines, as we will describe in Sect. 3.

3   The Elementary Gesture of a Pianist

The elementary gesture of the pianist is the movement of a finger that presses a key. In general, elementary gestures in music are conceptually similar, following the general idea of *arsis - thesis*, passing through the attack gesture of a conductor [14], or the breath with inspiration-exhalation of singers. It is the general

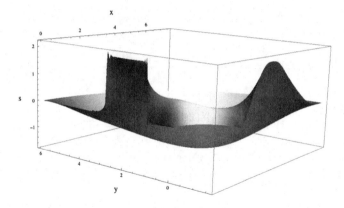

Fig. 5. The world-sheet for a simple up-down movement of one finger.

idea of a preparation by an ascending movement, and the accomplishment of a gesture by a descending movement. The completion of gesture is signaled by a third part, again an ascending movement, that prepares for a new gesture, in a chain of gestures—the simplest hypergesture.

Here we focus on the formal description of the elementary piano gesture, representing velocity by spline interpolation to describe variable velocity. Initially the finger—which we schematize as a massive point—is at rest on a pressed key. It then accelerates moving up, moves at constant speed for a time interval, decelerates, reaches a distance from the keyboard (*arsis* movement) where is at rest. It then starts to move again, with identical motion, but in opposite direction, until the key is completely pressed.

We choose cubic polynomial splines. For each part of the movement we use three different interpolations: cubic, constant velocity, and cubic again. If we consider the cubic polynomial $v(t) = at^3 + bt^2 + ct + d$, to find the first spline, representing the increasing velocity from zero to the maximal value, we have to solve the following system of equations:

- velocity zero at time $t_0 \Rightarrow at_0^3 + bt_0^2 + ct_0 + d = 0$
- acceleration zero at time $t_0 \Rightarrow 3at_0^2 + 2bt_0 + c = 0$
- maximal velocity at $t_1 \Rightarrow W = at_1^3 + bt_1^2 + ct_1 + d$
- zero acceleration at $t_1 \Rightarrow 0 = 3at_1^2 + 2bt_1 + c$.

The absolute value of acceleration is maximal at the center of the first and the third spline. Solving the system we obtain $v_y^1(t_0, t_1, t, W) = \frac{(t-t_0)^2(2t+t_0-3t_1)W}{(t_0-t_1)^3}$. Proceeding in an analogous way, we obtain the velocity for the first part of the motion, a massive point that starts from the key and reaches height H:

$$v_y^1(t_0, t_1, t, W) = \frac{(t - t_0)^2(2t + t_0 - 3t_1)W}{(t_0 - t_1)^3} \quad \text{if } t_0 < t < t_1,$$

$$v_y^2(t, W) = W \quad \text{if } t_1 < t < t_2,$$

$$v_y^3(t_2, t_3, t, W) = -\frac{(t - t_3)^2(2t - 3t_2 + t_3)W}{(t_2 - t_3)^3} \quad \text{if } t_2 < t < t_3.$$

To complete the vertical gesture with the *thesis* part, we also consider the descending motion:

$$v_y^4(t_3, t_4, t, W) = -\frac{(t - t_3)^2(2t + t_3 - 3t_4)W}{(t_3 - t_4)^3} \quad \text{if } t_3 < t < t_4,$$

$$v_y^5(t_4, t_5, t, W) = -W \quad \text{if } t_4 < t < t_5,$$

$$v_y^6(t_5, t_6, t, W) = \frac{(t - t_6)^2(2t + t_6 - 3t_5)W}{(t_5 - t_6)^3} \quad \text{if } t_5 < t < t_6.$$

where W is the maximal vertical velocity. The complete graph of velocity is shown in Fig. 6.

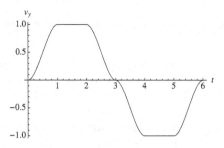

Fig. 6. Vertical velocity of piano primitive gesture, in the raising and lowering part.

Artistic parameters are related to these results. For example, acceleration is related to force via the well-known $F = ma$, and force determines the loudness of the sound. The maximum values of acceleration have a superior physical limit given by the following equation (with $t' = \frac{t_0 + t_1}{2}$):

$$\frac{d}{dt}\left[v_y^1(t_0, t_1, t, W)\right]_{t'} = \frac{d}{dt}\left[\frac{(t - t_0)^2(2t + t_0 - 3t_1)W}{(t_0 - t_1)^3}\right]_{t'} = \frac{3}{2}\frac{W}{(t_0 - t_1)} \leq \frac{F}{m}.$$

Summarizing, even in this simple case, we have the variety of parameters:

$$t_0, t_1, \ldots t_6, W, H, m.$$

The vertical displacement is obtained by integration of the previous speed functions:

$$Ascending\,Position[t_0, t_1, t_2, t_3, t, W, H_0] = H_0 +$$

$$\text{If } t < t_1, (t - t_0)^3 \frac{(t + t_0 - 2t_1)W}{2(t_0 - t_1)^3},$$

$$\text{If } t < t_2, -(t_0 W)/2 + (t_1 W)/2 + (t - t_1)W,$$

$$\text{If } t < t_3, -(t_0 W)/2 + (t_1 W)/2 + (-t_1 + t_2)W -$$

$$\frac{1}{2(t_2 - t_3)^3}(t - t_2)(t^3 - t_2^3 - tt_2(t_2 - 4t_3) +$$

$$4t_2^2 t_3 - 6t_2 t_3^2 + 2t_3^3 - t^2(t_2 + 2t_3))W,$$

and the following one for the descending movement (H_f is the highest vertical position):

$$DescendingPosition[t_0, t_1, t_2, t_3, t_4, t_5, t_6, t, W, H_0, H_f] = H_f +$$

$$\text{If } t < t_4, -(t - t_3)^3 \frac{(t + t_3 - 2t_4)W}{2(t_3 - t_4)^3},$$

$$\text{If } t < t_5, (t_3 W)/2 - (t_4 W)/2 - (t - t_4)W,$$

$$\text{If } t < t_6, (t_3 W)/2 - (t_4 W)/2 - (-t_4 + t_5)W +$$

$$\frac{1}{2(t_5 - t_6)^3}(t - t_5)(t^3 - t_5^3 - tt_5(t_5 - 4t_6) +$$

$$4t_5^2 t_6 - 6t_5 t_6^2 + 2t_6^3 - t^2(t_5 + 2t_6))W.$$

We then define the complete function Moto (in Mathematica) for the vertical movement depending on time (Fig. 7).

```
Moto[t0_,t1_,t2_,t3_,t4_,t5_,t5_,W_,H0_,Hf_] :=
  If[t < t3, AscendingPosition[t0,t1,t2,t3,t,W,H0],
    If[t < t6,
    DescendingPosition[t0,t1,t2,t3,t4,t5,t6,t,W,H0,Hf]]] .
```

We make the following choice of parameters: $t_0 = 0, t_1 = 1, t_2 = 2, t_3 = 3,$ $t_4 = 4, t_5 = 5, t_6 = 6, W = 1, H_0 = 0, H_f = 2$, and therefore $a = 6, b = 6$ for the rectangular domain of Poisson equation. With this parameter choice we obtain the right curve on Fig. 5.

The complete graph is obtained, as said above, solving the Poisson equation with a test potential. By variation of these parameters, we can describe an huge amount of piano touches. These concepts are useful to understand not only general ideas about music performance, but in particular to be used to avoid vagaries in performance didactics.

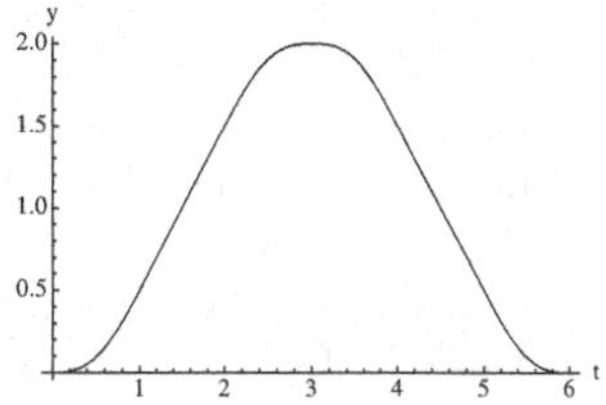

Fig. 7. Vertical movement of piano elementary gesture, in the raising and lowering part.

4   Opening the Aesthetic Question that Is Quantified in Lagrange Potentials

Regarding the artistic fight with the deaf material, let us recall with Dante's saying in his Divine Comedy [2, Paradiso, I Canto, vv. 127–132] that

Vero è che, come forma non s'accorda / molte fiate a l'intenzion de l'arte,
perch'a risponder la materia è sorda, / così da questo corso si diparte
talor la creatura, c'ha podere / di piegar, così pinta, in altra parte;

(It is true that, as form resists / many times to the intention of the artist,
because its matter in response is deaf, / so it moves away from this path
sometimes the creature with bending power / from innate attitude, in another way;)

To approach this struggle in a more quantitative and precise way, we have demonstrated that the complex time approach yields a dynamical framework where least Lagrange action can be solved by Poisson equation methods using Green functions for the corresponding Euler-Lagrange equations. This is a satisfactory result, it is the necessary initial study to approach the musical problems:

- How can potentials be defined from the aesthetic point of view?
- What is the variety of physical gestures for given symbolic gestures and potentials?
- How can the minimal action be understood as an effort of thinking in the making of musical creativity?
- What are the neurophysiological correlates of our model, e.g. with regard to a possible role of mirror neurons for gestural processing?

References

1. Agustin-Aquino, O.A., Junod, J., Mazzola, G.: Computational Counterpoint Worlds. Springer Series Computational Music Science, Heidelberg (2015)
2. Alighieri, D.: La Divina Commedia (1321). CreateSpace Independent Publishing Platform, USA (2014)
3. Châtelet, G.: Figuring Space. Kluwer 2000 (original French edition: Les Enjeux du mobile, Éditions du Seuil, Paris 1993)
4. Dahlhaus, C., et al.: Neues Handbuch der Musikwissenschaft, Bd. 1–13: Athenaion and Laaber, Laaber 1980–1993
5. Hawking, S.: A Brief History of Time: From the Big Bang to Black Holes. Bantam Books, New York (1988)
6. Kinderman, W.: Artaria 195. University of Illinois Press, Urbana (2003)
7. Lewin, D.: Generalized Musical Intervals and Transformations. Yale University Press, New Haven (1987)
8. Mazzola, G.: The Topos of Music. Birkhäuser, Basel-Boston (2002)
9. Mazzola, G.: Categorical gestures, the diamond conjecture, Lewin's question, and the Hammerklavier sonata. J. Math. Music **3**(1), 31–58 (2009)
10. Mazzola, G.: Musical Performance. Springer, Heidelberg (2011)
11. Mazzola, G., Park, J., Thalmann, F.: Musical Creativity. Springer, Heidelberg (2011)
12. Mazzola, G.: Hypergesture homology for performance stemmata with lie operators. In: Yust, J., Wild, J., Burgoyne, J.A. (eds.) MCM 2013. LNCS, vol. 7937, pp. 138–150. Springer, Heidelberg (2013)
13. Mazzola, G: Gestural dynamics in modulation–a musical string theory. In: Peyron, G. et al. (eds.): Proceedings of the International Congress on Music and Mathematics. Springer, Heidelberg (2015, to appear)
14. Nicotra, E.: Tecnica della direzione d'orchestra. Edizioni Curci, Milano (2007)
15. Penrose, R.: The Road to Reality: A Complete Guide to the Laws of the Universe. Vintage Books, New York (2004)
16. Valéry, P: Cahiers I-IV (1894–1914). Celeyrette-Pietri, N., Robinson-Valéry, J., (eds.), Gallimard, Paris (1987)
17. Zwiebach, B.: A First Course in String Theory. Cambridge University Press, Cambridge (2004)

Generating Fingerings for Polyphonic Piano Music with a Tabu Search Algorithm

Matteo Balliauw[(✉)], Dorien Herremans, Daniel Palhazi Cuervo,
and Kenneth Sörensen

Faculty of Applied Economics, University of Antwerp, Prinsstraat 13,
2000 Antwerp, Belgium
{matteo.balliauw,dorien.herremans,daniel.palhazicuervo,
kenneth.sorensen}@uantwerpen.be
http://www.uantwerpen.be

Abstract. A piano fingering is an indication of which finger is to be used to play each note in a piano composition. Good piano fingerings enable pianists to study, remember and play pieces in an optimal way. In this paper, we propose a tabu search algorithm to find a good piano fingering automatically and in a short amount of time. An innovative feature of the proposed algorithm is that it implements an objective function that takes into account the characteristics of the pianist's hand and that it can be used for complex polyphonic music.

Keywords: Piano fingering · Tabu search · Metaheuristics · OR in Music · Combinatorial optimisation

1 Introduction

Both pianists and composers often add a fingering to sheet music in order to indicate the appropriate finger that should be used to play each note. *Piano fingerings* are important as not every finger is equally suited to play each note, and some combinations of fingers are better suited to play certain note sequences than others [1]. Additionally, having a well thought-out piano fingering can help a pianist to study and remember a piece. It can also enhance the interpretation and the musicality of a performance [2].

Developing a high-quality fingering requires a considerable amount of time and expertise. For that reason, in this paper we develop an algorithm to automatically determine a good piano fingering. The purpose of the algorithm is to help pianists with little experience in deciding on a good fingering, as well as all pianists who quickly want to obtain a fingering they can use as a starting point.

In this paper, we model the generation of a good piano fingering as a combinatorial optimisation problem, in which a finger has to be assigned to every note in the piece. The quality of a fingering is evaluated by means of an objective function that measures playability. This objective function takes into account the characteristics of the player's hands. The proposed algorithm is a tabu search

© Springer International Publishing Switzerland 2015
T. Collins et al. (Eds.): MCM 2015, LNAI 9110, pp. 149–160, 2015.
DOI: 10.1007/978-3-319-20603-5_15

(TS) heuristic, capable to find a good fingering solution for a complex polyphonic piano piece in a reasonable amount of execution time. This technique has been successfully used in the field of computer-aided composing [3].

This paper is divided into six sections. Section 2 gives a short overview of the literature concerning the generation of piano fingerings and algorithms developed to solve this problem. In Sect. 3, a mathematical formulation of the problem is introduced. Section 4 explains the tabu search algorithm in detail. In Sect. 5, we show and discuss an example of a fingering generated by the TS for an existing piece of music. The final section gives some conclusions and suggestions for future research.

2   Literature Review

In the 18th century, exercises to learn frequently used fingering combinations fluently were published for the first time. Carl Philipp Emanuel Bach was among the first musicians to write down an entire set of rules for piano fingering. Previously, such rules were only transmitted orally through lessons. In the 19th century, many composers followed his example [4]. At the end of the 20th century, the first attempts to generate a mathematical formulation that could model a piano fingering and develop a suitable algorithm were published, as we show in the next subsection. Similar research has been carried out among others for string instruments [5].

2.1   Fingering Quality

In order to evaluate a piano fingering, it is necessary to quantify it quality. Three important dimensions of a piano fingering add to this quality. The main element is the ease of playing and this is where this research focuses on. Additional dimensions that add to the quality of a fingering are the ease of memorisation and the facilitation of the interpretation [6].

The dominant approach for evaluating the quality of a piano fingering using an objective function, which is also followed in this research, is based on hand-made rules [6–8]. Other strategies include using a machine learning approach such as Markov Models based on transition matrices [2,9–11] or Hidden Markov Models (HMM) [12].

In order to evaluate the playability of a fingering, the set of rules proposed by Parncutt et al. [6] and Parncutt [7] serve as the basis for this research. Every source of difficulty in a fingering, both monophonic and polyphonic, is assigned a cost. Every cost factor is attributed a weight. The playability measure is an objective function that consists of the weighted sum of costs and is to be minimised. This approach has the advantage that pianists can express their trade-offs between the sources of difficulty. This can be done by assigning different weights to each source of difficulty in the objective function. Parncutt's original cost factors for piano fingering have been expanded by Jacobs [13] and Lin and Liu [14]. More recently, some improvements and additions were made by the authors of this paper [15].

This latter set of rules is used in this research. Issues such as personal preferences, the use of the left hand and the difference between playing a black and white key have been ignored in the past [16], but are taken into account in this paper.

2.2 Algorithms for Automatic Generation of Fingerings

In many papers, dynamic programming is used to find optimal piano fingerings according to the selected objective function. Dijkstra's algorithm is used by Parncutt et al. [6] on monophonic music and by Al Kasimi et al. [11] on monophonic pieces with some simple polyphonic chords. Similar dynamic programming algorithms have been used in literature by Robine [2] and Hart et al. [9] on monophonic music. A rule-based expert system that optimises a fitness function for similar music was developed by Viana et al. [8] to implement their Intelligent System for Piano Fingering Learning Aid (SIEDP). For the HMM-model in monophonic music, Yonebayashi et al. used a Viterbi algorithm [12]. Although some improvements have been made to reduce the computing time of the aforementioned algorithms, it is often impossible for them to deal with complex polyphonic pieces (where simultaneous notes can have different starting and ending times and where hence a graphical representation would no longer work). For this reason, we develop an algorithm that can deal with complex polyphony in an effective and efficient way. This algorithm is explained in Sect. 4. In the next section, we mathematically formulate the problem of finding a piano fingering.

3 Problem Description

When generating a piano fingering, it is necessary to decide which finger should play each note. Each finger is represented using the traditional coding from 1 (thumb) to 5 (little finger).

In order to consider piano fingering as an optimisation problem, we define an objective function that measures the quality of a solution by looking at the playability of the piece. This objective function was described by Balliauw [15] as an adaptation of the work of Parncutt et al. [6] and Jacobs [13] and allows to work with both the right and the left hand.

The objective function takes into account a distance matrix, displayed in Table 1. This contains information for each finger pair about the distances that are easy and difficult to play, respectively called Rel (relaxed range), Comf (comfortable range) and Prac (practically playable range). These allowed distances can be adapted by the user of the algorithm according to the biomechanics of his or her hand.

The objective function consists of three sets of rules. Using the distance matrix, a first set of rules compares the actual distance (calculated by subtracting the corresponding values of the keys in Fig. 1) and the allowed distance between two simultaneous or consecutive notes for the proposed pair of fingers to play these two notes. A penalty score is applied when the actual distance is larger than the

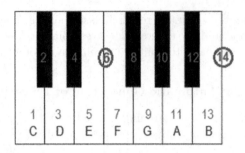

Fig. 1. Piano keyboard with additional imaginary black keys [15].

different types of allowed distances. As Jacobs [13] argued, the distances in the previous research of Parncutt et al. [6] were not accurate. To calculate these penalties more accurately, the authors increased in previous work [15] the distances between E and F and between B and C to two half notes, to equal the distances between other adjacent white keys on a piano keyboard, as illustrated in Fig. 1. The second set of rules is implemented to prevent unnecessary and inconvenient hand changes. The third group of rules prevents difficult finger movements in monophonic music. In this third set, two additional rules proposed by Balliauw [15] promote the choice of logical, commonly used fingering patterns. For a more detailed discussion of the rules and distances, we refer to the actual references.

The weighted penalty scores of all rules together, displayed in Table 2, are summed to obtain the objective function value. As this value indicates the difficulty of the fingering, it has to be minimised.

The problem description also takes into account one hard constraint. This constraint enforces that a single finger cannot be used to play two different, simultaneous notes, as this is not feasible to execute on a real piano.

Table 1. Example distance matrix that describes the pianist's right hand [15].

Finger pair	MinPrac	MinComf	MinRel	MaxRel	MaxComf	MaxPrac
1–2	−10	−8	1	6	9	11
1–3	−8	−6	3	9	13	15
1–4	−6	−4	5	11	14	16
1–5	−2	0	7	12	16	18
2–3	1	1	1	2	5	7
2–4	1	1	3	4	6	8
2–5	2	2	5	6	10	12
3–4	1	1	1	2	2	4
3–5	1	1	3	4	6	8
4–5	1	1	1	2	4	6

Table 2. Set of rules composing the objective function [15].

Rule	Application	Description	Score
1	All	For every unit the distance between two consecutive notes is below `MinComf` or exceeds `MaxComf`	+2
2	All	For every unit the distance between two consecutive notes is below `MinRel` or exceeds `MaxRel`	+1
3	Monophonic	If the distance between a first and third note is below `MinComf` or exceeds `MaxComf`: add one point. In addition, if the pitch of the second note is the middle one, is played by the thumb and the distance between the first and third note is below `MinPrac` or exceeds `MaxPrac`: add another point. Finally, if the first and third note have the same pitch, but are played by a different finger: add another point	+1 +1 +1
4	Monophonic	For every unit the distance between a first and third note is below `MinComf` or exceeds `MaxComf`	+1
5	Monophonic	For every use of the fourth finger	+1
6	Monophonic	For the use of the third and the fourth finger (in any consecutive order)	+1
7	Monophonic	For the use of the third finger on a white key and the fourth finger on a black key (in any consecutive order)	+1
8	Monophonic	When the thumb plays a black key: add a half point. Add one more point for a different finger used on a white key just before and one extra for one just after the thumb	+0.5 +1 +1
9	Monophonic	When the fifth finger plays a black key: add zero points. Add one more point for a different finger used on a white key just before and one extra for one just after the fifth finger	+1 +1
10	Monophonic	For a thumb crossing on the same level (white-white or black-black)	+1
11	Monophonic	For a thumb on a black key crossed by a different finger on a white key	+2
12	Monophonic	For a different first and third note, played by the same finger, and the second pitch being the middle one	+1
13	All	For every unit the distance between two following notes is below `MinPrac` or exceeds `MaxPrac`	+10
14	Polyphonic	Apply rules 1, 2 (both with doubled scores) and 13 within one chord	
15	All	For consecutive slices containing exactly the same notes (with identical pitches), played by a different finger, for each different finger	+1

4   Tabu Search Algorithm

A common approach to solve combinatorial optimisation problems is the use of exact algorithms, that ensure finding the optimal solution. The problem with this approach is that in the worst case the execution time can grow exponentially with the size of the instance treated. For this reason, they are often suited to deal with short or simple instances only. Metaheuristics form an alternative approach to generate good solutions for complex problems (often involving large and complex instances) in a reasonable amount of execution time. Several metaheuristic frameworks have been proposed in the literature, all of which offer guidelines to build heuristic algorithms [17].

Different categories of metaheuristics exist, such as constructive, population-based and local search [17]. In this research we develop a local search heuristic to generate piano fingerings for complex polyphonic music, as this class of heuristics better allows to take the characteristics of the problem into account [18]. Local search heuristics start from a *current solution*, to which small, incremental changes are made, called *moves*. All moves performing the same changes are part of the same *move type*. The set of solutions that can be reached from the current solution by a certain move type is called the *neighbourhood* of the solution. A *local optimum* is reached when the neighbourhood contains no improvement for the current solution [17].

An algorithm can escape from such a local optimum or avoid getting trapped into cycles by using a *tabu list*. This list contains a number of moves that were performed right before the current move and that are excluded from the neighbourhood of the current solution. These moves are *tabu active*. The length of the tabu list is called *tabu tenure*. The move with the best objective function value from this neighbourhood is performed, even if it worsens the current solution. In this way, the algorithm avoids getting trapped into cycles and can explore the solution space around the local optimum to eventually arrive at a better solution. The tabu tenure and the number of allowed iterations without improvement are the two parameters that define a *tabu search* [19].

When no more improvements can be made to a local optimum within one neighbourhood, the algorithm switches to another neighbourhood, defined by a different move type. This strategy is similar to that implemented by the *Variable Neighbourhood Search (VNS)* [20].

In this paper, finding a good piano fingering is modelled as a combinatorial optimisation problem. When this problem becomes larger (as is often the case), the number of possible solutions grows exponentially with the number of notes in the piece (5^n). To this end, we chose to develop a tabu search algorithm (TS), as it has already been applied efficiently to many other combinatorial optimisation problems such as the NP-hard travelling salesman problem [21] and vehicle routing problem [22]. More recently, it has also been used in the field of music [3]. As a result, developing a TS algorithm was considered as a viable choice in this paper.

The TS algorithm, displayed schematically in Algorithm 1, starts from a random initial solution, and optimises the fingering for the right hand.

The optimisation processes for the left hand is identical and executed subsequently. First, a preprocessing step Swap is applied to the initial solution. This step makes significant improvements (i.e., reducing the objective function value $f(\mathcal{S})$) by swapping two fingers throughout the entire piece. This preprocessing step has a positive impact on the solution quality and reduces the required execution time.

Algorithm 1. TS algorithm for each hand

Input : File $\mathcal{F}$ containing the piece
Output: File $\mathcal{F}'$ with the generated fingering included

1 $\mathcal{P} \leftarrow$ Parse($\mathcal{F}$)
2 $\mathcal{P} \leftarrow \mathcal{P}/\{$non-used hand$\}$
3 $\mathcal{S}_{best} \leftarrow$ Rand_Sol($\mathcal{P}$)
4 $\mathcal{S}_{best} \leftarrow$ Swap($\mathcal{S}_{cur}$)
5 **tabutenure** = $0.5 \cdot$Number_Notes($\mathcal{P}$)
6 **maxiters** = $5 \cdot$**tabutenure**
7 Init_Tabu_List(**tabutenure**)
8 $\mathcal{T} \leftarrow$ True
9 **while** $\mathcal{T} = True$ **do**
10 $\quad$ $\mathcal{T} \leftarrow$ False
11 $\quad$ **for** $o \in \mathcal{O}$ **do**
12 $\quad\quad$ $\mathcal{S}_{nbh} \leftarrow \mathcal{S}_{best}$
13 $\quad\quad$ $\mathcal{S}_{cur} \leftarrow \mathcal{S}_{best}$
14 $\quad\quad$ $i = 0$
15 $\quad\quad$ **while** $i <$**maxiters** **do**
16 $\quad\quad\quad$ $\mathcal{N} \leftarrow$ Neighbourhood$_o$($\mathcal{S}_{cur}$)
17 $\quad\quad\quad$ $\mathcal{N} \leftarrow \mathcal{N}/\{s : s$ involve $t_{ij} \in$ Tabu_List$\}$
18 $\quad\quad\quad$ $\mathcal{S}_{cur} \leftarrow$ Best_Sol($\mathcal{N}$)
19 $\quad\quad\quad$ Update_Tabu_List()
20 $\quad\quad\quad$ **if** $f(\mathcal{S}_{cur}) < f(\mathcal{S}_{nbh})$ **then**
21 $\quad\quad\quad\quad$ $\mathcal{S}_{nbh} \leftarrow \mathcal{S}_{cur}$
22 $\quad\quad\quad\quad$ $i = 0$
23 $\quad\quad\quad$ **else**
24 $\quad\quad\quad\quad$ $i \leftarrow i + 1$
25 $\quad\quad$ **if** $f(\mathcal{S}_{nbh}) < f(\mathcal{S}_{best})$ **then**
26 $\quad\quad\quad$ $\mathcal{S}_{best} \leftarrow \mathcal{S}_{nbh}$
27 $\quad\quad\quad$ $\mathcal{T} \leftarrow$ True
28 $\quad\quad$ Clear_Tabu_List()
29 $\mathcal{F}' \leftarrow$ Write($\mathcal{S}_{best}$)

The main loop of the algorithm uses three neighbourhood operators, each defined by a move type and executed consecutively. For each operator, the best solution from the neighbourhood is selected with a steepest descent strategy. As a result, the move that leads to the best fingering is chosen at each iteration.

(a) Change1   (b) Change2   (c) SwapPart

Fig. 2. Examples of each move explored in every neighbourhood considered by the tabu search algorithm.

A first neighbourhood, called Change1, is defined by a move type that changes the finger of a note to any other possible finger. The move type Change2 is similar to Change1, but it is expanded to two adjacent or simultaneous notes. This move is useful in situations in which changing two adjacent notes simultaneously improves the solution but changing each note separately has a detrimental effect on the fingering and is therefore discarded by Change1. To increase the interchangeability of two fingers in polyphonic music, moves of the third type SwapPart change the fingering of a note a from finger k to l, where the fingering of all notes b (played with finger l, starting before the end of note a and ending after the start of note a) are changed from l to k. An illustrative example of each neighbourhood operator $o \in \mathcal{O} = \{$Change1, Change2, SwapPart$\}$ can be found in Fig. 2.

To perform tabu search using a neighbourhood, a *tabu list* is initialized after the preprocessing step. The parameter tabutenure is defined as a percentage of the number of notes played by a given hand. The tabu list considers changing or swapping the fingering of a specific note i to fingering j as a forbidden move if the couple t_{ij} is *tabu active*, i.e., it is on the tabu list. A couple t_{ij} becomes tabu active after the fingering of note i has been moved to finger j and remains active for a number of moves, equal to the tabu tenure. As a result, it is possible that moving to the best neighbouring solution (that is not tabu) reduces the quality of the current solution. The number of allowed iterations without improvement is defined as a parameter, maxiters. This enables the algorithm to explore the solution space around the current solution and escape from a local optimum, given that enough iterations without improvement are allowed (here defined as a percentage of the tabu tenure). A path to arrive in a different area of the solution space and escape the local optimum can thus be pursued. This can be observed in Fig. 3.

When the number of non-improving iterations using a certain neighbourhood reaches maxiters, the content of the tabu list is cleared and in order to escape from the local optimum, the algorithm switches to the next neighbourhood. When a loop through all neighbourhoods $o \in \mathcal{O}$ successfully improves the solution $\mathcal{S}$, a new iteration over all these neighbourhoods is executed. Otherwise, the stopping criterion is met and the algorithm returns the best solution found. This solution is outputted to a MusicXML file, which can be processed by open source music sheet software, like MuseScore[1].

---

[1] Available from www.musescore.org.

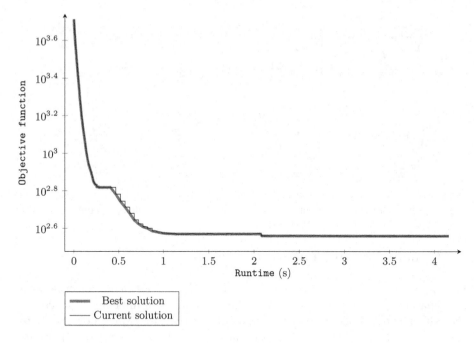

Fig. 3. Evolution of the right hand score over time in the first variation on the Saraband from Suite in D minor (HWV 437) by G.F. Händel.

5   Results

Figure 4 shows an example output of the described TS for the first variation on the Saraband from G.F. Händels Suite in D minor (HWV 437). The algorithm was run with all neighbourhood operators $o \in \mathcal{O}$ activated, the tabu tenure set as 50 % of the number of notes in the hand and the allowed iterations without improvement set as 5 times the tabu tenure (i.e., 500 %). These parameter settings were chosen during a limited pilot study. The fingerings were generated using the hand data shown in Table 1 and with the weights of the rules in the objective function set to 1.

The output of the algorithm shows the first eight bars of the piece. The execution time is very short (5 s in total and 4 s for the right hand) and the objective function value was 712 in total (362 for the right hand). The evolution of the right hand score over time is displayed in Fig. 3. Experts (researchers and pianists) confirmed that the solution displayed in the output in Fig. 4 is easily playable and thus forms a good fingering. However it is not perfect, as for example can be seen in the fifth bar where the first note (A) is played with finger 1, and the second (G), lower note with a 4. The algorithm might be improved to reach even better solutions in future research.

Fig. 4. Output for the first eight bars of the first variation on the Saraband from Suite in D minor (HWV 437) by G.F. Händel.

6   Conclusion and Future Research

In this paper, we described the problem of finding a good piano fingering as a combinatorial optimisation problem. Different sources of difficulty modelled in the implemented objective function allow to deal with complex polyphony, to analyse the left and right hand, and to have the option to adapt some parameters to personal preferences and biomechanics of the hand. We proposed a tabu search algorithm for the generation of piano fingerings, minimising the objective function value. By generating a piano fingering for an existing piece, we showed that the algorithm can find a good solution in a relatively short amount of execution time.

In the future, a possible enhancement of the objective function could be an even more accurate or detailed keyboard distance definition, accounting for the asymmetrical positioning of the black keys. The algorithm could also be expanded by including rules in the objective function that specify the interpretation and memorisation aspects of a piano fingering. Examples of such rules could be the use of a strong finger (e.g., the thumb) on the first beat of a bar, and using the same fingering patterns for identical note sequences. The objective function could also be improved by integrating models built by applying machine learning techniques to a database of existing piano fingerings. In this way, new rules could be defined and the weights of the different penalty scores in the objective function could be optimised. Further adaptations of the musical rules and interpretation of the outputs should always be verified by existing sheet music and expert piano professionals. The algorithm could also benefit from the inclusion of extra, more sophisticated neighbourhoods.

By taking into account musical sentences, a further improvement in both solution quality and execution time could be attained. The algorithm now improves an entire piece at once. When the piece would be split up into smaller parts based on musical sentences, the decreased size of the neighbourhoods would significantly speed up the execution time of the algorithm. Rules based on these musical sentences might equally improve the solution quality.

Acknowledgments. This research is supported by the Interuniversity Attraction Poles (IAP) Programme initiated by the Belgian Science Policy Office (COMEX project).

References

1. Sloboda, J.A., Clarke, E.F., Parncutt, R., Raekallio, M.: Determinants of finger choice in piano sight-reading. J. Exp. Psychol. Hum. Percept. Perform. **24**, 185–203 (1998)
2. Robine, M.: Analyse automatique du doigté au piano. In: Proceedings of the Journées d'Informatique Musicale, pp. 106–112 (2009)
3. Herremans, D.: Tabu search voor de optimalisatie van muzikale fragmenten. Master's thesis at University of Antwerp, Faculty of Applied Economics, Antwerp (2005)
4. Gellrich, M., Parncutt, R.: Piano technique and fingering in the eighteenth and nineteenth centuries: bringing a forgotten method back to life. Br. J. Music Educ. **15**, 5–23 (1998)
5. Sayegh, S.I.: Fingering for string instruments with the optimum path paradigm. Comput. Music J. **13**, 76–84 (1989)
6. Parncutt, R., Sloboda, J.A., Clarke, E.F., Raekallio, M., Desain, P.: An ergonomic model of keyboard fingering for melodic fragments. Music Percept. **14**, 341–382 (1997)
7. Parncutt, R.: Modeling piano performance: physics and cognition of a virtual pianist. In: Proceedings of International Computer Music Conference, pp. 15–18 (1997)
8. Viana, A.B., de Morais Júnior, A.C.: Technological improvements in the SIEDP. In: IX Brazilian Symposium on Computer Music, Campinas, Brazil (2003)
9. Hart, M., Bosch, R., Tsai, E.: Finding optimal piano fingerings. UMAP J. **2**, 167–177 (2000)
10. Radicioni, D.P., Anselma, L., Lombardo, V.: An algorithm to compute fingering for string instruments. In: Proceedings of the National Congress of the Associazione Italiana di Scienze Cognitive, Ivrea, Italy (2004)
11. Al Kasimi, A., Nichols, E., Raphael, C.: A simple algorithm for automatic generation of polyphonic piano fingerings. In: 8th International Conference on Music Information Retrieval, Vienna (2007)
12. Yonebayashi, Y., Kameoka, H., Sagayama, S. Automatic decision of piano fingering based on Hidden Markov models. In: IJCAI, pp. 2915–2921 (2007)
13. Jacobs, J.P.: Refinements to the Ergonomic model for keyboard fingering of parncutt, Sloboda, Clarke, Raekalliio desain. Music Percept. **18**, 505–511 (2001)
14. Lin, C.-C., Liu, D.S.-M.: An intelligent virtual piano tutor. In: Proceedings of the 2006 ACM International Conference on Virtual Reality Continuum and Its Applications, pp. 353–356 (2006)
15. Balliauw, M.: A variable neighbourhood search algorithm to generate piano fingerings for polyphonic sheet music. Master's thesis at University of Antwerp, Faculty of Applied Economics, Antwerp (2014)
16. Sébastien, V., Ralambondrainy, H., Sébastien, O., Conruyt, N.: Score analyzer: automatically determining scores difficulty level for instrumental e-learning. In: ISMIR, pp. 571–576 (2012)

17. Sörensen, K., Glover, F.: Metaheuristics. In: Glass, S.I., Fu, M.C. (eds.) Encyclopedia of Operations Research and Management Science, pp. 960–970. Wiley, New York (2013)
18. Herremans, D., Sörensen, K.: Composing first species counterpoint with a variable neighbourhood search algorithm. J. Math. Arts **6**, 169–189 (2012)
19. Glover, F., Laguna, M.: Tabu Search. Kluwer Academic Publishers, Boston (1993)
20. Mladenović, N., Hansen, P.: Variable neighbourhood search. Comput. Oper. Res. **24**, 1097–1100 (1997)
21. Fiechter, C.-N.: A parallel tabu search algorithm for large travelling salesman problems. Discrete Appl. Math. **51**, 243–267 (1994)
22. Gendreau, M., Hertz, A., Laporte, G.: A tabu search heuristic for the vehicle routing problem. Manag. sci. **40**, 1276–1290 (1994)

Logistic Modeling of Note Transitions

Luwei Yang[1]([✉]), Elaine Chew[1], and Khalid Z. Rajab[2]

[1] Centre for Digital Music, Queen Mary University of London, London, UK
{l.yang,elaine.chew}@qmul.ac.uk
[2] Antennas and Electromagnetics, Queen Mary University of London, London, UK
k.rajab@qmul.ac.uk

Abstract. Note transitions form an essential part of expressive performances on continuous-pitch instruments. Their existence and precise characteristics are not captured in conventional music notation. This paper focuses on the modeling and representation of note transitions. We compare models of excerpted pitch contours of performed portamenti fitted using a Logistic function, a Polynomial, a Gaussian, and Fourier Series, each constrained to six coefficients. The Logistic Model is shown to have the lowest root mean squared error and the highest adjusted R-squared value; an ANOVA shows the difference to be significant. Furthermore, the Logistic Model produces musically meaningful outputs: transition slope, duration, and interval; and, time and pitch of the inflection point. A case study comparing portamenti between erhu and violin on the same musical phrase shows transition intervals to be piece-specific (as it is constrained by the notes in the score) but transition slopes, durations, and inflection points to be performer-specific.

Keywords: Logistic model · Note transition · Portamento · Expressive music analysis · Performance

1 Introduction

Transitions between musical objects are extremely important in expressive musical performance and also in music composition. Musical expressivity lies in the way in which the performer plays the notes and connects them. In his book on violin teaching, Constantakos says, "There were connections and when you put notes together, you would start to think of it in connection with music, making phrases [4]".

Note transitions can be classified into two types. The first is a discrete note transition, which is the default mode in piano playing. In discrete note transitions, the player is unable to or does not wish to alter the pitch in the process of moving from one note to another. The other is the continuous note transition, which is prevalent in string, voice, and other instruments. In continuous note transitions, the player adjusts the pitch continuously. This type of note transition is usually referred to as portamento. Portamento is sometimes referred to as "glissando", "glide", or "slide". Here, we use "portamento" and

© Springer International Publishing Switzerland 2015
T. Collins et al. (Eds.): MCM 2015, LNAI 9110, pp. 161–172, 2015.
DOI: 10.1007/978-3-319-20603-5_16

"continuous note transition" interchangeably. A large range of expressivity exists in continuous note transitions.

Portamento is frequently used in violin playing. The sound of the violin is considered to be second only to the human voice in expressive beauty [1]. The violinist Joseph Joachim (1831–1907) considers portamento as being more important and indispensable than vibrato; and, portamento is ranked first among the vocal effects that can be recreated on the violin [2]. Aside from Western music, portamento has also been employed extensively in erhu, a key instrument in Chinese traditional music [21].

Next, we consider the technical and scientific literature on note transitions. Upon examining violin portamenti from eight master violinists, Lee [9] found that the violinists tend to use portamenti as a highly personalized device to exhibit their musicianship. Liu investigated violin glide differences between cadential and noncadential sequences, comparing their proportional duration and the notes' intonation [10]. Maher [11] focussed on vibrato synthesis over portamento transitions, and found that the vibrato rate should be in phase with the note onset so that the note duration is an integer multiple of the vibrato period. Krishnaswamy explored pitch perception, including vibrato and portamento, in South Indian classical music [8].

To the authors' knowledge, there is yet no mathematical model designed or tested for note transitions. The aim of this study is to shed light on the mathematical and computational modeling of note transitions. We observe that the S shape is prevalent in many note transitions, especially in string playing and vocal music. Practically speaking, the execution of a portamento consists of an accelerating process followed by a decelerating process. In a portamento, the player's finger will start to accelerate to a target speed, then decelerate to arrive at the target note position. This usually results an S shape in the spectrogram and pitch contour as shown in Fig. 1.

Inspired by the model for population growth, we propose to use the Logistic Model to fit the S shape of the portamento. We will show that the Logistic Model fits the shape of note transitions very well, and that it has the distinct advantage that its coefficients have direct musical meanings and interpretations.

This study seeks to achieve the following aims:

1. to model portamenti quantitatively using a mathematical model;
2. to provide a tool for investigating and comparing note transition; and,
3. to provide a note transition model that can be used for synthesizing natural-sounding music.

The remainder of this paper is organized as follows: we first propose a number of candidate models followed by the evaluation of these methods; next, a case study on note transitions based on the Logistic Model is presented; finally, the conclusions are presented.

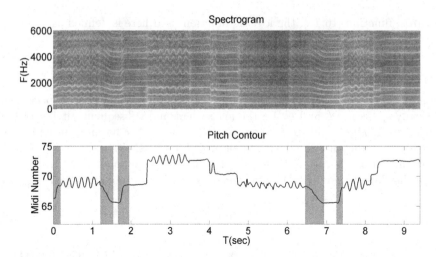

Fig. 1. Spectrogram and the corresponding pitch contour from a passage of erhu. The portamenti are highlighted by grey area in the lower plot.

2 Modeling of Note Transitions

The primary goal of this section is to introduce the Logistic Model for note transitions. At the same time, we offer some alternative modeling methods for comparison, namely, the Polynomial Model, Gaussian Model, and Fourier Series Model. We show that, in general, the Logistic Model has better explanatory value than the other methods. Moreover, other methods are not able to provide direct outputs with meaningful musical interpretations. All models mentioned are used to fit the pitch curve of note transitions.

2.1 Logistic Model

The Logistic Model was originally proposed to solve problems in population dynamics [15,19]. It has been applied successfully to the physical growth of organisms and to forestry growth [14]. Moreover, the Logistic Model has been extended to other fields: in [12], Marchetti and Nakicenovic applied the Logistic Model to energy usage and source substitution; in [5], Herman and Montroll presented the industrial revolution as modeled by the Logistic Model.

In string playing and singing, the players' portamento pitch curve (the log of the fundamental frequency) tends to exhibit an exponential start and an exponential end. In other words, the start and the end of portamenti have similar exponential-style increasing and decreasing shapes. The Logistic Model is especially well-suited to model such features.

To the best of the authors' knowledge, the Logistic Model has yet to be applied to note transitions or other relevant music areas. Inspired by the

Richards' function [16,18], the logistic function used here is defined as

$$P(t) = L + \frac{(U - L)}{\left(1 + Ae^{-G(t-M)}\right)^{1/B}},$$ (1)

where L and U are the lower and upper horizontal asymptotes, respectively. Musically speaking, L and U are the antecedent and subsequent pitches of the transition. A, B, G, and M are constants. G can further be interpreted as the growth rate, indicating the degree of slope of the transition.

An important characteristic to model is the inflection point of the transition, where the slope is maximized. The time of the inflection point is given by

$$t_R = -\frac{1}{G} \ln\left(\frac{B}{A}\right) + M.$$ (2)

This value is obtained by setting the second derivative of (1) to zero. Since (1) is monotonically increasing, the second order derivative has only one zero point. In other words, the zero point of the second derivative is the maximum of the first derivative, where the slope changes. The inflection point in pitch is calculated by substituting t_R into (1).

2.2 Polynomial Model

The Polynomial Model is given by

$$p(t) = a_n t^n + a_{n-1} t^{n-1} + \ldots + a_2 t^2 + a_1 t + a_0,$$ (3)

where n is the degree of the polynomial. The model then requires $n + 1$ coefficients. Although the Polynomial Model is widely used in many applications for curve fitting, this model performs poorly, especially outside the immediate range of the transition. It cannot model data having asymptotic lines. There is also a trade off between performance and polynomial degree. The larger the number of coefficients, the better the performance; however, the complexity and computational cost would also increase.

2.3 Gaussian Model

The Gaussian Model is given by

$$P(t) = \sum_{n=1}^{N} a_n e^{\left[-\left(\frac{t-b_n}{c_n}\right)^2\right]},$$ (4)

where a_n is the height of the model, b_n the location of the peak, and c_n controls the width of the Gaussian shape. The constant N denotes the number of Gaussian peaks, giving $3 \times N$ coefficients.

2.4   Fourier Series Model

The Fourier Series Model is given by

$$P(t) = a_0 + \sum_{n=1}^{N} a_n \cos(n\omega t) + b_n \sin(n\omega t). \tag{5}$$

Here, a_0 is the constant term, and ω is the fundamental frequency. The parameters a_n and b_n are amplitudes of the cosine and sine terms, respectively. The constant N is the number of the sinusoids used to fit the data, which results in $2 + 2 \times N$ coefficients.

Figure 2 shows the above modeling methods fitting a continuous note transition. For the purpose of unifying the comparison, each method is modeled by six coefficients. Note that the Logistic Model fits the transition better than the other three methods. A further statistical evaluation will follow.

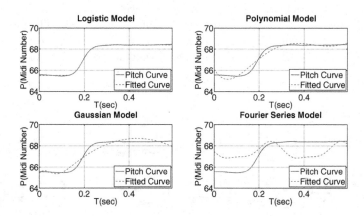

Fig. 2. Modeling of a note transition using the Logistic Model, Polynomial Model, Gaussian Model, and Fourier Series Model.

3   Evaluation

The Logistic Model in (1) has six coefficients. For comparison, the other three modeling methods were also constrained to the same number of coefficients version. As a result, we choose a 5-degree Polynomial Model (i.e., $n = 5$ in (3)), a 2-degree Gaussian Model was selected ($N = 2$ in (4)), and a 2-degree Fourier Series Model ($N = 2$ in (5)).

3.1   Data Annotation

First, pitch contours were extracted from audio files. As there is no prior note transition detection method, there also is no existing database for evaluation.

Portamenti can take place over an extremely short period of time, and it can be challenging to annotate the transition duration accurately. This is one of the reasons we choose to annotate a transition from the midpoint (of the note's duration) of the antecedent note to the midpoint of the consequent note. We create a note transition database using the following rules.

1. The portamento starts from the midpoint of the antecedent note and ends at the midpoint of the subsequent note.
2. If the portamento starts from an intermediate note[1], then the start point is the beginning of the intermediate note.
3. If the subsequent note is not the target of the portamento, then the end point is the end of the subsequent note.
4. If either of the two notes contains a vibrato, the vibrato is flattened to the fundamental frequency of the note.

Following the annotation rules above, we manually annotated portamenti for erhu and violin performances of a phrase in a well known Chinese piece *The Moon Reflected on the Second Spring* using Sonic Visualiser [3]. The violin score of this phrase is shown in Fig. 3. This phrase forms the backbone of the entire piece, and is the phrase that is least changed when adapting the score from erhu to violin.

Fig. 3. A phrase of *The Moon Reflected on the Second Spring* [6].

The two erhu performances used are from recordings by Jiangqin Huang [7] and Guotong Wang [20], and the two violin performances used are from solo recordings provided by Jian Yang and Laurel Pardue. Details about the excerpted portamenti can be seen in Table 1. The numbers in Table 1 show that erhu players tend to use more portamenti than violin players. This may be due to the fact that the erhu has only two strings while the violin has four. Thus,

---

[1] A portamento can be played with two fingers, in sequence, with the first finger sliding to an intermediate note or the second finger starting from an intermediate note.

erhu players have to initiate more slides to reach the target pitches while violin players are able to change strings to reach the target pitch without sliding. Except for the cases where portamenti are indicated in the score, the physical form of the instrument may be an important factor influencing the number of portamenti the player employs.

Table 1. Note transition dataset (corresponding to phrase shown in Fig. 3).

Instrument	Player	Duration	No. of Transitions
Erhu	Jiangqin Huang	55.649	31
	Guotong Wang	42.043	36
Violin	Jiang Yang	37.783	20
	Laurel Pardue	35.732	24
Total	N/A	171.207	111

This dataset is used in both this and the next sections. For the purposes of the study in this paper, we focus only on the continuous note transitions. It is worth pointing out that discrete note transitions can also be modeled by a Logistic Model by giving the slope an extremely high value.

3.2 Model Fitting

We use the Curve Fitting Toolbox in Matlab [17] to perform the note transition modeling. In this package, the non-linear least squares method was used.

Setting the correct search ranges and initial solutions can have a high impact on curve fitting performance. Unlike the Logistic Model, the Polynomial, Gaussian, and Fourier Series models do not have coefficients having direct musical meanings relating to note transitions. As a result, the search ranges of these methods were set to $(-\infty, +\infty)$, and the initial points were decided (randomly) by Matlab. For the Logistic Model, we found that its performance improves when we set the initial value of L to be the lowest pitch in the note transition, and the initial value of U to be the highest pitch in the note transition. The search ranges and initial points for the Logistic Model coefficients are given in Table 2, where x_{min} and x_{max} are the lowest and highest pitches in the note transition, respectively.

Table 2. Search ranges and initial points for coefficients of Logistic Model.

Coefficient	A	B	G	L	M	U
Search Range	$(0, +\infty)$	$(0, +\infty)$	$(-\infty, +\infty)$	$[1, 128]$	$[0, +\infty)$	$[1, 128]$
Start Point	0.8763	1	0	x_{min}	0.1	x_{max}

For each note transition (a finite pitch time series), the Root Mean Squared Error (RMSE) and Adjusted R-Squared values were calculated for each model. The performance of the four modeling methods is presented in Fig. 4, which shows the average Root Mean Squared Error (RMSE) and Adjusted R-Squared values. Note that the Logistic Model has the lowest RMSE and the highest Adjusted R-Squared value, showing that the Logistic Model performed better in the note transition modeling than any other methods. The Polynomial Model has the second best performance. While the Fourier Series Model gives the poorest modeling performance. The superiority of the Logistic Model is confirmed by an ANOVA (Analysis of Variance) [13] analysis. We performed the ANOVA analysis between the Logistic Model and other modeling methods to confirm that the mean values given in Fig. 4 are significant. From Table 3, all p-values are lower than the significant level of 0.01.

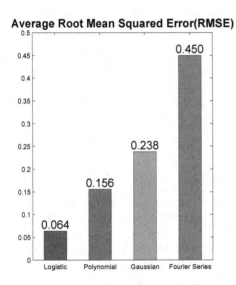
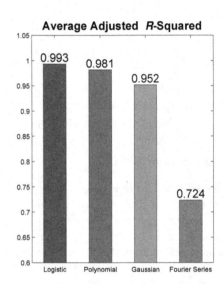

Fig. 4. Modeling performance of Logistic Model, Polynomial Model, Gaussian Model and Fourier Series Model.

Table 3. ANOVA Analysis (p-value) of Root Mean Squared Error and Adjusted R-Squared between Logistic Model and other three model methods.

Root Mean Squared Error			
p-value	Polynomial	Gaussian	Fourier
Logistic	1.11×10^{-9}	2.13×10^{-17}	5.81×10^{-12}
Adjusted R-Squared			
p-value	Polynomial	Gaussian	Fourier
Logistic	2.03×10^{-6}	3.94×10^{-15}	1.64×10^{-11}

4   A Case Study of Erhu and Violin Music

We present here the results of a case study investigating the behavior of portamenti as performed by erhu vs. violin players. The Logistic Model is employed
here to show the feasibility of such expressive performance analyses.

4.1   Parameters of Interest

Using the Logistic Model as defined in (1), we examine the following characteristic of the note transitions.

1. The slope of the transition, which is the coefficient G in (1).
2. The transition duration. Once the Logistic Model is set up, the first derivative
   of the Logistic curve that is larger than a threshold value can be employed to
   identify the transition duration. Empirically, this threshold is 0.861 semitones
   per sec.
3. The transition interval. The interval is obtained by calculating the absolute
   semitone difference between the lower and upper asymptotes.
4. The normalized inflection time. The actual time of the inflection point is given
   by (2). As transition durations are different one from another, this time is
   normalized to lie between 0 and 1, where 0 marks the beginning and 1 the
   end of the transition duration.
5. The normalized inflection pitch. This is similar to the normalized inflection
   time; this parameter is also normalized to lie between 0 and 1, where 0
   indicates the lower asymptote and 1 the higher asymptote in the transition
   interval.

An example of a note transition is given in Fig. 5, where a slope of 42.251 can
be observed. The transition duration and interval are 0.186 seconds and 2.914
semitones, respectively. The inflection point appears to lie in the first half of the
transition duration and interval; this is confirmed by the normalized inflection
time of 0.316 and pitch of 0.428.

4.2   Results

The slope, transition duration, and interval statistics are shown in Fig. 6. The
middle bar in the box indicates the median value. The lower and upper edges
mark the 25th and 75th percentiles, $Q1$ and $Q3$, respectively. The dotted lines
extend from $(Q1 - 1.5 \times (Q3 - Q1))$ to $(Q3 + 1.5 \times (Q3 - Q1))$, while dots beyond
these boundaries mark the outliers.

Consider the transition interval. All four players' transition intervals are on
the order of three semitones wide, with insignificant differences. A reason could
be that the pitches are constrained by the musical score, which limits the range
of the transition interval. Wang and Pardue exhibit wider variabilities in their
transition intervals, as indicated by the taller boxes, but it is likely that the
transition intervals may vary more widely across musical pieces than between
players. This hypothesis warrants further experiment and exploration.

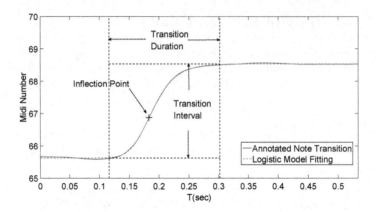

Fig. 5. Illustration of transition duration, transition interval, and inflection time and pitch from an erhu excerpt.

Wang has the largest average slope value. Since the four players have similar transition intervals, it is expected that Wang, due to the high slope, has the lowest average transition duration, as confirmed by Fig. 6. As expected, the slope and the transition duration are negatively correlated.

Figure 7 shows boxplots of the normalized time and pitch of the inflection point. Both erhu players tended to time their inflection points in the first half of the transition duration, while the violin players chose to put their inflection points around the middle of the transition duration. In contrast, the inflection pitch of the erhu players tended to lie around the middle of the transition interval, while that of violin players are located lower in the interval.

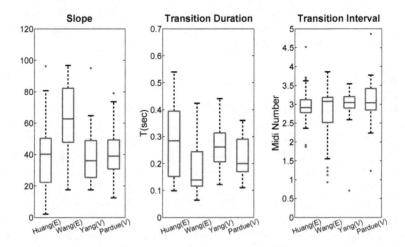

Fig. 6. Boxplots of slope, transition duration, and transition interval for all four players of *The Moon Reflected on the Second Spring*. E: Erhu, V: Violin.

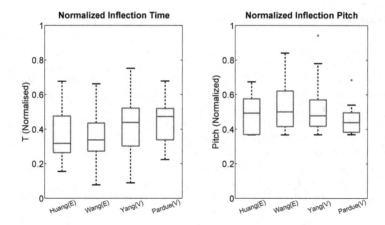

Fig. 7. Boxplots of normalized inflection time and normalized inflection pitch for all four players of *The Moon Reflected on the Second Spring*. E: Erhu, V: Violin.

5   Conclusion

In this study, we proposed a computational model of note transitions employing the Logistic Model. This model is able to fit discrete and continuous note transitions. The Logistic Model is shown to have better performance than other methods, namely the Polynomial Model, Gaussian Model, and Fourier Series Model. Moreover, parameters that convey musically meaningful information make the Logistic Model stand out from other methods. A case study on erhu and violin data was presented to demonstrate the feasibility of the Logistic Model in expressive music analyses.

For the four players analyzed, the transition interval was found to be largely constrained by the score, the transition duration varied by player, with the duration being inversely related to the slope. The two erhu players tended to place the inflection time in the first half of the duration, while the two violin players tended to put it around the middle; the two violin players tended to place the inflection pitch in the lower half of the interval, while erhu players tended to put it around the middle.

This study represents a first effort towards the modeling of note transitions. We plan to test this model on further datasets of continuous note transitions featuring other instruments, for example, the singing voice. Another direction would be to test the usability of this model for the synthesis of natural sounding music. Future work can also consider other models for note transitions such as a spline or the integral of a Gaussian function.

Acknowledgments. This research is supported in part by the China Scholarship Council. The authors would like to thank Jian Yang and Laurel Pardue for the violin recordings.

References

1. Applebaum, S.: The Way They Play. Paganiniana Publications, Neptune City (1972)
2. Brown, C.: The decline of the 19th-century German school of violin playing. http://chase.leeds.ac.uk/article/the-decline-of-the-19th-century-german-school-of-violin-playing-clive-brown/. Accessed in Jan 2015
3. Cannam, C., Landone, C., Sandler, M.: Sonic visualiser: An open source application for viewing, analysing, and annotating music audio files. In: Proceedings of the International Conference on Multimedia, pp. 1467–1468. ACM (2010)
4. Costantakos, C.A.: Demetrios Constantine Dounis: His Method in Teaching the Violin. Peter Lang Publishing Inc., New York (1997)
5. Herman, R., Montroll, E.W.: A manner of characterizing the development of countries. Nat. Acad. Sci. **69**, 3019–3023 (1972)
6. Hua, Y.: Erquanyingyue. Zhiruo Ding and Zhanhao He, violin edn. (1958), musical Score
7. Huang, J.: The Moon Reflected on the Second Spring, on The Ditty of the South of the Jiangsu. CD (2006). ISBN: 9787885180706
8. Krishnaswamy, A.: Pitch measurements versus perception of south indian classical music. In: Proceedings of the Stockholm Music Acoustics Conference (SMAC) (2003)
9. Lee, H.: Violin portamento: An analysis of its use by master violinists in selected nineteenth-century concerti. In: ICMPC9 Proceedings of the 9th International Conference on Music Perception and Cognition, August 2006
10. Liu, J.: Properties of violin glides in the performance of cadential and noncadential sequences in solo works by bach. In: Proceedings of Meetings on Acoustics. vol. 19. Acoustical Society of America (2013)
11. Maher, R.C.: Control of synthesized vibrato during portamento musical pitch transitions. J. Audio Eng. Soc. **56**(1/2), 18–27 (2008)
12. Marchetti, C., Nakicenovic, N.: The dynamics of energy systems and the logistic substitution model. Technical report. PRE-24360 (1979)
13. Ott, R.L., Longnecker, M.: An Introduction to Statistical Methods and Data Analysis, 6th edn. Brooks/Cole, Belmont (2010)
14. Payandeh, B.: Some applications of nonlinear regression models in forestry research. For. Chronicle **59**(5), 244–248 (1983)
15. Pearl, R.: The growth of populations. Q. Rev. Biol. **2**, 532 (1927)
16. Richards, F.J.: A flexible growth function for empirical use. J. Exp. Bot. **10**(2), 290–301 (1959)
17. The MathWorks, Inc., N.: Matlab r2013b (2013)
18. Tsoularis, A., Wallace, J.: Analysis of logistic growth models. Math. Biosci. **179**(1), 21–55 (2002)
19. Verhulst, P.F.: Notice sur la loi que la population suit dans son accroissement. correspondance mathématique et physique publiée par a. Quetelet **10**, 113–121 (1838)
20. Wang, G.: Track 4, disk 2, an anthology of chinese traditional and folk music a collection of music played on the erhu. CD (2009). ISBN: 9787799919928
21. Zhao, H.: Erhu yanzouzhong huayin de yunyong (the application of portamento in erhu playing). Chin. Music **4**, 020 (1987). (in Chinese)

Evaluating Singer Consistency and Uniqueness in Vocal Performances

Johanna Devaney[✉]

School of Music, The Ohio State University, Columbus, USA
`devaney.12@osu.edu`

Abstract. Identifying consistent and unique aspects of performances is an important aspect of modeling performance style. This paper presents a detailed analysis of inter-singer differences and intra-singer similarities using support vector machines to predict singer identity of a performance from pitch, timing, and dynamics performance parameters. The analysis was performed on a dataset of 72 recordings of the first verse of Schubert's "Ave Maria", the dataset consists of 3 *a cappella* and 3 accompanied performances by 6 professional and 6 non-professional singers.

Keywords: Singing · Music performance · Intra-performer similarity · Inter-performer differences · Classification · Discrimination

1 Introduction

Modeling inter-performer deviations in pitch, timing, dynamics, and timbre allows for a particular performer's style to be quantified. This is complicated, however, by intra-performer difference, i.e., how much variation there is when the same performer performs the same piece. If the amount of variance within a musician's performance of particular parameter exceeds the variance across the values for different musicians, then one cannot create a discriminative model between musicians based on that parameter. Likewise, if the amount of intra-performer variability at a particular point in time exceeds the variability across two recordings made at different times by the same musician, then the temporally divergent recordings cannot be considered developments in style strictly on the basis of the performance data. What perhaps is being modeled, in the case of professional recordings, is the aesthetic decision of the musician, and/or the producer, as to which recording was best representative of the musician's style.

This paper examines the amount of variation across six performances of the first verse of Schubert's "Ave Maria" by 12 different singers. The performances were all made in the same room, one after another, and so provide an opportunity to explore both inter- and intra-singer similarity. This is done through the task of singer identification using a support vector machine trained on pitch-, timing-, and dynamics-related features. Timbre is not considered because it is influenced not only by the singer's performance style but also their unique vocal physiology. Task is formulated in two different ways, both focusing on the opening and closing

© Springer International Publishing Switzerland 2015
T. Collins et al. (Eds.): MCM 2015, LNAI 9110, pp. 173–178, 2015.
DOI: 10.1007/978-3-319-20603-5_17

statements of the text "Ave Maria" in the performances of the first verse of the song: the first using a leave-one-out cross validation paradigm to examine consistency within either the opening or closing statements and the second one training on the opening statement data and testing on the closing statement data. The singer identification is also evaluated in terms of experience level (professional versus non-professional) and the presence or absence of accompaniment.

2 Background

Psychological studies have shown that even highly trained performers exhibit certain amounts of variability across performances [1]. In performance studies, however, researchers typically look at a single recording by each performer or treat multiple recordings by the same person (generally recorded at different points in time) as independent examples. An example of this is the research done with the Chopin Mazurka data set [3, 12]. When recordings of multiple performances are available, some researchers have used the mean values across the performances for analysis, such as [8, 9]. The issue with this, particularly when only a small number of performances is being considered, is that if one value is particularly high and another is low, the resulting medium-level value is neither representative of the performance parameter's values or its variance. Expressive performance modeling, where researchers develop machine learning models from large number of recordings of different pieces by multiple performers, e.g. [13], sidestep the issue of intra-performer consistency to a certain extent because of the sheer size of the dataset. This approach, however, can only model very general trends.

The evaluation task used in this paper relates to work in the area of artist identification from audio. For singing, this work has been dominated by timbral features, namely MFCCs (e.g., [10]). Thus this current project is more closely related to work that has focused on note-level performance parameters in instrumental performance rather than timbre, such as the work done on the Celtic violin in [11].

3 Method

3.1 Recordings

The data set consists of 72 recordings of the first verse of Schubert's "Ave Maria". Six professional and six non-professional singers each performed the verse six times: three times *a cappella* and three times listening to a piano accompaniment through one ear of a pair of headphones. All of the recordings were made in the same acoustically treated $4.85\,\text{m} \times 4.50\,\text{m} \times 3.30\,\text{m}$ room, chosen to minimize the effects of room acoustics on the performances. The "Ave Maria" was chosen because the singers were already familiar with and had previously rehearsed the piece. After the recordings were made, the singers verified that the performances were a good representations of their performance style.

3.2 Performance Data

A list of the performance features is provided is Table 1. The onset and offset times in the recordings were automatically estimated using a MIDI-audio alignment algorithm that was specifically developed to work robustly with recordings of the singing voice [5]. Two timing features were derived for each note from the onset and offset features: interonset interval (T1) and note duration (T2). Interonset interval provides information about the overall pacing of the performance while note duration provides information about the length of the notes themselves. Loudness estimates were made using the implementation of Glasberg and Moore's loudness model of time-varying tones [6] in the Genesis Loudness Toolbox[1]. The long-term loudness maximum was for each note was used as a feature (L).

Fundamental frequency (F0) estimates were made with a MATLAB implementation[2] of de Cheveign and Kawahara's YIN algorithm [2]. From the framewise F0 estimates, six note-level features were calculated. Perceived pitch estimates were made using Gockel, Moore, and Carlyon's model for time-varying tones [7]. From these calculations the interval size in cents between the current note and the previous note (P1) was calculated and the distance in cents between the current note and the opening note (P2). The interval size gives an indication of local tuning and the distance from the first note gives an indication of the amount of drift. The slope (P3) and curvature (P4) of the F0 trajectory over the duration of the note were estimated by taking the Discrete Cosine Transform of the F0 trace. In the resulting coefficients, the 0th corresponds to the mean of the F0 values, the 1st approximates slope, and the 2nd approximates curvature [4]. Vibrato rate and depth were estimated by taking fast Fourier transform (FFT) of the F0 trace, which assumes the vibrato to be sinusoid. The maximum absolute value of the FFT corresponds to half of the extent of the vibrato (P5) and the position of this maximum corresponds to the rate of the vibrato (P6).

3.3 Evaluating Similarity Through Classification

A support vector machine (SVM) with L1 regularization was used to predict singer identity with the normalized versions of the performance features for each

Table 1. Summary of note-level performance features

Timing	Loudness	Pitch
T1: Inter-onset interval	L:Long-term loudness	P1: Interval size
T2: Duration		P2: Distance from opening note
		P3: Slope of F0 trajectory
		P4: Curvature of F0 trajectory
		P5: Vibrato extent
		P6: Vibrato rate

---

[1] http://www.genesis-acoustics.com/.

[2] http://audition.ens.fr/adc/sw/yin.zip.

note detailed in Table 1. Four versions of the feature sets were used, one with all of the features, one just with the timing parameters (T1?2), one with just the loudness parameter (L), and one with just the pitch parameters (P1–6). One of the useful features of the "Ave Maria" is that the melody of the opening and closing phrases are the same (see Fig. 1). This analysis focuses on the opening and closing statements of the text "Ave Maria", as they provide the opportunity to evaluate both the consistency within the exact same musical material and across highly similar, but not identical, material. The first is achieved by using either the opening and closing data to predict other instances of the opening or closing using a leave-one-out cross validation paradigm. The second is achieved by using the data from opening predicting the identity of the data from the closing, which has a slightly different harmonic setting. The SVM was run separately on the data from the *a cappella* and accompanied recordings in order to observe trends about the relative contributions of the different features across contexts.

A - ve Ma - ri - - - - a

Fig. 1. Opening and closing phrases of the first verse of Schubert's "Ave Maria".

4   Results

The results of the two experiments are shown in Fig. 2. Overall there were higher singer identification rates within the *a cappella* versions than within the accompanied versions. Generally, there were also higher singer identification rates when all of the performance data was used than when timing, loudness, or pitch features were used exclusively, although an exception is noted below. In the *a cappella* versions, the individual groups of features were more or less equally effective for both the professional and non-professional singers. There was more variability in the accompanied versions, namely that loudness was a better predictor of singer identity on its own than all of the features combined in both experiments and that the pitch and timing predictors were considerably less accurate on their own for the professionals in experiment 1 than the non-professional. This trend also held for the professional's pitch parameters in Experiment 2. The weights matrix (built from the SVM's weight vectors) indicates how much each feature is used by the SVM in making predictions and was examined to determine the whether any individual performance parameter was considered irrelevant in the various conditions. No individual features were completely ignored by the SVM in either the version of the experiment or for any combination of features.

5   Discussion

The lower accuracy for the accompanied versions compared to the *a cappella* ones suggests that either the singers were more variable or that they were converging to the accompanied values. The latter is more likely to be true for pitch and

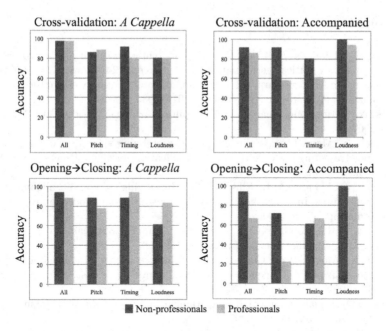

Fig. 2. Accuracy of singer identity prediction. The blue lines represent the non-professional singers and the red lines represent the professional singers. The top row of plots shows the mean accuracy for the first experiment: predicating singer identity within either the opening and closing data. The bottom row of plots shows the accuracy for the second experiment: using the data from opening to predict the identity of the singer from the closing data. The plots on the left show the results for the *a cappella* renditions and the plots on the right show the results for the accompanied renditions. Within each plot, the y-axis shows the accuracy of the prediction and the x-axis is grouped by features used: all, pitch-only, timing-only, and loudness-only (Color figure online).

timing parameters rather than loudness, since the accompaniment directly cues pitch and timing. This is confirmed by the differences in accuracy rates for the accompanied pitch and timing versus loudness in the accompanied experiment. There is also a similar trend within the pitch and timing values between the professionals and non-professionals, suggesting that the professionals are converging with the accompaniment, particularly for pitch parameters. Overall, these results indicate that even in the constrained case of predicting singer identity on the same melodic material these parameters are not complete accurate, demonstrating the presence of intra-singer variability that is comparable to the inter-singer variability. Plans for future work include listening experiments with both the original recordings digital re-creations using a uniform timbre with various combinations of these performance data in order to evaluate how important the timing, loudness, and timing descriptors are to people's assessment of similarity, which is arguably the gold standard in style modeling. Another avenue of inquiry is to assess whether mean measurements or range of variances are more useful for modelling for particular parameters.

References

1. Chaffin, R., Lemieux, A., Chen, C.: "It is different each time I play": variability in highly prepared musical performance. Music Percept. **24**(5), 455–472 (2007)
2. de Cheveigné, A., Kawahara, H.: Yin, a fundamental frequency estimator for speech and music. J. Acoust. Soc. Am. **111**(4), 1917–1930 (2002)
3. Cook, N.: Performance analysis and Chopin's mazurkas. Musicae Sci. **11**(2), 183–207 (2007)
4. Devaney, J., Mandel, M., Fujinaga, I.: Characterizing singing voice fundamental frequency trajectories. In: Workshop on Applications of Signal Processing to Audio and Acoustics, pp. 73–76 (2011)
5. Devaney, J., Mandel, M.I., Ellis, D.P.W.: Improving midi-audio alignment with acoustic features. In: Workshop on Applications of Signal Processing to Acoustics and Audio, pp. 45–48 (2009)
6. Glasberg, B.R., Moore, B.C.J.: A model of loudness applicable to time-varying sounds. J. Audio Eng. Soc. **50**(5), 331–342 (2002)
7. Gockel, H., Moore, B.C., Carlyon, R.P.: Influence of rate of change of frequency on the overall pitch of frequency-modulated tones. J. Acoust. Soc. Am. **109**(2), 701–712 (2001)
8. Howard, D.: Equal or non-equal temperament in a cappella SATB singing. Logop. Phoniatr. Vocol. **32**, 87–94 (2007)
9. Koren, R., Gingras, B.: Perceiving individuality in harpsichord performance. Front. Psychol. **5**(141), 1–13 (2014)
10. Lagrange, M., Ozerov, A., Vincent, E.: Robust singer identication in polyphonic music using melody enhancement and uncertainty-based learning. In: International Society for Music Information Retrieval Conference, pp. 595–600 (2012)
11. Ramirez, R., Maestre, E., Perez, A., Serra, X.: Automatic performer identification in celtic violin audio recordings. J. New Music Res. **40**(2), 165–174 (2011)
12. Spiro, N., Gold, N., Rink, J.: The form of performance: analyzing pattern distribution in select recordings of Chopin's Mazurka Op. 24 No. 2. Musicae Sci. **14**, 23–55 (2010)
13. Widmer, G., Dixon, S., Goebl, W., Pampalk, E., Tobudic, A.: In search of the Horowitz factor. AI Mag. **24**, 111–130 (2003)

A Change-Point Approach Towards Representing Musical Dynamics

Katerina Kosta[1](✉), Oscar F. Bandtlow[2], and Elaine Chew[1]

[1] Centre for Digital Music, School of Electronic Engineering and Computer Science,
London, UK
{katerina.kosta,elaine.chew}@qmul.ac.uk
[2] School of Mathematical Sciences, Queen Mary University of London, Mile End Rd,
London E1 4NS, UK
o.bandtlow@qmul.ac.uk

Abstract. This study proposes a novel application of change-point techniques to the question of how dynamic markings in a score correspond to performed loudness. We apply and compare two change-point algorithms–Killick, Fearnhead, and Eckley's Pruned Exact Linear Time (PELT) method, and Scott and Knott's Binary Segmentation (BS) approach–to detecting changes in dynamics in recorded performances of Chopin's Mazurkas. Dynamic markings in the score, assumed to correspond to change points, serve as ground truth. The PELT algorithm has a higher average best F-measure (15.78 % for 0 tolerance threshold; 29 % for one-beat tolerance threshold) compared to the BS algorithm (10.94 % and 19.74 %, respectively), it also results in a smaller average Hausdorff distance–32.8 vs. 77 score beats for 0 tolerance and 32 vs. 52.2 score beats for one-beat tolerance. Applications of loudness change-point detection include audio-to-score transcription.

Keywords: Dynamic markings · Dynamic level representation · Loudness categories · Musical prosody · Change points

1 Introduction

Information abstracted from music signals can serve as an important source of data for multimedia content analysis. A big challenge in creating these data resources lies in the creation of meaningful and salient high-level features for performed music, which describes almost all the music that we hear. The extraction of performance features, especially those related to expressivity, is one of today's most important unsolved problems in the description of music audio.

Focussing on concepts related to dynamic levels in musical expressivity, we introduce change-point detection as a means toward extracting conceptual dynamic levels, generally represented in the score by markings such as p (piano, meaning soft) and f (forte, meaning loud). Change-point algorithms have been applied to domains such as climatology, bioinformatics, finance, oceanography, and medical imaging (see details in [5]). To our knowledge, audio dynamics

© Springer International Publishing Switzerland 2015
T. Collins et al. (Eds.): MCM 2015, LNAI 9110, pp. 179–184, 2015.
DOI: 10.1007/978-3-319-20603-5_18

represents a novel application area. Dynamic markings are interpreted in performance and communicated through varying loudness levels. The aim of our study is to understand the connection between absolute loudness values and intended dynamics, which includes but are not only the notated markings.

In [4], Khoo describes dynamics as existing on two levels. "Primary dynamic shadings" are the main dynamic levels associated with notated marking, and "inner shadings" are dynamic variations associated with the foreground level, which may not be notated. This is analogous to Volk's [11] analysis of musical meter, where the notated time signature indicates the outer meter, while the local grouping of beats defines the inner meter. Inner and outer meters sometimes conflict; a notated p might be louder than a f, as shown in our previous study [8]. The analysis of the formation of different dynamic levels then depends on accurate models for detecting changes in the absolute dynamic information.

We analyze the dynamic values time series extracted from music audio. Using a change-point detection method, we estimate the position where the statistical properties of the sequence change. We compare and contrast the outputs of two change-point algorithms: the Pruned Exact Linear Time (PELT) algorithm by Killick, Fearnhead, and Eckley [6] and the Binary Segmentation (BS) algorithm by Scott and Knott [10]. We evaluate the algorithms by comparing the change points detected from the inner dynamics data (derived from the music signal) and dynamic markings (obtained from the score).

This research will enable the building of automatic methods that can more accurately and meaningfully map musical expression in recorded music to an ontology for music dynamics, beyond simply extracting a dynamic value. The analysis of context-dependent dynamic changes can benefit music structure analysis. These techniques will also be valuable for music transcription. On the synthesis side, a model based on these techniques will be a useful tool for generating expressive renderings of notated scores.

The paper is organized as follows: Sect. 2 describes the data used in the study and a heuristic for aligning multiple performances for analysis; Sect. 3 describes the two change-point algorithms we investigate; Sect. 4 presents the results and Sect. 5 the conclusions and discuss future work.

2   Data Acquisition

The dataset for this study is based on recordings of Chopin's Mazurkas by different pianists as provided by the CHARM project [12] (see details in Table 2). We consider the following markings: pp, p, f, ff and their positions. We did not consider the ones that were located on the first beat of the piece as they do not correspond to a loudness change. We included recordings that follow the repetitions in the score, and excluded the ones that do not and noisy recordings Table 1.

Multiple editions of Chopin's Mazurkas exist, and the particular edition used in each recording is not known. For the purposes of obtaining score-based dynamic markings as a basis for this study, we used the edition by Paderewski, Bronarski and Turczynski, as it is one of the more popular and readily available

Table 1. Details about the dataset used in this study.

Mazurka index	M1	M2	M3	M4	M5	M6
Opus/number	6/2	17/4	24/2	30/2	63/3	68/3
# Recordings	42	68	56	50	62	42
# Markings	12	6	12	13	3	7

editions. We created an XML version of each score, and automatically extracted the location of each dynamic marking using the Music21 software package [1].

In order to obtain a dynamic value time series from each recording, we applied the *ma_sone* function from the MA toolbox [9]. The loudness model estimates the strength of the loudness sensation per frequency band (see details in [3]). We used a hop size of 256 samples and window size of 0.002902 s. The dynamic level of a beat in the score (which covers numerous windows) is given by the median of the sone values within the beat.

To speed the labour-intensive process of annotating beat positions for each recording, only one recording (called the "reference recording") for a given Mazurka was annotated manually, and the beat positions transferred automatically to the remaining recordings using a multiple performance alignment heuristic. The alignment uses Ewert et al.'s algorithm, which is based on Dynamic Time Warping (DTW) and chroma features [2], and extends previous synchronization methods by incorporating features that indicate onset positions for each chroma separately. The heuristic optimises the choice of the reference recording by estimating the minimum match distance between the reference candidates, computing their Euclidean distance in pairwise manner, so as to reduce alignment error.

3   Change-Point Detection

Using the loudness time series obtained as described in Sect. 2, our aim is to detect points where changes to the underlying statistical properties of the data occur. The null hypothesis H_0 corresponds to the non-existence of a change point. Detection of a change point negates the null hypothesis H_0. The general likelihood ratio-based approach to change-point detection provides the asymptotic distribution of the likelihood ratio test statistic for a change in the mean and in the variance of normally distributed observations [7].

We use Killick and Eckley's R package, "changepoint" [5], to investigate the application of two multiple change-point search methods, BS and PELT, to the loudness time series. For both algorithms, the "meanvar" function was implemented, meaning that we detect the changes for both mean and variance for the datasets.

In both the BS and PELT algorithms, the aim is to minimize

$$\sum_{i=1}^{m+1} [C(y_{(\tau_{i-1}+1):\tau_i})] + \beta f(m), \tag{1}$$

where τ is a time point which separates two groups of data $\{y_1, ...y_\tau\}$ and $\{y_{\tau+1}, ..., y_n\}$ having different statistical properties, C is the log-likelihood cost function for a segment, and $\beta f(m)$ is a penalty, in sones, to avoid over fitting, where $f(m) = m$.

The BS algorithm applies a binary divide and conquer approach. It divides the data each time a change point is found, and applies the same process to the two new subsets. It is computational efficient; however, because the location of a change point depends on that of other change points, the BS algorithm gives an approximate solution as the whole data set is not scanned each time.

The PELT algorithm, on the other hand, provides an exact solution as it is based on dynamic programming. The dynamic programming approach is further helped by a pruning step which makes its computational cost linear in the data [6]. As the data set grows, the number of change points increases, meaning that the change points thus obtained also tend not to be closely bunched.

4   Results

In this section, we describe and report the results of the tests we conducted using the PELT and BS algorithms in answering the question of whether change points are located on the same score beat locations as dynamic markings.

For each recording's loudness time series, we perform multiple tests using a variety of penalty function values, so that each test returns a different number of change points. Assuming that change points correspond to dynamic markings in the score, the positions of the score markings then serve as ground truth for the change points. Comparing the change points detected and the positions of the score dynamic markings, we identify the test that returns the highest F-measure:

$$F = \frac{2PR}{P + R}, \tag{2}$$

where precision, P, is the number of true positives divided by the total number of change points, and recall, R, is the number of true positives divided by the total number of markings that do not coincide with change points. A value of 1 indicates that all change points are markings.

Table 2 reports the average best F-measure values for the loudness time series corresponding to each Mazurka (M1, M2, M3, M4, M5, or M6). It also reports the same values when we introduce a tolerance window threshold of one beat on either side of the beat location of the score marking. We also present the maximum F-measure obtained for each piece.

In all pieces, the average best F-measure value for the PELT algorithm (15.78 % for 0 tolerance threshold; 29 % for one-beat tolerance threshold) is better than that of the BS algorithm (10.94 % and 19.74 %, respectively). The result for the maximum F-measure is more mixed; bold numbers indicate when the BS algorithm has a higher maximum F-measure than the PELT algorithm.

To determine how closely the change points estimate the dynamic marking positions, for the recordings having the maximum F-measure, we compute the

Hausdorff distance between the change points and the marking positions:

$$d_H(M,C) = \max \left\{ \sup_{m\in M} \inf_{c\in C} d(m,c), \sup_{c\in C} \inf_{m\in M} d(m,c) \right\}, \qquad (3)$$

where M and C are the sets of score-beat positions of the markings and the change points, respectively. The distance is equal to zero when there is a perfect match, and it is small when the maximum distance from $m \in M$ to the closest point $c \in C$ is small.

For each threshold in BS and PELT, we report the minimum Hausdorff distance in Table 2. For the PELT algorithm, the average Hausdorff distance is 32.8 score beats for 0 tolerance threshold and 32 score beats for one-beat tolerance threshold; for BS, the values are 77 score beats and 52.2, respectively.

Table 2. Average best F-measure, maximum F-measure, and Hausdorff distance values for change point detection results using 0 ("Win 0") and one-beat ("Win 1") tolerance threshold.

Win 0	M1			M2			M3		
	Avg.	Max.	d_H	Avg.	Max.	d_H	Avg.	Max.	d_H
BS	0.2024	0.5217	21	0.0337	0.2857	85	0.1415	**0.3478**	144
PELT	0.2274	0.5217	21	0.0792	0.2857	51	0.1754	**0.3333**	38
	M4			M5			M6		
BS	0.0375	0.1600	33	0.0349	**0.5000**	157	0.2063	**0.4286**	22
PELT	0.1902	0.3111	14	0.0584	**0.2000**	52	0.2162	**0.3750**	21
Win 1	M1			M2			M3		
BS	0.2825	0.5217	21	0.0877	**0.4286**	45	0.2316	0.4615	47
PELT	0.3378	0.5600	21	0.1493	**0.3529**	52	0.3114	0.5926	52
	M4			M5			M6		
BS	0.1332	0.3158	20	0.0424	**0.5000**	157	0.4069	0.7692	23
PELT	0.3820	0.5333	11	0.1183	**0.2857**	36	0.4412	0.7692	20

5   Conclusions and Discussion

In this study we have introduced the use of change-point detection methods to determining changes in performed dynamic levels. We have shown that the PELT change-point algorithm performs better on average for identifying the positions of dynamic markings than the BS method. The Hausdorff distance shows that even though the BS algorithm sometimes has a maximum F-measure higher than the PELT algorithm, the PELT algorithm results in change points that are closer to the dynamic marking positions.

A common phenomenon was the presence of change points that appear in many recordings but which are not near to the markings that were tested. We found that a number of these popular change points corresponded to events such as the start of a *crescendo*, a *calando* or a *poco rit.* While this study focussed on step changes in dynamics, more gradual changes could be a step-forward for future contributions.

In order to analyse more music pieces, we need more sophisticated methods for selecting the appropriate penalty value for the algorithms as this directly impacts the number and quality of the change points detected. Future work will consider other change-point methods and extended set of dynamic-related markings, including *crescendo* and *decrescendo*, which define gradually changing dynamics and may necessitate a different approach. A challenge to be addressed include the detection of changes near (and not on) the score beat of a marking and changes on positions where other type of markings are present (e.g. markings that indicate change in tempo) or where there is no marking. Another next step would be to consider the impact of the sequence of the dynamic markings on the performed dynamic levels.

References

1. Cuthbert, M.S., Ariza, C.: Music21: a toolkit for computer-aided musicology and symbolic music data. In: 9th International Conference on Music Information Retrieval, pp. 637–642 (2010)
2. Ewert, S., Müller, M., Grosche, P.: High resolution audio synchronization using chroma onset features. In: IEEE International Conference on Acoustics, Speech, and Signal Processing (ICASSP), Taipei, Taiwan, pp. 1869–1872 (2009)
3. Hartmann, W.M.: Signals, Sound, and Sensation. Springer, USA (1997)
4. Khoo, H.C.: Playing with dynamics in the music of Chopin. Ph.D. thesis. Royal Holloway, University of London (2007)
5. Killick, R., Eckley, I.A.: Changepoint: an R package for changepoint analysis. J. Stat. Softw. **58**(3), 1–19 (2014)
6. Killick, R., Fearnhead, P., Eckley, I.A.: Optimal detection of changepoints with a linear computational cost. J. Am. Stat. Assoc. **107**(500), 1590–1598 (2012)
7. Killick, R., Eckley, I.A., Jonathan, P.: Efficient detection of multiple changepoints within an oceanographic time series. In: 58th World Statistical Congress, pp. 4137–4142 (2011)
8. Kosta, K., Bandtlow, O.F., Chew, E.: Practical implications of dynamic markings in the score: is piano always piano?. In: 53th Conference on Semantic Audio (2014)
9. Pampalk, E.: A Matlab toolbox to compute similarity from audio. In: 3rd International Conference on Music Information Retrieval (2004)
10. Scott, A.J., Knott, M.: A cluster analysis method for grouping means in the analysis of variance. J. Biometrics **30**(3), 507–512 (1974)
11. Volk, A.: Persistence and change: local and global components of metre induction using inner metric analysis. J. Math. Comput. Music **2**(2), 99–115 (2008)
12. Mazurka Project. www.mazurka.org.uk
13. Sonic Visualiser. www.sonicvisualiser.org

Similarity and Contrast

Structural Similarity Based on Time-Span Sub-Trees

Masatoshi Hamanaka[1][(✉)], Keiji Hirata[2], and Satoshi Tojo[3]

[1] Kyoto University, Kyoto, Japan
masatosh@kuhp.kyoto-u.ac.jp
[2] Future University Hakodate, Hakodate, Japan
hirata@fun.ac.jp
[3] JAIST, Nomi, Japan
tojo@jaist.ac.jp

Abstract. We propose structural similarity of two melodies based on sub-trees from the time-span tree provided by the Generative Theory of Tonal Music. The structural distance of the tree was previously defined and called the "maximal time-span distance," and experimental results showed to some extent a correspondence between the maximal time-span distance and psychological similarity. However, there is a big problem in that almost all pairs of melodies are not similar on the basis of the maximal time-span distance because the definition of the similarity is too strict. Therefore, we attempt to express a melodic structural similarity by using the coincidence rate of time-span sub-trees to weaken the condition for calculating similarity. We have set up three experimental conditions and compared their results.

Keywords: GTTM · Time-span tree · Similarity · Meet · Join

1 Introduction

We have been developing music analyzers based on the Generative Theory of Tonal Music (GTTM) [1,2]. The main advantage of analysis with GTTM is that it can acquire a tree structure called a "time-span tree" from a score, and this tree structure provides a method for manipulating a piece of music [3].

Before using such manipulation, we have to manually select a melody similar to the target melody. For example, the melody morphing method [3] that generates an intermediate melody between a melody and another melody requires similar melodies for input; otherwise, the two melodies cannot be aligned, and then, a melody cannot be made that has the flavor from both input melodies.

If the values of similarity are not distributed appropriately, it is difficult to represent the search or recommendation result. For example, when there are many pieces of music for which similarity is 1.0 between query pieces, as a result of search or recommendation, there are too many top ranking pieces, and we need a lot of time to preview the top ranking pieces. Therefore, it is desirable

© Springer International Publishing Switzerland 2015
T. Collins et al. (Eds.): MCM 2015, LNAI 9110, pp. 187–192, 2015.
DOI: 10.1007/978-3-319-20603-5_19

that the similarity value is appropriately distributed without duplication. It is desirable that the obtained subjective similarity results of listening experiments become close to similarity by calculation.

The maximal time-span distance is the sum of the lengths of time spans that match perfectly from root to leaf and is good for measuring the similarity between variations because these variations have a common structure [4]. However, the maximal time-span distance cannot measure the similarity between two different melodies because they usually do not have a common structure.

To measure this similarity, we attempted to weaken the condition for matching time-span trees by using the coincidence rate of time-span sub-trees. We compare three kinds of matching conditions: tight, middle, and weak. Experimental results show that the middle condition is better for the use tasks of searching or recommendation. In the maximal time-span, similarity is calculated by using the sum of matched numbers of notes weighted by the length of the maximal time-span. We also discuss whether this weight is proper or not by considering the subjective similarity and similarity by calculation.

2 Maximal Time-Span Similarity

The time-span tree is a binary tree and is a hierarchical structure describing the relative structural importance of notes that differentiate the essential parts of a melody from the ornamentation. In the tree, the essential notes are connected to a branch nearer to the root of the tree. In contrast, the ornamentation notes are connected to the leaves of the tree. In a separation, we hereafter call the branch "primary" and leaf "secondary" (Fig. 1a). The time-span tree can extract an abstracted melody by reducing ornamentation notes.

Distance between melodies before and after reduction can be expressed as the length of the maximal time-span m_i of a reduced note. Maximal time-span m_i is a sum of time spans t_is of the entire notes, which are recursively connected to the note as secondary. In Fig. 1b, note 2 is connected to note 1 as secondary. Therefore, m_2 is the same as t_2, and m_1 is the sum of t_1 and t_2. In the same way, m_3 is the sum of t_3 and m_1 because note 1 is connected to note 3 as secondary.

We equate melody A with its time-span tree A hereafter. The distance between A and B in Fig. 1b is the sum of maximal time-spans of reduced notes, which we express as $|B - A|$. The maximal time-span distance [4] between P

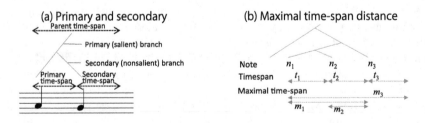

Fig. 1. Time-span trees

and Q can be expressed as the distance via meet $|P - P \sqcap Q| + |Q - P \sqcap Q|$ or distance via join $|P \sqcup Q - P| + |P \sqcup Q - Q|$. The meet operator $P \sqcap Q$ extracts the largest common part or the most common information of the time-span trees in a top-down manner (Fig. 2a). The join operator $P \sqcup Q$ unites two time-span trees in a top-down manner as long as the structures of the two time-span trees are consistent. In fact, the distances via join and meet are the same [4]. The distance between P and Q is maximized when P and Q do not have a common part, which means $P \sqcap Q$ is empty $\perp$ (bottom) as $|P - \perp| + |Q - \perp|$.

The maximal time-span similarity is calculated by normalizing the maximal time-span distance by dividing the maximized maximal time-span distances and subtracting them from 1.[1]

$$1 - \frac{|P - P \sqcap Q|}{2 \cdot |P - \perp|} - \frac{|Q - P \sqcap Q|}{2 \cdot |Q - \perp|} \qquad (1)$$

The maximal time-span similarity is higher when the common part of P and Q, which can be expressed as $P \sqcap Q$, is larger. However, the condition for matching is too strict because it compares perfectly from root to leaf through the separation of branches. This strict matching is very good at comparing very similar melodies such as themes and variations [4].

In comparison, the maximal time-span similarity is usually zero between two different melodies.[2] In the experiments, we use time-span trees of 300 8-bar-long monophonic classical music in the GTTM database [5]. In fact, 32 out of 300 pieces in the GTTM database can be interpreted as two kinds of time-span trees. Therefore, we use all the pairs of 332 time-span trees and 32 pairs of the maximal time-span similarity that are bigger than zero.

3   Time-Span Sub-trees Similarity

In the time-span tree, each note is connected to the root through the separations. The matching of maximal time-span similarity is done by comparing all the separations of each note that is a primary or secondary. If the primary/secondary separation was different at some level, all the branches below the separation were discarded as unmatched. In this sub-tree matching, we only pay attention to the final primary/secondary separation, i.e., the separation to leaf note. From now on, we call a pair of a primary branch and a secondary branch with no other separation a *sub-tree*. Figure 2b is an example of matching in the maximal time-span tree, where n_1 is the secondary of the first separation from the root and also the secondary of the second separation from the root, while n_2 is the primary of

---

[1] This definition of similarity is different from general definitions such as of the Jaccard coefficient, the Simpson coefficient, and Dice coefficient in that we independently normalize $P - P \sqcap Q$ and $Q - P \sqcap Q$; otherwise, only one melody is influential when one melody has a lot of notes and the other has little numbers of notes.

[2] The distance by maximal time-span has been improved in [6], however, we here omit its detail by our limited space. We will compare this improved distance by maximal time-span with our sub-tree similarity in future.

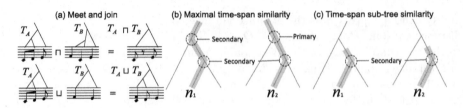

Fig. 2. Examples of meet ⊓, join ⊔, and matching of nodes

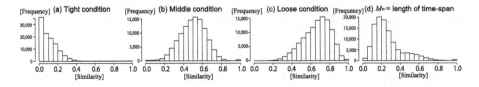

Fig. 3. Histograms of time-span sub-trees similarity

the first separation from the root and secondary of the second separation. Thus, n_1 and n_2 are unmatched. Figure 2c is an example of matching in time-span sub-tree similarity, where n_1 is the secondary of the last separation and also the secondary of last separation. Thus, n_1 and n_2 match.

The matching of maximal time-span similarity is done by comparing the pitches of each note exactly same or not. In our sub-tree matching, we ease the matching condition only focusing on whether two notes of the sub-trees are ascending or descending. Thus, we can regard two sub-trees are matching only if this ascending/descending feature is common between them. Then, we can consider the following three patterns: (i) the pitch of the primary note is higher than that of the secondary, (ii) the pitch of the primary note is lower than that of the secondary, or (iii) the primary and secondary notes have the same pitch. Now we name this classification $M_p = 3$. Furthermore, we can loosen the condition, as (i') the pitch of the primary note is higher than or equal to that of the secondary or (ii') the primary is lower than or equal to the secondary, and we call this classification $M_p = 2$. On the contrary, we can tighten the condition; the pitch of two primary notes must be the same when two sub-trees are matched. We will experiment all these conditions and will compare the results.

The matching of maximal time-spans is done by comparing the onset and length of the maximal time-spans in all separations from the root to leaf that are exactly same. Here we pay attention to the middle point of a maximal time span as there may exist rests at either end of the span. The relative location of the middle points of two time-spans is named *gap*. The matching of time-span sub-tree similarity is done by comparing the gap of the middle of the maximal time-span and the length ratio of the maximal time-span in the sub-trees that are in a certain range. In other words, if the gap of the middle of maximal time-spans is in M_g times of the length of a piece of music and the ratio of the longer maximal time-span and shorter maximal time-span is in M_r, the time-span sub-trees are matching. We will change M_g to 0.1, 0.2, and 0.3 and also change M_r to 1.1, 1.3, and 1.5, and discuss which condition is appropriate.

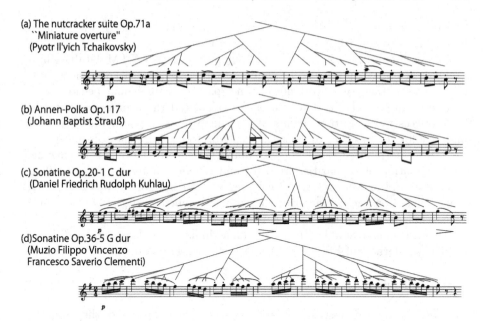

(a) The nutcracker suite Op.71a
   ``Miniature overture''
   (Pyotr Il'yich Tchaikovsky)

(b) Annen-Polka Op.117
   (Johann Baptist Strauß)

(c) Sonatine Op.20-1 C dur
   (Daniel Friedrich Rudolph Kuhlau)

(d)Sonatine Op.36-5 G dur
   (Muzio Filippo Vincenzo
   Francesco Saverio Clementi)

Fig. 4. Example of analysis

In maximal time-span similarity, notes close to the root branch are more important than notes far from the root branch because the weights for calculating maximal time-span similarity are the lengths of the maximal time-spans. In the experiment, we compare the calculation of time-span similarity has a weight (M_w = length of the maximal time-span) or has no weight ($M_w = 1$).

4   Experimental Results

The time-span sub-tree similarity proposed in Sect. 3 enables the condition for matching pitch, time, the nodes of the tree, and weight for matiching to be changed. We compare three kinds of matching conditions as follows: (a) Tight condition, with $M_p = 3$, $M_g = 0.1$, $M_r = 1.5$, $M_w = 1$ and exactly match of primary pitch; (b) Middle condition, with $M_p = 3$, $M_g = 0.2$, $M_r = 1.3$, $M_w = 1$; (c) Loose condition, with $M_p = 2$, $M_g = 0.3$, $M_r = 1.5$, $M_w = 1$. Figure 3 is histograms of similarities of each matching condition from the 300 pieces in the database. Compared with the original maximal time-span similarity, Fig. 3a is a decrease in the number of similarities, which is zero; however, most of the similarities are still zero, and all the similarities are under 0.5. In Fig. 3b, the similarities were distributed in a wide area, and the shape of the distribution was similar to the standard distribution. In comparison, the center of the distribution of Fig. 3c leaned to the right, and very few similarities were in the area of zero to 0.3. From the above results, (b) is appropriate in the three conditions.

We compared two kinds of weight for matching: (c) $M_w = 1$ and (d) $M_w =$ the length of maximal time-span. Here, for the values of M_p, M_g, and M_r, we use the condition of (b). Figure 3d is a histogram of similarity made by using the weight as (d), which has also a long distribution.

As described in Sect. 2, the 32 out of 300 pieces in the GTTM database have two kinds of time-span trees corresponding to interpretation from musicologists. It is preferable that each two kinds of time-span trees are similar because there are interpretations of the same piece. We calculated the average similarity of the 32 pieces by using the weights of (c) and (d). As a result, the average of similarity for (c) was 0.77 and for (d) was 0.6. To compare the values from (c) and (d), we normalized the average to 0 and variance to 1 because the average and distribution of similarities from (c) and (d) were different. As a result of normalization, the average similarity for (c) was 2.16 and for (d) was 2.65. These values depend on the character of the corpus. In current 300 piece the weight for matching is appropriately by using condition of (c) than (d).

We show some examples that show high similarities where the condition is (b). The similarities of Fig. 4a and b was 0.90, Fig. 4c and d was 0.88.

5   Conclusion

Although maximal time-span similarity was previously proposed, the similarity of two different pieces of music is usually zero. We proposed time-span sub-tree similarity, for which similarity is made to be more than zero in many cases by weakening the condition for matching maximal time-span similarity. In the experimental results obtained with three kinds of matching condition, the middle one was better than the other conditions with the GTTM database. We plan to construct applications for similarity with the time-span tree because the appropriate definition of similarity depends on the application.

Acknowledgments. This work was supported in part by JSPS KAKENHI Grant Numbers 23500145, 25330434, and 25700036 and PRESTO, JST.

References

1. Lerdahl, F., Jackendoff, R.: A Generative Theory of Tonal Music. MIT Press, Cambridge (1983)
2. Hamanaka, M., Hirata, K., Tojo, S.: Implementing "A generative theory of tonal music". J. New Music Res. **35**(4), 249–277 (2007)
3. Hamanaka, M., Hirata, K., Tojo, S.: Melody morphing method based on GTTM. In: Proceeding of ICMC 2008, Belfast, pp. 155–158 (2008)
4. Tojo, S., Hirata, K.: Structural similarity based on time-span tree. In: Aramaki, M., Barthet, M., Kronland-Martinet, R., Ystad, S. (eds.) CMMR 2012. LNCS, vol. 7900, pp. 400–421. Springer, Heidelberg (2013)
5. Hamanaka, M., Hirata, K., Tojo, S.: Music structural analysis database based on GTTM. In: Proceeding of ISMIR 2014, Taipei, pp. 325–330 (2014)
6. Hirata, K., Tojo, S., Hamanaka, M.: Cognitive similarity grounded by tree distance from the analysis of K.265/300e. In: Aramaki, M., Derrien, O., Kronland-Martinet, R., Ystad, S. (eds.) CMMR 2013. LNCS, vol. 8905, pp. 589–605. Springer, Heidelberg (2014)

Cross Entropy as a Measure of Musical Contrast

Robin Laney[1]([⊠]), Robert Samuels[1], and Emilie Capulet[2]

[1] Open University, Milton Keynes MK7 6AA, UK
robin.laney@open.ac.uk
[2] London College of Music, University of West London, London W5 5RF, UK

Abstract. We present a preliminary study of using the information theoretic concept of cross entropy to measure musical contrast in a symbolic
context, with a focus on melody. We measure cross entropy using the
Information Dynamics Of Music (IDyOM) framework. Whilst our long
term aim is to understand the use of contrast in sonata form, in this
paper we take a more general perspective and look at a broad spread of
Western art music of the common practice era. Our results suggest that
cross entropy has a useful role as an objective measure of contrast, but
that a fuller picture will require more work.

Keywords: Contrast · Similarity · Cross entropy · N gram model ·
Markov model

1   Introduction

In Western art music of the common practice era, similarity and contrast play
key roles. Similarity to a corpus is a way of establishing style, and similarity
within a work provides coherence. Contrast, on the other hand, not only adds
interest, but is essential to the æsthetic illusion of dramatic resolution of conflict.
In terms of empirical understanding, similarity has been widely studied in terms
of music information retrieval techniques, but contrast much less so.

Conklin and Witten [4], Pearce [8], and Whorley [12], in principled approaches
to generation from a corpus, use a low value for cross entropy between a generated
piece and a model of the corpus as a measure of stylistic coherence. Thus low cross
entropy is used as a model of similarity. In this paper we investigate whether high
entropy between motifs or themes might be a good measure of contrast, on the
basis that both similarity and contrast are related to listeners expectations being
respectively fulfilled or denied. We build on the Information Dynamics Of Music
(IDyOM) framework [8,9].

In this paper we focus on melody. Whilst it is clear that many other features
play a key role in establishing contrast, in particular dynamics, texture, harmony,
timbre, and articulation, there is a sense in which melody is central, for example
in analysing a theme a musicologist will typically look at the shape of the melody.

In the results section we take five pieces from composers of the common practice period and measure cross entropy between patterns representing themes,

© Springer International Publishing Switzerland 2015
T. Collins et al. (Eds.): MCM 2015, LNAI 9110, pp. 193–198, 2015.
DOI: 10.1007/978-3-319-20603-5_20

subjects or motifs that might be expected to contrast. As a comparison we also look at cross entropy between the pieces and the patterns. Our measurements are of intervals and durations, leaving the investigation of other features, including combinations, to later work. A long term motivation is to see whether musically meaningful contrast between themes and motifs, ascribed or assumed by musicological writing, can be correlated with objectively measurable properties.

2 Background

In this section we briefly introduce the use of entropy in the context of music, and consider the relationship of similarity and contrast, and the extent to which a study of the latter can draw on work on the former.

Shannon's notion of entropy in the context of his theory of information [10] has been influential in the field of computational musicology. The use of cross entropy in a probabilistic predictive/generative context was introduced by Conklin and Witten [4] who also note that Meyer [6] relates musical experience and entropy. Collins et al. [3] use entropy as part of a stylistic consistency metric. The IDyOM model of Pearce [8,9] combines the predictions of multiple length n gram models, using Prediction by Partial Match, a technique developed by Cleary and Witten [1]. The following equation follows Pearce [8].

Let a sequence $(a_1...a_n)$ be represented as a_1^n. For a given a value of n, an n gram model predicts the probability of the next event a_n given a previous sequence a_1^{n-1}. That is $p(a_n \mid a_1^{n-1})$. If we let m be a model, that can predict, using variable lengths of initial sequences, probabilities p_m, then the cross entropy $H(p_m, a_1^j)$ is:

$$H_m(p_m, a_i^j) = -\frac{1}{j}\log_2 p_m(a_1^j)$$

$$= -\frac{1}{j}\sum_{i=1}^{j}\log_2 p_m(a_i \mid a_1^{i-1}) \qquad (1)$$

Cross entropy, is a measure of how likely the sequence $a_1...a_n$ is relative to the model m. Low values suggest that if the model was built from a corpus, then the sequence is similar to the corpus, and we explore in this paper whether high values correspond to a sequence that contrasts with the corpus. In fact we will compare individual phrases by using one phrase as the corpus, and similarly use the piece as a corpus to measure the contrast of a phrase with the overall piece. We are interested in whether cross entropy is a useful model of contrast.

Similarity in music is non-trivial, as musical segments can be analysed from many perspectives, using a variety of models. Wiggins et al. [13] argue that music is a subjective experience that in a sense only exists in the mind of listeners. So what models and measures correspond to human experience is a non-trivial question requiring empirical work. A similar argument can be applied to contrast, but more initial work is required. Given the multi-dimensional nature of both similarity and contrast, and the subjective nature of music, we should not assume that contrast and similarity are simply opposites. Tversky [11] in seminal work taking a feature based approach to similarity makes a similar point, as well as showing that similarity (and by implication) contrast are context specific.

In a similar vein, Marsden [5] has questioned whether musical similarity is a definitive phenomenon or a product of interpretation. He reviews a number of approaches: the edit distance between melodies, possibly under Time-Warping; structural models based on reduction; and feature extraction combined with machine learning or statistical analysis. Whilst contrast is not simply the opposite of similarity, it is worth considering how similarity measures can be applied to contrast. Reduction approaches remove decorative notes leaving structural ones, though from the point of view of contrast such a dichotomy might be misleading in that deviance from a structure might be a way of establishing a contrast. Edit distance methods appear appropriate if we are dealing with segments of the same length or where one is an elongation of the other. The advantage of a cross entropy based approach is that it allows us to compare the character of two segments regardless of length considerations. We can ask for example, if a particular short motif provides contrast after a long passage.

3   Corpus

We used the contents of the Johannes Kepler University Patterns Development Database training corpus [2]: Orlando Gibbons' The Silver Swan (1612), Johann Sebastian Bach's Fugue in A minor, BWV 889, Wolfgang Amadeus Mozart's Minuet from Piano Sonata in E flat major, K. 282, Ludwig van Beethoven's Minuet and Trio from Piano Sonata in F minor, op. 2, no. 1, and Frédéric Chopin's Mazurka in B minor, op. 24, no. 4. This gives a broad spread across music of the common practice period. The training corpus contains a number of patterns (labelled A,B...), that according to accepted conventions would contrast as follows:

- A and C in the Gibbons (ascending v descending),
- A and B in the Bach (in the sense of subject and countersubjects),
- C and D in the Mozart (first and second themes in a minuet),
- C and D in the Beethoven (themes from a minuet and trio),
- B and D in the Chopin (themes from a mazurka).

The corpus database includes scores for the pieces and patterns.

4   Methodology

Using the IDyOM software we measured the cross entropy between a number of combinations of patterns and the whole piece. IDyOM constructs models that are multi order Markov models. We label the phrases under consideration a and b and the whole piece p, and denote the cross entropies of interest as:

- CE(a,b) Cross entropy for pattern a with respect to a model of pattern b

- CE(b,a) Cross entropy for pattern b with respect to a model of pattern a
- CE(a,p) Cross entropy for pattern a with respect to a model of the piece
- CE(b,p) Cross entropy for pattern b with respect to a model of the piece.

Additionally, for completeness and transparency we measured:

- CE(p,a) Cross entropy for piece with respect to a model of pattern a
- CE(p,b) Cross entropy for piece with respect to a model of pattern b.

We also take averages of CE(a,b) and CE(b,a) (labelled pattern in tables in Sect. 5) and averages of CE(a,p) and CE(b,p) (labelled pattern-piece in tables in Sect. 5) in order to give an overall idea of the cross entropy between patterns and the cross entropy between the patterns and the whole piece respectively. A working hypothesis is that if:

- (a) cross entropy is a good model of contrast (and similarity) and
- (b) pieces are coherent through similarity with interest generated by contrast between the identified themes.

Then we expect that the average of CE(a,b) and CE(b,a) to be greater than the averages of CE(a,p) and CE(b,p). However we also need to take into account that contrast might, for example, be established through pitch whilst keeping durational content more similar.

5   Results

In the following two Tables 1 and 2 we see our results for the cross entropy of pitch predicted using a model of intervals and for the cross entropy of durations predicted using a model of durations.

For both pitch and durations, the average entropies between patterns (labelled pattern) is larger than that between the patterns and the whole piece (labelled pattern-piece). So the data is consistent with our working hypothesis. However, an important confounding factor relates to differences in the sizes of the corpus used in the comparisons. Clearly, the corpus for a comparison between two patterns is smaller than when using a corpus of the whole piece and there will often be a fall in cross entropy as corpus size increases. We also need a clearer idea of what magnitude of ratio suggests a given strength of contrast. One approach might be to look at a range of patterns across a piece and use an average as a baseline.

Table 1. Cross entropy of pitch predicted using a model of intervals

Composer	CE(a,b)	CE(b,a)	CE(a,p)	CE(b,p)	CE(p,a)	CE(p,b)	Pattern	Pattern-piece
Gibbons	3.03	3.57	2.09	2.95	4.45	4.42	3.30	2.52
Bach	5.23	4.64	1.94	0.83	6.02	3.61	4.93	1.38
Mozart	3.10	3.00	1.24	2.79	4.48	4.26	3.05	2.02
Beethoven	3.13	3.16	2.19	0.80	4.17	3.87	3.14	1.50
Chopin	3.23	3.39	0.71	1.03	4.16	4.53	3.31	0.87

Table 2. Cross entropy of durations predicted using a model of durations

Composer	CE(a,b)	CE(b,a)	CE(a,p)	CE(b,p)	CE(p,a)	CE(p,b)	Pattern	Pattern-piece
Gibbons	1.83	2.23	1.23	1.34	3.11	1.98	2.03	1.28
Bach	3.87	4.23	1.43	0.70	4.16	2.32	4.05	1.07
Mozart	2.59	2.56	0.76	1.62	2.54	2.90	2.57	1.19
Beethoven	3.40	4.40	0.27	0.28	3.15	2.26	3.90	0.28
Chopin	2.61	2.40	0.77	1.01	3.49	3.91	2.50	0.89

6   Conclusions

There is much existing work on similarity, although that work is far from complete. However, contrast, has received little attention. Given the centrality of identifying contrasts when carrying out analysis, a better understanding of relevant objective features would enrich the analytical process, and widen the scope of digital tools. It would also support a fresh look at historical musicological writings on the role of contrast in a range of forms.

Why should we wish to provide digital tools? Currently the degree of training required for a competent analysis is very steep. Moreover, once one commits to a particular way of viewing a piece in terms of its structure, it is expensive time-wise to consider alternatives. Yet, often it is the tension between alternative interpretations of similarity and contrast that leads to the sublime nature of a work. See Marsden [5] for a highly pertinent example. Automated tools would widen the likelihood of working with alternative readings and make it feasible for a wider group to engage with analysis, such as instrumental teachers, conservatoire staff, and professional performers. Currently there is much interest in electronic editions such as the Online Chopin Variorum Edition [7] which allow a much more flexible attitude towards score editions, their alternatives, and blended versions. Easy to use tools would further open up and democratise an area of work which has sometimes been seen as the preserve of a small elite.

For music in the common practice period, coherence depends on adhering to the style, but interest is generated through contrast. From this assumption it is easy to see a potential weakness of generating a piece from a single Markov model. The problem is that contrasts in pieces within a corpus might effectively get averaged out in the model and when generating pieces this will be reflected.

The picture painted by our numerical data was consistent with our working hypothesis. However, differences in the sizes of corpus used in each comparison is a compounding issue that needs further work. The overall corpus, whilst being wide-ranging historically, was of limited size. Further work is needed to give a firmer idea of how cross entropy and contrast relate and also how the measure compares with others, or indeed could be meaningfully combined with them. Further work might also look at layers of contrast: If a motif A contrasts with a motif B, this might be perceived as a contrast at a micro-level. But if A and B recur, they are still contrasting, but the question is whether A and B together create a new level of contrast with another, more large-scale, element of the music, lets say C, and so forth.

Acknowledgements. We are grateful for discussions and advice to Byron Dueck, Tom Collins, Alan Marsden, Marcus Pearce, Raymond Whorley, Geraint Wiggins, and Alistair Willis.

References

1. Cleary, J.G., Witten, I.H.: Data compression using adaptive coding and partial string matching. IEEE Trans. Commun. **32**(4), 396–402 (1984)
2. Collins T., Böck S., Krebs, F., Widmer, G.: Bridging the audio-symbolic gap: the discovery of repeated note content directly from polyphonic music audio. In: Proceedings of the Audio Engineering Society's 53rd Conference on Semantic Audio, London, p. 12 (2014)
3. Collins, T., Laney, R., Willis, A., Garthwaite, P.H.: Developing and evaluating computational models of musical style. In: Jin, Y. (ed.) Artificial Intelligence for Engineering Design, Analysis and Manufacturing, vol. 9, p. 28. Cambridge University Press, Cambridge (1995)
4. Conklin, D., Witten, I.H.: Multiple viewpoint systems for music prediction. J. New Music Res. **24**(1), 51–73 (1995)
5. Marsden, A.: Interrogating melodic similarity: a definitive phenomenon or the product of interpretation? J. New Music Res. **41**(4), 323–335 (2012)
6. Meyer, L.B.: Meaning in music and information theory. J. Aesthet. Art Crit. **15**, 412–424 (1957)
7. Online Chopin Variorum Edition. http://www.ocve.org.uk/index.html
8. Pearce, M.: The construction and evaluation of statistical models of melodic structure in music perception and composition, M.T. Ph.D. thesis. City University London, Pearce (2005)
9. Pearce, M.T., GA Wiggins, G.A.: Auditory expectation: the information dynamics of music perception and cognition. Top. Cogn. Sci. **4**(4), 625–652 (2012)
10. Shannon, C.E.: A mathematical theory of communication. Bell Syst. Tech. J. **27**, 379–423 (1948)
11. Tversky, A.: Features of similarity. Psychol. Rev. **84**, 327–352 (1977)
12. Whorley, R.: The construction and evaluation of statistical models of melody and harmony. Whorley, R.P. Ph.D. thesis. Goldsmiths, University of London (2013)
13. Wiggins, G.A., Müllensiefen, D., Pearce, M.T.: On the non-existence of music: why music theory is a figment of the imagination. Musicae Sci. **5**, 231–255 (2010)

Symbolic Music Similarity Using Neuronal Periodicity and Dynamic Programming

Rafael Valle[✉] and Adrian Freed

Center for New Music and Audio Technologies (CNMAT), Berkeley, CA 94709, USA
rafaelvalle@berkeley.edu

Abstract. We introduce NP-MUS, a symbolic music similarity algorithm tailored for polyphonic music with continuous representations of pitch and duration. The algorithm uses dynamic programming and a cost function that relies on a mathematical model of tonal fusion based on neuronal periodicity detection mechanisms. This paper reviews the general requirements of melodic similarity and offers a similarity method that better addresses contemporary and non-traditional music. We provide experiments based on monophonic and polyphonic excerpts inspired by spectral music and Iannis Xenakis.

1 Introduction

Symbolic music similarity has been extensively investigated within the scope of query systems focused on popular music [5]. This focus on popular/traditional musical genres and equal-tempered schemes deeply influenced the formulation of MIR systems from the very beginning of melodic similarity [6]. We refer the reader to [1] for an overview of symbolic melodic similarity algorithms.

This paper evaluates the requirements of MIR systems for symbolic melodic similarity under the prism of non-traditional[1] music and expands the investigations on music similarity beyond discrete and equal-tempered systems.

The Neuronal Periodicity Music Similarity (NP-MUS) method uses dynamic programming with a cost function based on a mathematical model of tonal fusion developed by Martin Ebeling [4] that uses neuronal periodicity analysis methods.

In summary, the novel contributions of this paper are:

- a symbolic music similarity method for continuous pitch and duration
- a similarity measure based on tonal fusion and neuronal periodicity
- a method for similarity of Vertical Pitch Structures (chords)
- a method for polyphonic melodic similarity.

2 Symbolic Music Similarity Requirements

MIR researchers have enumerated the basic requirements of every MIR system, such as being able to handle polyphonic queries and evaluate at least the pitch (vertical) and rhythm (horizontal) domains [3,8,10].

---

[1] By non-traditional, we refer to musical practices that do not conform with western traditional music and its common discrete representations of pitch and rhythm.

© Springer International Publishing Switzerland 2015
T. Collins et al. (Eds.): MCM 2015, LNAI 9110, pp. 199–204, 2015.
DOI: 10.1007/978-3-319-20603-5_21

2.1   Pitch Requirements and Solutions

Transposition Invariance: Some of the pitch requirements include octave equivalence, degree equality and note equality. In his excellent publication about melodic similarity and in opposition to general practice, Hoffman-Engl [6] affirms that transposition is a similarity factor. In addition, we emphasize that transposition invariance ignores tonal fusion, roughness and harmonic similarity. Facing these issues, we itemize below possible modifications to those requirements:

- Transposition variance should be addressed.
- Non-tempered systems must be handled appropriately.
- Glissandi should be handled dependent on pitch range and duration.

Our solution to these requirements is a tonal fusion based similarity measure that is related to consonance and dissonance [2] and supports non-tempered pitch analysis and transposition variance. Details are provided in Sect. 5.

2.2   Rhythmic Requirements and Solutions

Tempo Equivalence: Some solutions include using timeless models, representing note durations as multiples of a base score duration and using only the duration ratio between two notes. It is known [6] that tempo changes cause impacts in melody recognition and, therefore, melodic similarity.

Duration Equality: This problem refers to small tempo changes and variations in rhythm and a general *solution* is to ignore these issues. Melodies played *rubato* and *swing*, for example, are not equal, therefore modifications to those requirements are necessary:

- Tempo equivalence should be substituted with a tempo similarity measure.
- Duration equality should be substituted with support for continuous time.
- The rhythm-pitch continuum must be appropriately handled.

Our method projects durations onto a continuous tonal fusion space, thus supporting tempo equivalence and duration equality. The rhythm-pitch continuum uses a hard threshold for repeated notes at 20 notes per second. For glissandi, the threshold is dependent on the pitch range and duration.

2.3   Pitch and Rhythm Features

Our method represents pitch and rhythm continuously using the conversion formula $P(f) = 69 + 12 * \log_2(f/440)$. Rhythm has a pre-processing step that transforms note durations into frequency by taking the inverse of the duration in seconds. The feature vector for pitch and rhythm includes intervals and absolute values. The algorithm accepts MIDI, XML, and numerical data[2].

---

[2] The examples used in this paper were generated by using equations to produce numerical data for pitch and rhythm. MIDI and XML are converted when necessary.

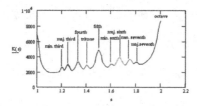

Fig. 1. GCF for intervals within an octave calculated on the basis of rectangular pulse trains of width $\epsilon = 0.8\,\text{ms}$ and $100\,\text{Hz}$ as lowest frequencies.

3   Neuronal Periodicity as Similarity Measure

The model developed by Martin Ebeling [4] is inspired in tonal fusion and neuronal periodicity. Tonal fusion is very relevant for melodic similarity because it is in agreement with roughness, consonance and pitch chroma and height [7].

A measure for the degree of overlapping, correlated with similarity, of an interval with vibration ratio $s = f_1/f_2$, illustrated in Fig. 1, is defined by squaring the autocorrelation function of s. After derivations of this concept, Ebeling defines the Generalized Coincidence Function (GCF) as $K(S) = \int_0^{50} \rho^2(s,\tau)d\tau$.

4   Generalized Coincidence Function and DTW Method

4.1   Perceptual Similarity Measure

Definition 1. *The perceptual similarity, $\mathcal{PS}$, between two frequencies f_1 and f_2, with $f_1 \leq f_2$ and interval $s = f_2/f_1$ is defined as follows*[3]:

$$\mathcal{PS}(f_1, f_2) = (1 + \mathcal{D}(s))^{-1} \tag{1}$$

where $\mathcal{D}(s)$ is a dissimilarity function based on the Generalized Coincidence Function (GCF):

$$\mathcal{D}(s) = -log(GCF(s)) \tag{2}$$

Our method uses the discrete GCF(dGCF), with values at equal-tempered intervals within the octave and an octave penalty. A linear interpolation of the dGCF is used to obtain the perceptual similarity for non-tempered intervals. Algorithmically, the $dGCF$ is a look-up table with normalized similarity values for equal-tempered intervals within the octave, from unison to major seventh:

$$dGCF = [1.0, 0.38, 0.28, 0.39, 0.44, 0.53, 0.43, 0.72, 0.45, 0.57, 0.53, 0.5] \tag{3}$$

Definition 2. *Based on the dGCF, the perceptual similarity, $\mathcal{PS}_d$, between pitches P_1 and P_2, where P_1 and P_2 are obtained using the conversion formula $P(f)$, with $P_1 \leq P_2$ and interval $z = P_2 - P_1$ is defined as follows:*

$$\mathcal{PS}_d(P_1, P_2) = (1 + \mathcal{D}_d(z))^{-1} \tag{4}$$

---

[3] We also consider using Shepard's universal law: $PS(f_1, f_2) = e^{-D(s)}$.

where $\mathcal{D}_d$ is a dissimilarity function with octave penalty π:

$$\mathcal{D}_d(z) = -log(\mathcal{C}(z) * \pi(z))$$
$$\pi(z) = (1 + z/12)^{-1/3}$$
$$\mathcal{C}(z) = dGCF(I(z)) + F(z)(dGCF((I(z) + 1) \bmod 12) - dGCF(I(z))) \quad (5)$$
$$I(z) = \lfloor z \rfloor \bmod 12$$
$$F(z) = z \bmod 12 - I(z)$$

4.2 Dynamic Time Warping

Dynamic Time Warping is an dynamic programming algorithm to compare two sequences $X := x_1, x_2, ..., x_n$ and $Y := y_1, y_2, ..., y_m$ using some distance function $d(X, Y)$. The minimal overall cost is defined by the shortest warping path in a cost matrix. For a thorough account of DTW, we refer the reader to [9]. Our method uses (5) as DTW's cost function to compute similarity.

4.3 Similarity of Musical Events

Definition 3. *The similarity between two musical events, $\mathcal{S}(M_1, M_2)$, is computed as the normalized weighted average of similarities for each feature $f \in F$, where W is a weight vector and F is the set of features. Formally:*

$$\mathcal{S}(M_1, M_2) = (\sum_{f \in F} W_f)^{-1} \sum_{f \in F} \mathcal{S}(M_f^1, M_f^2) * W_f \quad (6)$$

5 Experiments

5.1 Monophonic Similarity

Glissandi Similarity: Glissandi are treated with dynamic programming and a hard threshold is used for selecting the number of notes, dependent on the glissando's frequency range. Results in Fig. 2 shows that an increase in angle between C5 and a glissando yields a decrease in similarity, as expected.

(a) Glissandi Excerpts   (b) Similarity   (c) Glissandi Perception

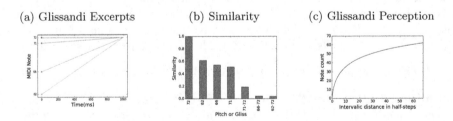

Fig. 2. Similarity between C5(72) and other notes/glissandi. Perception heuristics provides the notes-per-second rate at which a glissando is perceived.

Rhythmic Similarity: Figure 3 addresses tempo equivalence, duration equality and the rhythm-pitch continuum. The results provide evidence that the method handles time-stretching appropriately and is in consonance with the common use of *doppio tempo* and *hemiola*. Time-stretch coefficients very close to 1, 2, and 1.5 have the highest similarity. Pitch similarity is computed above 20 articulations per second.

(a) Impulse   (b) Similarity   (c) Subdivisions   (d) Similarity

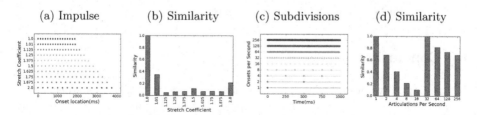

Fig. 3. Left: similarity between a impulse sequence and its time-streched versions. Right: similarity between note 32 Hz played every second and other equal-length articulations per second.

5.2 Polyphonic Similarity

Non-equal Length VPS Similarity: The example illustrated on Fig. 4 computes similarity by treating polyphonic music as a sequence of chords with different sizes. A pre-processing step is applied to each piece such that sustained notes over shorter notes are re-articulated at the onset of the shortest note and with the same duration of that shorter note. The similarity is computed by using DTW for chords and music of different length. Absolute values, contour (intervals)[4] and average are used.

(a) 1   (b) 2   (c) 3   (d) Similarity

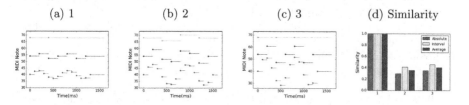

Fig. 4. Polyphonic VPS similarity. 1 is the base excerpt. Notice that the contour of voices between 1 and 3 are the most similar.

Voice Similarity: Figure 5 illustrates this similarity measure, which is based on similarity computed by averaging the similarity between all voice pairs between two musical events. This method provides the expected similarity rankings but must be used with caution: as expected, the self-similarity does not equal to 1. This method provides a useful tool for composition based graphical notation.

---

[4] Acknowledgments to composer Matt Sandahl for his insightful feedback on contour.

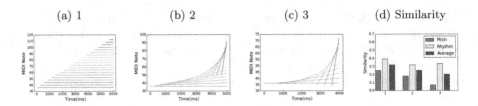

(a) 1   (b) 2   (c) 3   (d) Similarity

Fig. 5. Polyphonic voice similarity. 1 is the base excerpt.

6   Conclusion

In this paper, we described a symbolic music similarity method that specializes in polyphonic music with continuous representations of pitch and rhythm. The results obtained with the NP-MUS algorithm represent an important contribution to contemporary and non-traditional music, specially for musicology and composition. In addition to the traditional MIR query task, NP-MUS empowers the design of analytical and compositional processes based on similarity at different levels and prioritizing specific features.

Further developments include specifying the method's type of metric, comparing the results provided by the algorithm with results provided by human listeners, creating a GUI and ultimately a fast implementation for large datasets.

References

1. Aloupis, G., Fevens, T., Langerman, S., Matsui, T., Mesa, A., Nuñez, Y., Rappaport, D., Toussaint, G.: Algorithms for computing geometric measures of melodic similarity. Comput. Music J. **30**(3), 67–76 (2006)
2. Bidelman, G.M.: The role of the auditory brainstem in processing musically relevant pitch. Front. Psychol. **4**(264), 1–13 (2013)
3. Byrd, D., Crawford, T.: Problems of music information retrieval in the real world. Inf. Process. Manage. **38**(2), 249–272 (2002)
4. Ebeling, M.: Neuronal periodicity detection as a basis for the perception of consonance: a mathematical model of tonal fusion. J. Acoust. Soc. Am. **124**(4), 2320–2329 (2008)
5. Ghias, A., Logan, J., Chamberlin, D., Smith, B.C.: Query by humming: musical information retrieval in an audio database. In: Proceedings of the Third ACM International Conference on Multimedia, pp. 231–236. ACM (1995)
6. Hoffmann-Engl, L.: Melodic Similarity - Providing a Cognitive Groundwork. Chamaleon group: Online Publication, Essex (2004)
7. Lerdahl, F.: Tonal pitch space. Music Percept. **5**, 315–349 (1988)
8. Mongeau, M., Sankoff, D.: Comparison of musical sequences. Comput. Humanit. **24**(3), 161–175 (1990)
9. Rabiner, L.R., Juang, B.H.: Fundamentals of Speech Recognition, vol. 14. PTR Prentice Hall, Englewood Cliffs (1993)
10. Selfridge-Field, E.: Conceptual and representational issues in melodic comparison. Comput. Musicology: Directory Res. **11**, 3–64 (1998)

Post-Tonal Music Analysis

Applications of DFT to the Theory of Twentieth-Century Harmony

Jason Yust[(✉)]

Boston University, 855 Commonwealth Avenue,
Boston, MA 02215, USA
jason.yust@gmail.com
http://people.bu.edu/jyust

Abstract. Music theorists have only recently, following groundbreaking work by Quinn, recognized the potential for the DFT on pcsets, initially proposed by Lewin, to serve as the foundation of a theory of harmony for the twentieth century. This paper investigates pcset "arithmetic" – subset structure, transpositional combination, and interval content – through the lens of the DFT. It discusses relationships between interval classes and DFT magnitudes, considers special properties of dyads, pcset products, and generated collections, and suggest methods of using the DFT in analysis, including interpreting DFT magnitudes, using phase spaces to understand subset structure, and interpreting the DFT of Lewin's interval function. Webern's op. 5/4 and Bartok's String Quartet 4, iv, are discussed.

Keywords: Discrete fourier transform · Pitch-class set theory · Twentieth-century harmony · Posttonal theory · Webern · Bartok · Generated sets

1 Introduction

In American music theory of the 1960 s and 1970 s, the era of Allen Forte's ambitiously titled book *The Structure of Atonal Music* [12], a theory of harmony for the twentieth century seemed not only a possible but a natural goal of the discipline. In later years, the idea of pursuing a general theory for such an eclectic century would come to seem increasingly audacious. But recent advances in mathematical music theory should reignite this enterprise: in particular the application of the Fourier transform to pcsets [3,5,9,15,18,19].

Forte's project [11,12] to develop a theory based on interval content and subset structure was propitious in that he identified general properties would be relevant to a wide range of music despite great disparities in compositional aesthetic and technique. The DFT makes it possible to establish a more solid mathematical foundation for such a theory than was available to Forte.

© Springer International Publishing Switzerland 2015
T. Collins et al. (Eds.): MCM 2015, LNAI 9110, pp. 207–218, 2015.
DOI: 10.1007/978-3-319-20603-5_22

2   Preliminaries

Amiot and Sethares [5] define *scale vectors* as characteristic functions of pcsets:

Definition 1. *The* characteristic function *of a pcset is a vector with twelve places, one for each pc starting from C = 0, with a 1 indicating the presence of a pc and 0 indicating its absence.*

The characteristic function naturally generalizes to include pc-multisets (by allowing positive integers other than 1) and, more generally, *pc-distributions* (by allowing non-integers). I will refer to real-valued vectors corresponding to pc-distributions as *pc vectors*.

    Pc-distributions are best identified with equivalence classes of pc vectors under addition of a constant. In other words, it is the differences between pc values that define a pc-distribution, not the values themselves. As we will see, these equivalence classes can be neatly described using the DFT. They also bypass the potential conundrum of assessing the meaning of negative-valued pc vectors: negative values can always be eliminated by addition of a constant.

    The insight of Lewin [14,15], Quinn [18], and others is that reparameterizing such characteristic functions by means of the DFT reveals a wealth of musically significant information:

Definition 2. *Let $A = (a_0, a_1, a_2, \ldots, a_{11})$ be the characteristic function of a pc-distribution. Let $\hat{A} = (\hat{a}_0, \hat{a}_1, \hat{a}_2, \ldots, \hat{a}_{11})$ denote the* discrete Fourier transform (DFT) *of A. Then $\hat{A}$ is given by $\forall (0 \leq k \leq 11)$,*

$$\hat{a}_k = \sum_{j=0}^{11} a_j e^{-i2\pi kj/12} = \sum_{j=0}^{11} a_j(\cos(2\pi kj/12) + i\sin(2\pi kj/12)) \qquad (1)$$

The components of the Fourier transform, $\hat{A}$, as defined above, are complex numbers. They are most useful when viewed in polar form (magnitude and phase).

Definition 3. *Let $\hat{a}_k = re^{i\theta}$. Then the* magnitude *of $\hat{a}_k$ is $r = \sqrt{Re(\hat{a}_k)^2 + Im(\hat{a}_k)^2}$ and is denoted $|\hat{a}_k|$. The* phase *of $\hat{a}_k$, φ_{a_k}, is $\arg(\hat{a}_k) = \theta = \arctan(Re(\hat{a}_k), Im(\hat{a}_k))$. We will often normalize phases to a mod12 circle, denoted $_{12}\varphi_{a_k} = 6\theta/\pi$.*

Guerino Mazzola has pointed out (in informal response to [4]) that the DFT is one of many possible orthonormal bases for the space of pc-distributions. (See, e.g., [5].) Any of these would reflect the common-pc-content–based topology promoted by Yust [19] as a fundamental strength of this space. However, the DFT basis is of special music-theoretic value because it reflects evenness (i.e., periodicity) properties of fundamental musical importance. For instance, Amiot [3] and Yust [19] have shown that a space based on phases of the third and fifth components reflects many properties of tonal harmony by isolating evenness properties particular to triads and scales. Amiot [2] has also used the DFT to evaluate temperaments on the basis of the evenness of diatonic subsets. The analyses below consider the musical significance of other DFT components, especially the second and sixth.

Some basic properties of the DFT:

Remark 1. The components of the pc vector (pc magnitudes) are real valued. Therefore components 7–11 of the DFT have the same magnitude as their complementary components and multiply the phases by -1.

Remark 2. The zeroeth component of the DFT is always equal to the cardinality of a pcset or multiset.

Remark 3. Adding a constant to a pcset changes *only the zeroeth component* of the DFT. Therefore members of an equivalence class of pc vectors (as defined above) always have equivalent non-zero DFT components.

Remark 4. Negation preserves DFT magnitudes adds π to all well-defined phases. Adding a constant of 1 to all pcs of the negation produces the complement, meaning that these belong to an equivalence class, differing only in the zeroeth component of the DFT.

Remark 5. Transposition and inversion change the phases of DFT components but do not affect magnitudes.

3   Pcset Arithmetic in Fourier Coefficients

3.1   Sums of Pcsets

A sum of pcsets is the componentwise sum of their characteristic functions. This differs from the set-theoretic concept of pcset *union* in that the latter eliminates doublings, whereas pcset sums preserve doublings by allowing for multisets.

Pcset sums also correspond to the componentwise sum of their DFTs. This is straightforward when the components of the DFT are expressed in real and imaginary parts, but less so in the more meaningful polar representation.

Proposition 1. *Let pcset B be a sum of pcsets $A, A', A'', \ldots$. Then for all $0 \le k \le 11$,*

$$\varphi_{b_k} = \arg(|\hat{a}_k| \cos(\varphi_{a_k}) + |\hat{a}'_k| \cos(\varphi_{a'_k}) + |\hat{a}''_k| \cos(\varphi_{a''_k}) + \ldots,$$
$$|\hat{a}_k| \sin(\varphi_{a_k}) + |\hat{a}'_k| \sin(\varphi_{a'_k}) + |\hat{a}''_k| \sin(\varphi_{a''_k}) + \ldots) \quad (2)$$

If φ_{b_k} is undefined, then $|\hat{b}_k| = 0$. Otherwise,

$$|\hat{b}_k| = |\hat{a}_k| \cos(\varphi_{b_k} - \varphi_{a_k}) + |\hat{a}'_k| \cos(\varphi_{b_k} - \varphi_{a'_k}) + |\hat{a}''_k| \cos(\varphi_{b_k} - \varphi_{a''_k}) + \ldots \quad (3)$$

Equation (2) is derived simply by converting to rectangular coordinates, summing, and converting back to polar. Equation (3) is most easily demonstrated geometrically, by projecting each summand, as a vector in the complex plane, onto the sum.

From (3) we see that the contribution of each pcset to the sum is determined by its magnitude and its difference in phase from the sum. It maximally reinforces the sum when its phase is the same, contributes nothing when its phase is oblique (a difference of $\pi/2$ or 3 mod 12) and maximally reduces the sum when its phase is opposite (π or 6 mod 12). The contribution of each pcset to the phase of the sum is also weighted by magnitude, as (2) shows. Two pcsets with equal magnitude and opposite phases cancel one another out in the sum.

3.2   Product of Pcsets

"Multiplication" of pcsets was first defined by Pierre Boulez [7,13] in reference to his own compositional technique. Cohn [10] demonstrates the applicability of the operation, which he calls "transpositional combination," in analysis of twentieth-century music. Mathematically, Boulez and Cohn's operation is a variant of *convolution*. For pc vectors A and B, the convolution $C = A * B$ is given by:

$$c_k = \sum_{j=0}^{11} a_j b_{(k-j) \bmod 12} \tag{4}$$

The difference between convolution and Boulez's multiplication or Cohn's transpositional combination is that it allows for pc-multisets, whereas Boulez and Cohn take the additional step of eliminating doublings (replacing all positive integers in the pc vector with 1 s).

Boulez's term is fortuitous for present purposes, because according to one of the basic Fourier theorems, convolution of pc vectors corresponds to the termwise product of their DFTs.

$$\hat{c}_k = \hat{a}_k \hat{b}_k = |\hat{a}_k| e^{i\varphi_{a_k}} |\hat{b}_k| e^{i\varphi_{b_k}} = |\hat{a}_k| |\hat{b}_k| e^{i(\varphi_{a_k} + \varphi_{b_k})} \tag{5}$$

As this shows, convolution is particularly straightforward when viewed from the polar form of the DFT: it corresponds to simply multiplying the magnitudes and adding the phases of each component. It is therefore appropriate to refer to the convolution as a *product of pcsets*.

Lewin [15] noted that the convolution of one pcset with the inverse of another (or the *cross-correlation*) gives his *interval function*, a vector that lists the number of occurences of each pc interval from the first pcset to the second. The interval function of a pcset to itself gives Forte's *interval vector* (as components 1–6 of the twelve-place interval function). The DFT of the interval vector is purely real-valued (all well-defined phases are zero), as can be seen from (5) and the fact that inversion (about 0) negates the phases and does not affect magnitudes (see Remark 5):

$$\hat{a}_k(\hat{Ia})_k = |\hat{a}_k|^2 e^{i(\varphi_{a_k} - \varphi_{a_k})} = |\hat{a}_k|^2 \tag{6}$$

Singularities, zero-magnitude DFT components [5], are of special importance for pcset products in particular, because a singularity in one multiplicand leads to a singularity in the product. Note that phases are undefined when there is a singularity on a given component.

4 Fourier Components and Intervallic Content

4.1 Relating Fourier Components to Interval Classes

A motivating factor behind Forte's [11,12] focus on interval vectors is their invariance with respect to transposition and inversion (implied by (6) and Remark 5). An advantage of the DFT is that while distilling essentially the same intervallic information as the interval vector in the magnitudes of its components, it also preserves essential information in their phases. These are important, for instance, in understanding subset structure, as Eq. (3) shows.

Quinn [18] has emphasized the association of individual Fourier components with specific interval classes (ic1 $\leftrightarrow$ $\hat{a}_1$, ic2 $\leftrightarrow$ $\hat{a}_6$, ic3 $\leftrightarrow$ $\hat{a}_4$, ic4 $\leftrightarrow$ $\hat{a}_3$, ic5 $\leftrightarrow$ $\hat{a}_5$, ic6 $\leftrightarrow$ $\hat{a}_2$). The primary grounds for such associations are that Quinn's *generic prototypes* (set classes maximal with respect to a given component, often generated sets – see Sect. 5) have maximal representation of the associated interval class. The associations can be misleading in other respects, however.

For example, let $A = \{$C, F, F$\sharp\}$ and let $B = \{$C, D, E$\}$. Although A contains an instance of ic5 and B does not, $|\hat{a}_5|^2 = 1$, while $|\hat{b}_5|^2 = 4$. Component 5 does not indicate the "fifthy-ness" of a pcset so much as its diatonicity, and B is a more characteristic diatonic subset than A. Or, for another example, consider the set $A = \{$C, D, E, F$\sharp\}$. It has a relatively large number of ic4s for a tetrachord, but $|\hat{a}_3|^2 = 0$, because the two ic4s cancel one another out ($_{12}\varphi_{\{C,E\}_3} = 0$ while $_{12}\varphi_{\{D,F\sharp\}_3} = 6$). (See also the discussion in [9].)

The relationship of interval classes to Fourier components is best summarized by their own DFTs, as shown in Table 1.

Table 1. Squared DFT magnitudes for all twelve-tone interval classes

	0	1	2	3	4	5	6	7	8	9	10	11
ic1	4	3.73	3	2	1	0.27	0	0.27	1	2	3	3.73
ic2	4	3	1	0	1	3	4	3	1	0	1	3
ic3	4	2	0	2	4	2	0	2	4	2	4	2
ic4	4	1	1	4	1	1	4	1	1	4	1	1
ic5	4	0.27	3	2	1	3.73	0	3.73	1	2	3	0.27
ic6	4	0	4	0	4	0	4	0	4	0	4	0

This information can be summarized by defining *delta values* as minimal phase distances between the two pcs in the dyad:

Definition 1. *Let h be an interval in a u-ET universe. The* delta value *of h for each component k is the shortest* mod u *distance represented by $h \cdot k$,* $\delta = |((hk + u/2) \bmod u) - u/2|$.

For 12-tET, $\delta = |((hk + 6) \bmod 12) - 6|$, ranging from 0 to 6, and the squared DFT magnitude for any ic/component pair is $|\hat{a}_k| = 4cos^2(\delta\pi/12)$. (See also Sect. 5 below.)

Components are not neatly associated one-to-one with interval classes. Each has a maximum value, but the maximum could correspond to $\delta = 0$ (for components 2, 3, 4, and 6) or $\delta = 1$ (for 1 and 5). In the former situation a component might have maximum value for more than one ic, as is the case for components 4 and 6. Also, the maximum values do not tell the full story: at least as important are the *singularities* of each interval class, where $\delta = 6$ ($= u/2$).

4.2   Webern, Op. 5, No. 4

Forte [11], in his classic analysis of Webern's *Satz für Streichquartett*, op. 5 no. 4, uses interval vectors and abstract subset structure to demonstrate how the piece is sectionalized by harmonic content. A similar conclusion can be reached using the DFT. Figure 1 shows mm. 1–10, the first two sections of the ternary form, and labels some significant pcsets. Table 2 lists the squared magnitudes of the DFT components for each of these. From this we can make the following generalizations: Universe A (sets A, A', A'' and combinations involving them) is characterized by a high component 2 and low odd components, Universe B (sets D, E, F) by a high component 3 and low component 2. Intermediate between these are sets B and C, which have a high component 2 and moderate presence of 3. As the values for dyads show (Table 1), the interval classes that Forte associates with universes A and B (ic6 and ic4 respectively) manifest these properties only in part: component 2 is one of three maximum values for ic6, and one of four relatively low values in ic4. The reason for this is evident from the fact that many of these are products of dyads: $A = $ ic1×ic6 or ic5×ic6, $A + A' = $ ic1×ic1×ic6 or ic5×ic5×ic6 or ic1×ic5×ic6, $A \cup A' \cup A'' = $ ic1×ic2×ic6 or ic5×ic2×ic6, $E = $ ic3×ic4. Interval classes 1 and 5 have higher values of component 2 than 4 or 6, so they contribute this feature when multiplied by ic6. Similarly, ic3 has a singularity on component 2, making $E = $ ic3×ic4 a particularly appropriate antipode to A. Other sets from Universe B are more similar to ic4: D is generated by ic4 (see Sect. 5), and $D + E$ and $E + F$ are, like ic4, weighted towards component 6 as well as 3. The accompaniment by itself, F, gives the weakest contrast from the harmony previous section. B and C are also factorable: $B = $ ic1×ic5 and $C = (012)$×ic5.

Table 2. Squared DFT magnitudes for pcsets from Webern's Op. 5 No. 4

	1	2	3	4	5	6		1	2	3	4	5	6
A	0	12	0	4	0	0	D	0	0	9	0	0	9
$A \cup A'$	0	16	0	0	0	4	E	2	0	8	4	2	0
$A \cup A' \cup A''$	0	12	0	4	0	0	$D + E$	2	0	5	4	2	9
$A + A'$	0	36	0	4	0	0	$E + F$	3	3	9	3	3	9
$A + A' + A''$	0	48	0	0	0	0	F	0.27	3	5	1	3.73	9
$A \cup B$	1	13	1	1	1	1							
B	1	9	4	1	1	0	V	0.27	7	2	1	3.73	4
C	2	12	2	0	2	0	*Flyaway*	1	7	1	7	1	1

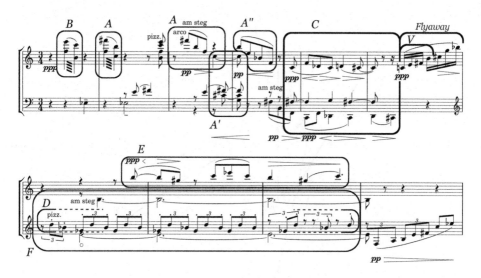

Fig. 1. Webern op. 5 no. 4, mm. 1–10: Some significant pcsets

Other authors (Perle [17] and Burkhart [8]) see more continuity in the piece by emphasizing V, which is a (literal) subset of $A \cup B$, $A \cup A' \cup A''$, and, most explicitly, *Flyaway* (borrowing Lewin's [16] nickname for this motive). Perle observes that the same set class is a subset of F ({EG♭B♭B}). As Table 2 shows, V and *Flyaway* are intermediate between the harmonic universes, like B and C. And the presence of component 2 in F suggests a link with Universe A.

We can clarify subset/superset relationships by using the kind of phase spaces proposed by Amiot [3] and Yust [19]. These authors construct toroidal spaces using phases of Fourier components 3 and 5 as axes to make *Tonnetz*-like maps for tonal harmony. For Webern's piece, a space based on phases of components 2 and 3, as seen in Fig. 2, is appropriate. Yust [19] demonstrates how the third component represents the triadic aspect of tonality; Webern's Universe B might accordingly be heard as a reference to tonal harmony, made hazy by a lack of diatonicity. The second DFT component does not feature in Amiot and Yust's treatment of tonal harmony, but its use can be identified with the "quartal" sonorities emblematic of early twentieth-century modernism – i.e., the second component comes into play specifically in a harmonic palette that avoids thirds and sixths (ics 3 and 4), as is evident from Table 1.

According to (3), the more spread out pcsets are in phase, the weaker their sums are on a given component. The pcsets from the first section of the piece are concentrated in a small zone of φ_2, but spread out in φ_3. Pcsets connected by lines in Fig. 2 have equal magnitude but opposite phase on φ_3, so their sums have component 3 singularities, except $F + E$ and $-F$ (the negation of F) which are opposite on φ_2. The position of F is opposite that of $-F$ (a difference of 6 in all dimensions, see Remark 4), so, like $F + E$, the phase of its second component is within the region defined by the pcsets of Universe A.

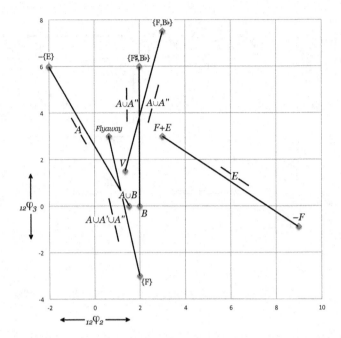

Fig. 2. Pcset sums from Webern op. 5 no. 4 shown in phase space

5   DFT of Generated Collections

As noted above, Quinn [18] and Amiot [1] place special emphasis on generated collections (and maximally even sets in particular) as pcsets most representative of a particular interval. The following formulas simplify the calculation of the DFT of generated collections and provide some insight into their properties.

Proposition 1. *Let A be a pcset of cardinality n generated by interval g/u, where u is the cardinality of the ET universe. Then,*

$$\varphi_{a_k} = -(n-1)gk\pi/u \tag{7}$$

$$|\hat{a}_k| = \begin{cases} n & \text{if } gk = 0 \bmod u, \\ \dfrac{\sin(ngk\pi/u)}{\sin(gk\pi/u)} & \text{otherwise} \end{cases} \tag{8}$$

Proof. The first case of (8) is evident from the fact that when $gk = 0 \bmod u$, the unit vectors all have the same phase, 0.

For $gk \neq 0$ the Fourier series (1) can be written out in the order of generation and simplified as a geometric series:

$$\hat{a}_k = \sum_{j=0}^{n-1} e^{-i2\pi kgj/u} = \frac{1 - e^{-i2ngk\pi/u}}{1 - e^{-i2gk\pi/u}}$$

$$= \left(\frac{e^{ingk\pi/u} - e^{-ingk\pi/u}}{e^{igk\pi/u} - e^{-igk\pi/u}}\right)e^{-(n-1)igk\pi/u} = \frac{\sin(ngk\pi/u)}{\sin(gk\pi/u)}e^{-(n-1)igk\pi/u} \tag{9}$$

The second step factors out the phase of the component, and the final step applies Euler's formula. The resulting magnitude function is a Dirichlet kernel.[1]

This result complements those of Amiot [1] and generalize his formulas for maximum values, which represent the special cases $\delta = 0$ and $\delta = 1$, giving a more general picture of the special status of generated collections viewed through the DFT. Equation (8) can be viewed as a function of n, so that the denominator, $sin^{-1}(gk\pi/u)$, is a constant, indicating the maximum value of the given component for the given generator. Note that (8) gives the same result for $-gk$ as for gk, so δ (Definition 1) can substitute for gk, making the maximum value $sin^{-1}(\delta\pi/u)$. $|\hat{a}_k|$ is a sinusoidal function of n with period u/δ, minimum value (0) at $n = 0 \bmod u/\delta$, and maximum at $n = \frac{1}{2}u/\delta \bmod u/\delta$. Amiot and Quinn focus on the cases $\delta = 0$, where $|\hat{a}_k|$ is unbounded, and $\delta = 1$, which maximizes $sin^{-1}(\delta\pi/u)$ and gives a period of u (for $|\hat{a}_k|$ as a function of n), and hence a unique maximum. However, this is one extreme of a range of possibilities, the other being $\delta = \frac{1}{2}u$, which minimizes $sin^{-1}(\delta\pi/u) = 1$ and gives a period of 2. (In other words, this component alternates between magnitudes 0 for n even and 1 for n odd). For D in the Webern analysis above (the augmented triad, $g = 4$), $\delta = 4$ for components 1, 2, 4, and 5, reaching a minimum at $n = 3$, while $\delta = 0$ for components 3 and 6.

As another example, compare B and C from the Webern analysis, which both involve the product of an ic1-generated collection with ic5. From $n = 2$ to $n = 3$, components with large δ values and short periods (3 and 4) decrease, while component 5, with a long period, increases incrementally. The result in C intensifies the strength of component 2 relative to 3 and 4.

6   Example: Bartók, "Allegro Pizzicato" from String Quartet No. 4 (iv)

The example from Webern demonstrated how contrasting harmonic profiles can operate as a means of formal delineation. In the pizzicato fourth movement of Bartók's Fourth String Quartet, we find similar contrasts being used for *stratification* of harmonic materials as well as formal delineation. The first section of the piece consists of fugal entries of a scalewise theme accompanied by ostinato-like patterns in the other instruments. Figure 3 shows the first entry and its accompaniment. The melody is written in the acoustic scale on A♭ (a collection favored by Bartók; see [6]), while the accompanimental collection is {DE♭GA♭}.

Table 3 shows the DFT magnitudes for these two collections. Remarkably, the largest component of the accompaniment ($\hat{a}_2$) is a singularity for the acoustic scale, while the largest component for the acoustic scale ($\hat{a}_6$) is a singularity for the accompaniment. The acoustic scale also has a relatively high value on component 5 while {DE♭GA♭} has a relatively low value (contrary to what subset relations suggest – the (0156) tetrachord is a subset of the diatonic, but contains

---

[1] I am indebted to Emmanuel Amiot for pointing this out and helping me improve upon a previous less elegant proof.

Fig. 3. The subject of Bartók's "Allegro Pizzicato"

Table 3. Squared DFT magnitudes of pcsets in Bartók's "Allegro Pizzicato"

	Accompaniment						Melody					
	1	2	3	4	5	6	1	2	3	4	5	6
Mm. 6–12	1	**9**	4	1	1	0	**Acoustic scale:**					
Mm. 13–19	1	**13**	1	1	1	1	0.54	0	1	4	**7.46**	9
Mm. 20–27	2.27	3	1	1	**5.73**	9	**Whole-tone pentachord:**					
							1	1	1	1	1	**25**

precisely the most marginal members of the diatonic on the circle of fifths). Note also that this accompanimental collection is the same set class as B from the Webern analysis above, and can be expressed as a product of dyads, ic1×ic5.

As previously noted, high component 2 values typify the sonorous landscape of modernism, and its role here may reflect upon the Fourth Quartet's reputation for reflecting a turn towards a modernist aesthetic. The second accompanimental collection intensifies the focus on component 2, while the third shifts towards a closer match to the harmonic properties of the acoustic scale. This shift occurs precisely at the point where contrapuntal writing begins.

Taken by itself, the melodic subject realizes a harmonic motion from the acoustic scale (dominated by components 5 and 6) to its five-note whole-tone subset, where the presence of component 5 (diatonicity) is overtaken by 6 (whole-tone). (See Table 3.)

Bartók's stratification of hamonic materials is perhaps best viewed through the lens of Lewin's interval function [15,16], which is a pcset product (see Sect. 3.2). The DFT magnitudes of this product are transposition-independent, just as they are for pcsets themselves, so phase is significant in determining the specific intervals between collections. Note that DFTs of interval functions in Table 4 shows are a product of the magnitudes and a difference of phases (as implied by (5)). In mm. 6–12, the singularities on components 2 and 6 annhilate these components in the product, leaving component 5 to predominate. This means that the most prevalent intervals tend to be fifth-related, which can be seen in the resulting interval function below. However, depending on the *phase* of component 5, these could be intervals of circle-of-fifths proximity (0, 5, 7, 10, 2, ...) or of circle-of-fifths remoteness (6, 1, 11, ...). The phase difference of component 5 ($\delta = 1$) is small but not minimal, shifting the interval function towards slightly more remote intervals. This small difference is most directly

Table 4. Interval functions between melody and accompaniment

Component	1		2		3		4		5		6	
	mag²	12φ	mag²	12φ	mag²	12φ	mag²	12φ	mag²	12φ	mag²	12φ
Mm. 6–12, melody	0.54	8	0	–	1	6	4	2	7.46	10	9	0
accompaniment	1	7	9	8	4	3	1	4	1	11	0	–
Interval function	0.54	1	0	–	4	3	4	10	7.46	11	0	–
Mm. 20–27, melody	0.54	6	0	–	1	12	4	6	7.46	0	9	0
accompaniment	2.27	2.2	3	9	1	3	1	6	5.73	11.7	9	0
Interval function	1.22	3.83	0	–	1	9	4	0	42.8	0.3	81	0

Interval Functions

	0	1	2	3	4	5	6	7	8	9	10	11
Mm. 6–12	3	2	2	3	2	2	2	3	2	2	3	2
Mm. 20–27	5	1	4	2	3	3	3	3	3	3	4	1

Fig. 4. Transpositional combination in the middle section, mm. 47–49

manifest in the mild polyscalar dissonance of the accompanimental G against the melodic G♭. In mm. 20–27 Bartók moves the accompanimental collection into near-perfect alignment with the melody, which is evident in the balance of the interval function around interval zero. The singularity in component 6 is also eliminated, leading to a strong imbalance in even versus odd intervals.

Bartók also, like Webern, uses harmonic contrasts for formal delineations in a three-part design. Figure 4) shows two multiplications that make up the principal accompanimental and melodic material of the section. The first is the product of an ic2-generated trichord and ic1. The trichord represents the whole-tone saturated harmonic universe of the first part, but, as Table 5 shows, the multiplication ironically annihilates its sixth component altogether. A similar point can be made about the melodic construction, the product of an ic1-generated trichord and ic2. Both combinations result in intensely chromatic pcsets, reflected in the dominance of component 1.

Table 5. Transpositional combinations in the middle section

	1	2	3	4	5	6		1	2	3	4	5	6
{ABC♯}×	4		0	1	0	4		9 {EFF♯}×	7.46	4	1	0	0.54	1
ic1 =	3.73		3	2	1	0.27		0 ic2 =	3	1	0	1	3	4
{AB♭BCC♯D}	14.9		0	2	0	1.07		0 {EFF♯²GA♭}	22.4	4	0	0	1.61	4

References

1. Amiot, E.: David Lewin and maximally even sets. J. Math. Mus. **1**, 157–172 (2007)
2. Amiot, E.: Discrete Fourier transform and Bach's good temperament. Mus. Theor. Online **15** (2009)
3. Amiot, E.: The Torii of phases. In: Yust, J., Wild, J., Burgoyne, J.A. (eds.) MCM 2013. LNCS, vol. 7937, pp. 1–18. Springer, Heidelberg (2013)
4. Amiot, E.: Viewing diverse musical features in Fourier space: a survey. Paper presented to the International Congress on Music and Mathematics, Puerto Vallarta, 28 November 2014
5. Amiot, E., Sethares, W.: An algebra for periodic rhythms and scales. J. Math. Mus. **5**, 149–169 (2011)
6. Antokoletz, E.: Transformations of a special non-diatonic mode in twentieth-century music: Bartók, Stravinsky, Scriabin and Albrecht. Mus. Anal. **12**, 25–45 (1993)
7. Boulez, P.: Boulez on Music Today. Translated by Bradshaw, S., Bennett, R.R., Harvard University Press, Cambridge (1971)
8. Burkhart, C.: The symmetrical source of Webern's Opus 5, No. 4. In: Salzer, F. (ed.) Music Forum V, pp. 317–334. Columbia University Press, New York (1980)
9. Callender, C.: Continuous harmonic spaces. J. Mus. Theor. **51**, 277–332 (2007)
10. Cohn, R.: Transpositional combination and inversional symmetry in Bartók. Mus. Theor. Spectr. **10**, 19–42 (1988)
11. Forte, A.: A theory of set-complexes for music. J. Mus. Theor. **8**, 136–183 (1964)
12. Forte, A.: The Structure of Atonal Music. Yale University Press, New Haven (1973)
13. Koblyakov, L.: Pierre Boulez: A World of Harmony. Harwood, New Haven (1990)
14. Lewin, D.: Re: intervallic relations between two collections of notes. J. Mus. Theor. **3**, 298–301 (1959)
15. Lewin, D.: Special cases of the interval function between pitch-class sets X and Y. J. Mus. Theor. **45**, 1–29 (2001)
16. Lewin, D.: Generalized Musical Intervals and Transformations, 2nd edn. Yale University Press, New Haven (2007)
17. Perle, G.: Serial Composition and Atonality: An Introduction to the Music of Schoenberg, Berg, and Webern, 6th edn. UC Press, Berkeley (1991)
18. Quinn, I.: General equal-tempered harmony (in two parts). Perspectives of New Mus. **44**, pp. 114–159 and **45**, pp. 4–63 (2006)
19. Yust, J.: Schubert's harmonic language and Fourier phase space. J. Mus. Theor. **59**, pp. 121–181 (2015)

Utilizing Computer Programming to Analyze Post-Tonal Music: Contour Analysis of Four Works for Solo Flute

Kate Sekula[✉]

University of Science and Arts of Oklahoma, 1727 West Alabama Avenue,
Chickasha, OK, USA
ksekula@usao.edu
http://kate.cooleysekula.net/java

Abstract. A computer application was written to complete the task of contour reduction. The application was used to complete analyses of twentieth-century post-tonal works for solo flute. The methodology of Rob Schultz's Contour Reduction Algorithm was chosen for implementation. While contour reduction is a useful analytical tool, it is a meticulous and time-consuming process. Computer implementation of this procedure produces quick and accurate results while reducing analyst fatigue and human error. Java computer programming language is used to create a contour reduction application. This implementation greatly reduces the time needed to analyze a melody. Computer programming is combined with music analysis to produce informed and expressive musical interpretations.

Keywords: Contour reduction · Flute · Kazuo Fukushima · Java · Robert Morris · Post-tonal · Tōru Takemitsu · Rob Schultz

1 Introduction

The study of contour was born out of ethnomusicological studies of folk melodies and focused primarily on melodic pitches.[1] More recently, it has been applied to post-tonal music and to parameters other than pitch, such as dynamics or duration. As contour theory has evolved, the vocabulary used to describe this feature has become more technical, pulling terminology and metaphors from other research areas such as linguistics, phenomenology, genealogy, mathematics, and biology. It is also an important structural feature in the music of many post-tonal composers such as Iannis Xenakis, Olivier Messiaen, Edgard Varèse, and György Ligeti. The fact that composers of post-tonal music specifically turned to contour as a formative ingredient has made the practice of contour analysis a necessity.

The work of Robert Morris and Rob Schultz was chosen for computer implementation [2,3]. Robert Morris applies set theory to contour, creating hierarchical

---

[1] For a complete discussion on the history of contour analysis, see [1].

© Springer International Publishing Switzerland 2015
T. Collins et al. (Eds.): MCM 2015, LNAI 9110, pp. 219–230, 2015.
DOI: 10.1007/978-3-319-20603-5_23

contour relationships, applies contour to any musical parameter, and determines a measure for contour equivalence. He is largely responsible for a generalized theory of contour with formal methodology. He combines set theory with contour theory and demonstrates how "pitch classes and their sets, brought out by contour hierarchies, are related to each other as well as to adjacent pitch-class sets by abstract intersection and complement relations". The end result is "a complete taxonomy of all contour types" [2, p. 206]. Morris's work with contour led to his development of the Contour Reduction Algorithm (CRA), a process which reduces any contour to a prime.

Rob Schultz reconfigures Morris's work to repair methodological errors. He then combines contour theory with phenomenology and genealogy to produce a diachronic analysis that reveals the transformational nature of a contour through time. This process will be referred to as Diachronic-Transformation Analysis (DTA).

A detailed historical account of computer assisted analysis through the year 2000 has been completed by Nico Schüler [4]. David Cope continued the cataloging of historical computation analysis through 2008 [5]. The music analysis computer programs written and/or ongoingly maintained since 2008 include the Melisma Music Analyzer,[2] RUBATO,[3] Tonalities,[4] Harmonia and Music Theory Workbench (MTW),[5] Music21,[6] and VisiMus.[7]

The majority of these programs focus on the analysis of tonal music. Music21 and VisiMus offer contour analysis options. Music21 only locates contours and does not offer the capability of contour reduction and further analysis or comparison between contours.

VisiMus is a software application for calculating and plotting musical contour operations [6]. It can process contours related to different parameters such as pitch, duration, dynamic level, and chord density. The application calculates various contour properties as defined in the publications of Michael Friedmann, Elizabeth Marvin West and Paul Laprade, and Robert Morris [2,7,8].

In both of these program, input must be performed by hand, as Music21 and VisiMus do not allow for file upload. VisiMus only allows for mod12 input of integers 0–11. Finally, none of these computer programs offer the ability to compute DTA, as described below.

The purpose of this research was to automate Rob Schultz's version of the CRA as well as the procedure for DTA. These tools were then used to determine hierarchical relationships in post-tonal music for solo flute by Japenese composers Kazuo Fukushima and Tōru Takemitsu.

---

[2] Version 1.0 www.link.cs.cmu.edu/music-analysis/ Version 2.0 http://theory.esm.
rochester.edu/temperley/melisma2/.

[3] http://www.rubato.org/.

[4] http://www.hud.ac.uk/tonalities/.

[5] http://camil.music.illinois.edu/software/harmonia/.

[6] http://web.mit.edu/music21/.

[7] http://visimus.com.

1.1 Definitions

Contour is the function that tracks a musical parameter over time. A *parameter* is any set of compositional or auditory properties whose values determine the characteristics or behavior of a piece of music. Examples of parameters are dynamics, articulation, timbre, and pitch.

Post-tonal refers to any type of musical composition in which there is no hierarchy of functional harmony, no distinction between consonance and dissonance, for which a traditional tonal analysis does not yield satisfying analytical results, or for which a pitch-class set analysis alone is not sufficient to illuminate the structure of the music.[8]

Octave identification standards are used as established by The Acoustical Society of America [10]. Middle C = C4.

Diachronic Transformation Analysis (DTA) is a procedure which recursively performs the CRA as each pitch is added to a melody. The contour data unfolds from the initial to the final pitch revealing linear contour prime data in the form of percentages of contours used.

A Linear Contour Prime (LCP) is a contour prime that results from application of the CRA and in which no simultaneities occur. A predominant LCP is a LCP which has a high rate of recurrence as DTA occurs.

1.2 Scope

Only post-tonal music for solo flute is analyzed, specifically works by Kazuo Fukushima and Tōru Takemitsu. However, the methods can be applied to any monophonic piece of music. While the author does not deny that composer intent and listener cognition are important for music analysis, particularly of works written in the twentieth and twenty-first centuries, this study focuses primarily on objective principles and data analysis.

2 Methods

2.1 Robert Morris and the Contour Reduction Algorithm

Robert Morris defines contour as an ordered set of *contour pitches* (cps). The cps of a *contour segment* (cseg) generate a *contour space* (c-space). C-space is the ranking of cps from low to high, ignoring the exact intervals between cps [11]. The lowest cp is renamed 0 and the highest is renamed $n - 1$, where n equals the number of distinct pitches in a musical unit. For example, in the melodic cseg (C4, G3, E4, C4), the c-space would be $<1\,0\,2\,1>$.

Further development of contour theory led to Morris's creation of the CRA. The full algorithm is reproduced in Fig. 1. Through implementation of the CRA, Morris delineates local high and low points of a cseg, which he calls "maxima" and "minima," and recursively eliminates all non-maxima and non-minima

---

[8] For a complete discussion of pitch-class sets, see [9].

Algorithm: Given a contour C and a variable N

0. Set N to 0.
1. Flag all maxima in C; call the resulting set the *max-list*.
2. Flag all minima in C; call the resulting set the *min-list*.
3. If all pitches in C are flagged, go to step 9.
4. Delete all non-flagged pitches in C.
5. N is incremented by 1 (i.e., N becomes N+1).
6. Flag all maxima in the max-list. For any string of equal and adjacent maxima in the max-list, either: (1) flag only one of them; or (2) if one pitch in the string is the first or last pitch of C, flag only it; or (3) if both the first and last pitch of C are in the string, flag (only) both the first and last pitch of C.
7. Flag all minima in the min-list. For any string of equal and adjacent minima in the min-list, either: (1) flag only one of them; or (2) if one pitch in the string is the first or last pitch of C, flag only it; or (3) if both the first and last pitch of C are in the string, flag (only) both the first and last pitch of C.
8. Go to step 3.
9. End. N is the "depth" of the original contour C.

Fig. 1. Robert Morris's CRA [2].

(a process which he calls "pruning") until a cseg's fundamental structure, or "prime," is revealed. He introduces the concept of "contour depth" which measures how many times one traverses through the algorithm before reaching the contour prime. Morris determines 25 basic contour prime classes. The contour prime generates a hierarchical level of contour pitch salience and the contour depth number gives a rough measure of the complexity of a contour. Out of these 25 basic contour prime classes there are seven LCP classes: <0>, <0 1>, <0 1 0>, <0 2 1>, <1 0 2 1>, <1 0 3 2>, and <1 3 0 2>.

2.2 Rob Schultz and the Contour Reduction Algorithm

Rob Schultz refines Morris's CRA [3,12]. He discusses two flaws. First, it fails to reduce "wedge-shaped" contours; contours in which every pitch is a maximum or minimum. Second, the manner in which it deals with repeated cps provides no indication of which repeated cp should be pruned. Arbitrary pruning of a repeating maximum or minimum may result in different cps at the deepest level.

Schultz modifies the algorithm so that Step 3 (see Fig. 1) no longer prematurely ends the algorithm before every cp has been subjected to the pruning procedure at least once. This solves the issue pertaining to wedge-shaped contours, since in Morris's CRA none of the cps would be flagged in step 1 and step 2, leading to no cps being pruned and no reduction of the contour. Schultz modifies the algorithm further to include additional instructions for the flagging

Algorithm: Given a contour C and a variable N:

0. Set N to 0.
1. Flag all maxima in C upwards; call the resulting set the *max-list*.
2. Flag all minima in C downwards; call the resulting set the *min-list*.
3. If all c-pitches are flagged, go to step 6.
4. Delete all non-flagged c-pitches in C.
5. N is incremented by 1 (i.e., N becomes N + 1).
6. Flag all maxima in the max-list upward. For any string of equal and adjacent maxima in the max-list, flag all of them, unless: (1) one c-pitch in the string is the first or last c-pitch of C, then flag only it; or (2) both the first and last c-pitches of C are in the string, then flag (only) both the first and last c-pitches of C.
7. Flag all minima in the min-list downward. For any string of equal and adjacent minima in the min-list, flag all of them, unless: (1) one c-pitch in the string is the first or last c-pitch of C, then flag only it; or (2) both the first and last c-pitches of C are in the string, then flag (only) both the first and last c-pitches of C.
8. For any string of equal and adjacent maxima in the max-list in which no minima intervene, remove the flag from all but (any) one c-pitch in the string.
9. For any string of equal and adjacent minima in the min-list in which no maxima intervene, remove the flag from all but (any) one c-pitch in the string.
10. If all c-pitches are flagged, and no more than one c-pitch repetition in the max-list and min-list (combined) exists, not including the first and last c-pitches of C, proceed directly to step 17.
11. If more than one c-pitch repetition in the max-list and/or min-list (combined) exists, not including the first and last c-pitches of C, remove the flags on all repeated c-pitches except those closest to the first and last c-pitch of C.
12. If both flagged c-pitches remaining from step 11 are members of the max-list, flag any one (and only one) former member of the min-list whose flag was removed in step 11; if both c-pitches are members of the min-list, flag any one (and only one) former member of the max-list whose flag was removed in step 11.
13. Delete all non-flagged c-pitched in C.
14. If N != 0, N is incremented by 1 (i.e., N becomes N + 1).
15. If N = 0, N is incremented by 2 (i.e., N becomes N + 2).
16. Go to step 6.
17. End. N is the "depth" of the original contour C.

Fig. 2. Rob Schultz's reinterpretation of the CRA [3, p. 130].

of adjacent maximum or minimum. If two adjacent maxima have an intervening minimum (or if two adjacent minima have an intervening maximum) then the adjacent maximum (or minimum) is flagged and retained. If there is no intervening cp, then only one of the adjacent cps is kept. Finally, he proposes the addition of two new contour prime classes, which are linear, to Morris's basic contour prime classes: $<1\,0\,2\,0\,1>$ and $<1\,0\,3\,0\,2>$. These classes allow for the irreducible adjacent minima which are a product of the new version of his algorithm. The final version of Schultz's algorithm is shown in Fig. 2.

Rob Schultz combines contour theory, phenomenology, and genealogy into a general theory of temporally ordered contour relationships [3]. From a phenomenological perspective, Schultz uses the work of Edmund Husserl to explain that contours should be heard as a series of "now-points." A listener does not conceive that a contour instantaneously exists. Rather, they hear it unfold through time and contour analysis should account for this phenomenon. His dissertation focuses on a *diachronic*, rather than a *synchronic*, view of contour. Diachronic relates to how something evolves through time. Synchronic means that elements, such as cps, exist at the same time; they are concurrent.

In a diachronic analysis, a contour begins with an initial cp. From this point it could proceed to a cp $=$, $>$, or $<$ itself. These possibilities produce three contour options: $<0\,0>$, $<0\,1>$, or $<1\,0>$. With the addition of a third cp, each of these options must also be evaluated to determine how the contour will proceed, generating thirteen possible three-note contours. The number of possibilities increases greatly as the cardinality of the contour increases [3, pp. 17–18].

Every contour is viewed as a unique transformational path and contour similarity is based on how similar two contours' paths are. The CRA passes over the contour as the cardinality changes. As a result the first, last, highest, and lowest cps and the depth number are all subject to continued revision as the contour unfolds from the initial cp to its final length. The result is a transformational path whose various csegs contain contours of differing maxima/minima and depth due to the reiteration of the CRA.

The analyses presented here utilize the idea of DTA in a different manner, determining how often a contour occurs as DTA unfolds from the initial to final pitch of a work. The frequency of certain diachronic LCPs is then compared to instances of surface and synchronic LCPs to determine embedding/nesting relationships.

2.3   Computing

A Java web-application has been developed that can perform Rob Schultz's version of the CRA as well as DTA. The application is named: Contour Analysis Tools (CAT) and is available for use at kate.cooleysekula.net/java. The application also provides information such as Forte set class designation, normal form, prime form, contour depth, contour normal form, and LCP.

Pitches are entered as integers with spaces in between. The program allows for decimals, so microtones are easily accounted for. A user has the option to manually enter integers or to upload a MIDI file. Parameter data may be entered in any numerical format: mod12, MIDI number, pitch frequency, or any numerical system chosen by the analyst. The application can give a snapshot of the synchronic contour of a given cseg or provide repetitive DTA as each new pitch is added to a melody (or any parameter which the analyst chooses that can be represented as an integer). Pairing the usage of CAT with software like Finale, Sibelius, or Photoscore can further expedite the analytical process.

3 Results and Discussion

3.1 *Mei* for Solo Flute by Kazuo Fukushima

Mei is composed in ABA' form. Section B is divided into four subsections, referred to as B1–B4.[9] The opening and closing A-sections show a chromatic ascent to the pitch B5 (see Fig. 3). B5 is highlighted in the score in several ways: it is the highest pitch in section A, it is followed by a quarter-rest, and it is part of a *rallentando* which proceeds to an *a tempo* in the next measure (see Fig. 4).

Fig. 3. Section A pitches of Kazuo Fukushima's *Mei*. Arrows represent microtones. Duration is removed.

Fig. 4. Section A, measures 11–12 of Kazuo Fukushima's *Mei*.

DTA was performed on the pitches of section A. As demonstrated across the top of Fig. 3, a reduced contour was determined from the initial pitch of the section to every other pitch of the section, creating 37 contours in total. The results were conflicting to those discussed above. DTA predominantly produced ascending LCPs <0 1> (19 of 37 csegs, or roughly 51 %) and <0 2 1> (15 of 37 csegs, or

---

[9] For a complete discussion of the form of *Mei*, see [1].

roughly 40 %).[10] LCP <0 1> is obviously ascending. LCP <0 2 1> is referred to here as *overall ascending*, since its first and last cp demonstrate an overall rise. These LCPs reveal that the diachronic contour of section A continually ascends until the final pitch, C4. At this point the ascending diachronic contour <0 2 1> retrogrades to the descending contour <1 2 0> unveiling a facet of pitch contour previously overlooked without the use of DTA: the diachronic contour is ever-ascending beyond B5 and does not relax to descent until the last pitch of the section. This downplaying of B5 is also reflected in Fukushima's change of dynamic from *fff* to *mp* and his indication to slur, rather than re-articulate, the pitch.

Analysis of contour nesting reveals previously unseen relationships within the piece. The synchronic contour, that is, the result of application of the CRA to the entire piece, of *Mei* is <1 2 0>. This is also the synchronic contour of sections A and A' as well as subsection B3. In section A and A', <1 2 0> is representative of the pitches E♭4, B5, and C4, again centering the discussion on the interpretation of B5. This pitch is obviously the goal, but the anti-climax that results after the chromatic ascent requires a high amount of nuance and sensitivity on the part of the performer.

<1 2 0> is the retrograde of LCP <0 2 1>, which is the synchronic contour of subsection B1. It is also prevalent as a diachronic contour in both sections A and A'. Subsection B4 has a synchronic contour of <0 1>, a LCP which is prevalent diachronically in sections A and A'. Finally, subsection B2 has a synchronic contour of <1 0 3 2>, which is also the LCP for the whole of section B. It is also the most predominant diachronic contour as the piece progresses from the initial to the final pitch, representing 60 % of the contours generated by DTA. There is a revelation of contour relationships within and between formal strata of the piece.

3.2   *Requiem* for Solo Flute by Kazuo Fukushima

The formal structure of *Requiem* is ABC [13, p. 43].[11] Again, analysis of contour nesting highlights or reveals hierarchical structures of the piece. For example, the contour <2 3 0 1> is initially observed as a melodic surface contour (i.e., a contour apparent on the musical surface without reduction), as shown in Fig. 5. This is the retrograde of LCP <1 0 3 2>.

The same contour appears synchronically at higher levels of segmentation.[12] There are 15 *sequences*. Of these, five are four-pitch synchronic contours and of those five, three are the prime or retrograde of LCP <1 0 3 2>. The piece is built from the tone-row <2 5 6 4 3 0 7 e 1 t 9 8>, which is repeated/transformed seven

---

[10] It is worth noting that of the possible LCPs, these analyses contain predominantly LCPs <0 1>, <0 2 1>, <1 0 3 2>, and <1 3 0 2>. The nature of these LCPs only allows for one transformation, generally the retrograde, to be created from the prime. The exception is LCP <0 2 1>, for which a prime, retrograde, inversion, and retrograde inversion can exist.

[11] Change in predominant diachronic LCP demarcates formal divisions of the work. For space limitations, the discussion of form is limited here. See [1].

[12] Segmentation of this work refers to the segmentation process developed by James Tenney and Larry Polansky, see [14].

Fig. 5. Example of surface occurrences of contour <2 3 0 1> in Kazuo Fukushima's *Requiem*.

times in total. Two of the tone-row statements produce synchronic occurrences of <1 0 3 2>.

In terms of diachronic analysis, relationships exist between overall synchronic contour and predominant diachronic contour. The contours which occur most often as the piece unfolds from beginning to end are found in places of synchronic prominence. In *Requiem*, section C has the synchronic contour <2 3 0 1>, another instance of retrograde. At the highest level, the entire piece, the synchronic contour is, again, the retrograde <2 3 0 1>. Not only is this contour seen at the various hierarchical formal segmentations of this piece, but it is also shown as an important diachronic contour, representing the highest percentage, 37 %, of the diachronic contours as the piece unfolds from its initial to final pitch.

3.3 *Itinerant* for Solo Flute by Tōru Takemitsu

Itinerant was composed in memory of Takemitsu's friend, sculptor Isama Noguchi, who traveled extensively throughout his lifetime: hence the title [15]. There are 25 phrases in *Itinerant*, each demarcated by a combination of rests and *fermatas*.[13] Contour combines with other musical aspects of the work to reveal a permeation of gestural material focused on the idea of ascent.[14]

---

[13] For a complete discussion of the form of *Itinerant*, see [1].

[14] Roger Graybill uses the term "gesture" to describe a grouping structure with a distinguishing internal dynamic shape [16].

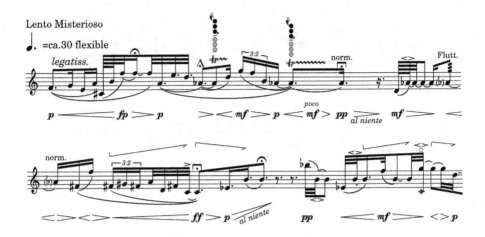

Fig. 6. First two lines of *Itinerant*.

The piece is full of short, melodically ascending motives which are combined into ascending phrase ideas. While the piece does not contain the contour nesting functions demonstrated in *Mei* and *Requiem*, its synchronic contours reflect the ascending nature of the melodic gestures. 75 % of the phrases have synchronic contours which are overall ascending. Furthermore, the piece's synchronic contour is <1 3 0 2>, another instance of overall ascent. At a diachronic level, the highest contour percentage belongs to two overall ascending contours: <1 0 3 2> (28 %) and <1 3 0 2> (31 %). In terms of interpretive meaning, the piece not only serves to represent Noguchi's itinerant nature, but also demonstrates the greater metaphor of the souls' departure from the body.

3.4   *Air* for Solo Flute by Tōru Takemitsu

Contour reduction analysis of *Air* proved to be the least successful. While the use of small-scale, unreduced, surface contours is integral to the identification of musical gestures, nesting relationships or hierarchical relationships to LCPs were not apparent. It was revealed, though, that the piece's overall synchronic contour, LCP <1 0 3 2>, is also the most predominant diachronic contour. However, the main compositional processes of this piece are the combination of small musical gestures with pitch-class set analysis and the use of overlapping modes of limited transposition.[15] The lack of embedded contour relationships demonstrates how contour can be a useful analytical tool for post-tonal analysis, but it must be added to the theorist's analytical toolbox and applied when appropriate (Fig. 7).

---

[15] For the complete analytical discussion of *Air*, see [1].

Fig. 7. First two lines of *Air*.

4   Conclusion

The CRA proved to be an insightful mechanism for analysis, providing useful and previously unseen information about the music discussed herein. In terms of synchronic contour, reduction, repetition, inversion, and retrograde exist just as they would for melodic motives. In the works of Fukushima, contour nesting/embedding was apparent in multiple levels of formal strata.

In terms of diachronic analysis, relationships exist between overall synchronic contour and predominant diachronic contour. The contours which occur most often as a piece unfolds from beginning to end are found in places of synchronic prominence.

Contour reduction analysis offers insight for a performer. Post-tonal music is often difficult to perform with continuity due to the lack of melodic relationships. The parameter of contour is a useful focal point for a performer to bring continuity to a performance.

Next steps would be continued development of open source applications for contour analysis. Further analysis of all post-tonal works for solo flute could reveal large-scale relationships between the pieces discussed in this paper and other works. This extended analytical activity could then be paired with performance practice to help teachers and performers in their understanding of such difficult repertoire.

References

1. Sekula, K.: Utilizing computer programming to analyze post-tonal music: a segmentation and contour analysis of twentieth-century works for solo flute. Ph.D. dissertation. University of Connecticut, Storrs, CT (2014)
2. Morris, R.: New directions in the theory and analysis of musical contour. Music Theor. Spectr. **15**(2), 205–228 (1993)
3. Schultz, R.D.: A diachronic-transformational theory of musical contour relations. Ph.D. dissertation. University of Washington, Seattle, WA (2009)
4. Schüler, N.: Methods of computer-assisted analysis: history, classification, and evaluation. Ph.D. dissertation. Michigan State University, East Lansing, MI (2000)

5. Cope, D.: Hidden Structure: Music Analysis Using Computers. AR Editions, Madison (2008)
6. Sampaio, M.S., Kröger, P.: Goiaba: a software to process musical contours. In: Proceedings of the 12th Brazilian Symposium on Computer Music, pp. 203–206 (2009)
7. Friedmann, M.L.: A methodology for the discussion of contour: its application to schoenberg's music. J. Music Theor. **29**(2), 223–248 (1985)
8. Marvin, E.W., Laprade, P.A.: Relating musical contours: extensions of a theory for contour. J. Music Theor. **31**(2), 225–267 (1987)
9. Forte, A.: The Structure of Atonal Music. Yale University Press, New Haven (1973)
10. Young, R.W.: Terminology for logarithmic frequency units. J. Acoust. Soc. Am. **11**(1), 134–139 (1939)
11. Morris, R.: Composition with Pitch-Classes: A Theory of Compositional Design. Yale University Press, New Haven (1987)
12. Schultz, R.D.: Melodic contour and nonretrogradable structure in the birdsong of Olivier Messiaen. Music Theor. Spectr. **30**(1), 89–137 (2008)
13. Lee, C.-L.: Analysis and interpretation of Kazuo Fukushima's solo flute music. Ph.D. dissertation. University of Washington, Seattle, WA (2010)
14. Tenney, J., Polansky, L.: Temporal gestalt perception in music. J. Music Theor. **24**(2), 205–241 (1980)
15. Burt, P.: The Music of Tōru Takemitsu. Cambridge University Press, Cambridge (2006)
16. Graybill, R.: Prolongation, gesture and musical motion. In: Raphael, A., Michael, C. (eds.) Eleven Essays in Honor of Davis Lewin, Musical Transformation and Musical Intuition, pp. 199–224. Ovenbird Press, Dedham (1994)

A Statistical Approach to the Global Structure of John Cage's Number Piece $Five^5$

Alexandre Popoff[(✉)]

Paris Dauphine University, 119 Rue de Montreuil, 75011 Paris, France
al.popoff@free.fr

Abstract. The Number Pieces are a body of late works composed by John Cage using particular temporal structures called time-brackets. In a Number Piece with multiple performers, the superposition of the various parts, each containing time-brackets, creates an everchanging polyphonic work whose structure cannot be determined in advance. We provide here a statistical study of the global structure of the Number Piece $Five^5$ in terms of probabilities of occurence of the various possible set classes, completed by the use of various measures from information theory.

Keywords: John Cage · Number Piece · Time-bracket · Statistics · Information theory · Global structure

1 Introduction

From 1987 to 1992, John Cage wrote a series of scores named the "Number Pieces". In almost all the Number Pieces, a particular time-structure, called "time-bracket", is used for determining the temporal location of sounds. A typical time-bracket is shown in Fig. 1.

A time-bracket is performed as follows: the performer decides to start playing the written sounds anywhen inside the first time interval on the left, and chooses to end them anywhen inside the second one. In the case of Number Pieces written for multiple performers, the superposition of different voices each performing time-brackets creates a polyphonic landscape in constant evolution. The description of this landscape is complex, owing to the freedom given to the performer over the starting and ending times, along with the possible overlaps between the time intervals of different time-brackets. Previous authors [1–3] have pointed out the difficulty of analyzing the Number Pieces. In his dissertation, Haskins notices that "...the brackets offer a flexibility that creates many possibilities", and taking the specific example of $Five^2$, notes that "(...) coping with the myriad possibilities of pitch combinations - partially ordered subsets - within each time-bracket of Five² remains an important issue" ([3], p. 207).

Based on previous works [4,5], we provide here a statistical approach of the global structure of the Number Piece $Five^5$, a short piece with a reduced number of players which permits a convenient computer implementation.

© Springer International Publishing Switzerland 2015
T. Collins et al. (Eds.): MCM 2015, LNAI 9110, pp. 231–236, 2015.
DOI: 10.1007/978-3-319-20603-5_24

Fig. 1. A time-bracket containing a single sound.

2   Methodology

The Number Piece $Five^5$ is scored for flute, two clarinets, bass clarinet, and percussion for a total duration of five minutes [6]. All the time-brackets contain only a single sound. Given the reduced pitch content per bracket, the analysis focuses on set-classes, which can provide a concise description of the harmonic content and sonorities of the piece [7]. We acknowledge however that the analysis of Number Pieces whose time-brackets contain multiple pitches, such as *Four* or *Five*, would certainly benefit from a more detailed analysis at the pitch-class set level. The percussion part does not specify any particular instruments: in this analysis of set classes, we will therefore discard it. We denote by S the set of all 49 set classes which can be heard from the four remaining parts, including silence, single tones and dyads. The temporal structure of $Five^5$ is given on Fig. 2(b) using the graphical representation of time-brackets presented in Fig. 2(a).

For the needs of the computer implementation, time is discretized using a time step of 0.1 s. This may be considered at odds with Cage's notion of time as a continuous and progressive phenomenon, which has been discussed in [1] and [4]. Nevertheless, our results remain similar with smaller time steps, used to approximate continuous time. In addition, such a discretization is a reasonable choice for analysis since performers would probably rely on a stopwatch or a clockwatch (whose use has been cited by Cage [8]), which usually have a resolution of 0.01 s.

In order to study a Number Piece as a whole from a statistical point of view, we consider the starting times and ending times for each time-bracket as random variables with chosen probability distributions. This turns the Number Piece into a stochastic process and allows us to define, for each time t, a random variable X_t taking its values in S, which represents the possible set classes occuring at time t during a realization. The discrete probability distributions $P(X_t = i)$ are estimated by a frequentist approach, using a computer program which generates a large number (typically 10^5–10^6) of realizations of $Five^5$. For each realization, the parts are independently generated, and for a given part, time-brackets are processed successively in the order given on the score. The selection of starting and ending times follows a similar procedure to the one exposed in [4].

Given the knowledge of pitch classes in each part at each time t, we determine the corresponding set class using Daniel Starr's algorithm [9,10]. The set classes are given using Forte's notation [11], which we extend to take into account silence (notated by 0–1), single sounds (1–1) and dyads (2–1 to 2–6). Note that the algorithm does not differentiate a pitch class set from its inverted form.

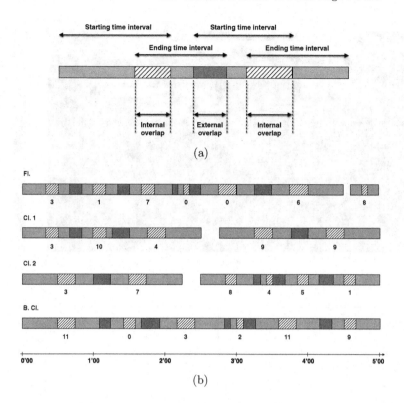

Fig. 2. (a) Graphical representation of two consecutive time-brackets. The internal overlaps are represented by the striped interval, whereas the external overlap is represented by the darker interval. (b) Temporal structure of the flute, clarinet and bass clarinet parts in $Five^5$, using the representation of (a). The pitch class of each time bracket is given under the internal overlap, using the usual circle of semitone encoding.

Inspired by the work of Plumbley [12] and Crutchfield [13] for the study of time-series, useful information measures can be derived from the random variables X_t. The *instantaneous entropy* (in bits) is the entropy of the random variable X_t, given by

$$H(X_t) = -\sum_{i \in S} P(X_t = i) \log_2 P(X_t = i).\qquad(1)$$

Since the random variables X_t are clearly not independent, we also study the *partial conditional entropy* $H(X_{t+\tau}|X_t = i)$, and the *conditional entropy* $H(X_{t+\tau}|X_t)$, as well as the *mutual entropy* $I(X_{t+\tau}, X_t) = H(X_{t+\tau}) - H(X_{t+\tau}|X_t)$, which represents the amount of information shared between the present and the future.

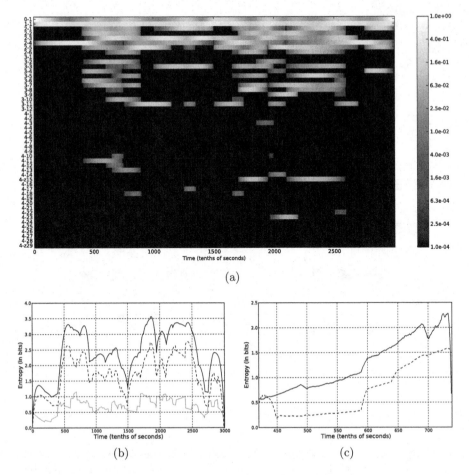

(a)

(b)   (c)

Fig. 3. (a) Heatmap of the probabilities $P(X_t = i)$ over the 49 possible set classes at each time t in $Five^5$. 3(b) Plots of the entropy measures $H(X_t)$ (in black), $H(X_{t+\tau}|X_t)$ (in gray), and $I(X_{t+\tau}, X_t)$ (dashed line) for $\tau = 1\,\text{s}$, with respect to time t. 3(c) Plots of the partial conditional entropy $H(X_{t+\tau}|X_t = \text{'3–4'})$ (in black), and $H(X_{t+\tau}|X_t = \text{'3–7'})$ (dashed line) for $\tau = 1\,\text{s}$, between 42.0 and 75.0 s.

3   Analysis of $Five^5$

The probabilities $P(X_t = i)$ are presented in a heatmap plot on Fig. 3(a). We also present the plot of entropy measures $H(X_t)$, $H(X_{t+\tau}|X_t)$, and $I(X_{t+\tau}, X_t)$ as a function of the time t, for a delay τ of one second, on Fig. 3(b).

It can be noted that the sonic possibilities vary greatly throughout the duration of the piece. We can roughly segment this Number Piece in five sections. This observation is readily captured by the evolution of the instantaneous entropy $H(X_t)$, which is a good indicator of the complexity of this Number Piece throughout time. From $t = 0$ to $t = 40\,\text{s}$, the only possible set classes are silence, single sounds and the dyad 2–4. At $t = 40$ s, the second time-bracket of the flute and first clarinet parts are accessible, leading to a second section which extends to $t = 90$ s, in which 9 out of the 12 possible triads, as well as tetrachords (in particular 4–11) may be heard. From $t = 90$ s to $t = 165$ s, the third section is characterized by the reduced sonic possibilities it offers, and by the central role of set class 3–3 at $t = 125$ s. This corresponds to the beginning of a gap in the clarinet parts wherein the successive time-brackets do not overlap. At $t = 165$ s, new time-brackets are accessible to the clarinets which mark the beginning of a fourth section extending approximately to $t = 265$ s, and which is characterized by the greater variety of possible set classes. New tetrachords appear, with 4z–15 and 4–23 as their main representatives. The last section extends from $t = 265$ s to the end and has a reduced content, with only one possible triad, since the first clarinet and the bass clarinet play the same pitch class. Overall, this Number Piece shows a structure of the form A-B-C-B'-A' which is not readily apparent from the score.

The set class heard at a time t is not necessarily subject to random changes for every subsequent times (a case which would happen if the random variables X_t were truly independent), hence the use of the information measures $H(X_{t+\tau}|X_t)$ and $I(X_{t+\tau}, X_t)$. As can be seen from Fig. 3(b), the values taken by the conditional entropy $H(X_{t+\tau}|X_t)$ are small and vary little. As a direct consequence, the plot of the mutual information $I(X_{t+\tau}, X_t)$ indicates that the amount of information shared between t and $t + \tau$ is high.

We can also observe that for two different set classes i and j having comparable probabilities $P(X_t = i)$ and $P(X_t = j)$ in a certain period of time, the evolution of $H(X_{t+\tau}|X_t = i)$ and $H(X_{t+\tau}|X_t = j)$ may actually be very different. We provide here a particular example by considering the set classes 3–4 and 3–7 between $t = 42$ and $t = 75$ s as shown on Fig. 3(c). In the case of set class 3–4, the evolution of $H(X_{t+\tau}|X_t = \text{'3–4'})$ is almost monotonous throughout the time interval [42, 75]. Its values are high and reach a maximum of 2.3 bits. From the structure of the score given in Fig. 2, it can be seen that the set class 3–4 can only be generated by the set $\{3, 10, 11\}$ of pitch classes. This set is rather unstable: in the time period considered, the flute may begin playing the pitch class 1, the first clarinet may end playing the pitch class 10, and the second clarinet as well as the bass clarinet are about to finish playing. On the other hand, the partial conditional entropy $H(X_{t+\tau}|X_t = \text{'3–7'})$ is almost constant with low values in the interval [45, 59], before increasing monotonously. Its values are lower than those of $H(X_{t+\tau}|X_t = \text{'3–4'})$, with a maximum of 1.55 bits. This is explained by the fact that the set class 3–7 has a more predictable evolution. Consider for example the set $\{1, 3, 10\}$ in the time interval [45, 59]. Since the pitch class 11 is not heard, the bass clarinet is therefore silent during this entire time interval. The pitch classes 1 and 10 have to be sustained since their ending time interval has not been reached yet. The second clarinet is about to finish playing the pitch class 3. Therefore the set $\{1, 3, 10\}$ has only two possibilities of evolution: the same set if the pitch class 3 is sustained, or the set $\{1, 10\}$ if it stops. We thus see that the partial conditional entropies provide useful information about the stability and the evolution of each set class in time.

4   Conclusions

The statistical approach presented in this paper allows one to analyze at once the distribution of the possible set classes during a performance of a Number Piece. This approach may be extended to any other Number Piece, in particular those with a higher pitch content per bracket, such as *Four*, *Seven*², *Thirteen*, etc. This analysis however relies on a number of assumptions, for example the random selection procedure for each time-bracket, or the independency of the different parts. The possible shortcomings of this approach have been discussed in [5], among which the fact that it may not represent human behavior accurately. The modelling of the Number Pieces would thus benefit from a more thorough investigation of actual musicians' behavior. The application of a similar approach to other works of Cage and the investigation of the potential value of information theory measures in such analyses may also be envisioned for future research.

References

1. Weisser, B.J.: Notational Practice in Contemporary Music: A Critique of Three Compositional Models (Luciano Berio, John Cage and Brian Ferneyrough), Ph.D. dissertation, City University of New York, pp. 82–83 (1998)
2. Weisser, B.: John Cage: '.. the whole paper would potentially be sound': time-brackets and the number pieces (1981–92). Perspect. New Music **41**(2), 176–225 (2003)
3. Haskins, R.: An Anarchic Society of Sounds: The Number Pieces of John Cage, Ph.D. dissertation, University of Rochester, New York, p. 245 (2004)
4. Popoff, A.: John Cage's number pieces: the meta-structure of time-brackets and the notion of time. Perspect. New Music **48**(1), 65–84 (2010)
5. Popoff, A.: Indeterminate music and probability spaces: the case of John Cage's number pieces. In: Agon, C., Andreatta, M., Assayag, G., Amiot, E., Bresson, J., Mandereau, J. (eds.) MCM 2011. LNCS (LNAI), vol. 6726, pp. 220–229. Springer, Heidelberg (2011)
6. Cage, J.: Five⁵, EP 67431. C.F. Peters, New York (1991)
7. Martorell, A., Gómez, E.: Hierarchical multi-scale set-class analysis. J. Math. Music **9**(1), 95–108 (2015)
8. Cage, J.: Je n'ai jamais écouté aucun son sans l'aimer; le seul problème avec les sons, c'est la musique. La Souterraine: La Main Courante, France (2002)
9. Starr, D.: Sets, invariance and partitions. J. Music Theor. **22**(1), 1–42 (1978)
10. Brinkman, A.R.: Pascal Programming for Music Research, p. 629. University of Chicago Press, Chicago (1990)
11. Forte, A.: The Structure of Atonal Music. Yale University Press, New Haven and London (1973)
12. Abdallah, S.A., Plumbley, M.D.: Anatomy of a bit: information in a time series observation. Chaos **21**, 037109 (2011)
13. James, R.G., Ellison, C.J., Crutchfield, J.P.: Information dynamics: patterns of expectation and surprise in the perception of music. Connection Sci. **21**(2-3), 89–117 (2009)

Exact Cover Problem in Milton Babbitt's All-Partition Array

Brian Bemman and David Meredith[✉]

Department of Architecture, Design and Media Technology, Aalborg University,
Rendsburggade 14, Aalborg, Denmark
{bb,dave}@create.aau.dk
http://www.create.aau.dk

Abstract. One aspect of analyzing Milton Babbitt's (1916–2011) all-partition arrays requires finding a sequence of distinct, non-overlapping aggregate regions that completely and exactly covers an irregular matrix of pitch class integers. This is an example of the so-called *exact cover* problem. Given a set, A, and a collection of distinct subsets of this set, S, then a subset of S is an exact cover of A if it exhaustively and exclusively partitions A. We provide a backtracking algorithm for solving this problem in an all-partition array and compare the output of this algorithm with an analysis produced manually.

Keywords: Babbitt · Knuth · All-partition array · Exact cover · Computational music analysis

1 Introduction

The *exact cover problem* is a constraint satisfaction problem known to be NP-complete [5, p. 2]. It is defined as follows: given a collection of subsets, S, of a set, A, an *exact cover* of A is a sub-collection, s, of S that exhaustively and exclusively partitions A. The classic example of such a problem was provided by Scott and Trotter in 1958 [7]. They found all ways to cover a chessboard with the 12 distinct, non-overlapping pentaminoes while leaving the center four squares uncovered (see Fig. 1).

Following our definition above, the chessboard in Fig. 1 would be A, the collection of 63 distinct pentaminoes would be S, and the 12 distinct pentaminoes selected to cover the chessboard would be s.

The exact cover problem is typically solved using a greedy backtracking algorithm that performs a depth-first search of the solution space [5, p. 2]. The backtracking process finds a complete solution to a problem by accumulating partial solutions to a set of constraints. It selects the first of these partial solutions until a complete solution is found, or, in the event that the constraints are no longer satisfied by the currently selected partial solution, it returns to the previous point and selects the next partial solution. It continues this process until either a solution is found or it fails.

© Springer International Publishing Switzerland 2015
T. Collins et al. (Eds.): MCM 2015, LNAI 9110, pp. 237–242, 2015.
DOI: 10.1007/978-3-319-20603-5_25

Fig. 1. An exact covering of part of a chessboard using 12 pentaminoes while leaving the center four squares uncovered. As taken from [5, p. 2].

```
11  4  3  5  9 10 10  1  1  8  2  0  7  6  5  5  4  4 11  9  3 10  1  2  6  8  7  0  9 10  3  5 11  4  4  1  0  0  8
 6  7  7  0  2  2  8  1 10  9  5  5  3  4  4 11 11  0  7  8  8  6  2  1 10  3  9 11  4  5  8  1  0  2  6  7  7 10  5
 5  6 11  1  7  0  9  9  8  4  2  2  3 10  1  1  0  7  5 11  6  9 10  2  4  3  8  5  0  0  1 11  7  6  3  8  8  2  4
 2  9 10  8  4  3  3  3  0  5 11  1  6  7  7  4  9  8 10  2  3  6  1  7  5  0 11  2  3  8 10  4  9  6  5  1 11 11  0
 0  0  5  4  6  6  6 10 11  2  9  3  1  8  7  4  5 10  0  6 11  8  7  3  1  2  9  4 11  0 10  6  5  2  7  1  3  3  8
 1  8  8  9  7  3  2 11 11  4 10 10  0  5  6  3  2  9  9  7  1  8 11  0  4  6  5 10  9  2  1  3  7  8 11  6  0 10 10
```

Fig. 2. The beginning of the projection of Babbitt's *Sheer Pluck*.

2   All-Partition Array as Exact Cover

A significant number of Milton Babbitt's (1916–2011) works are based on the *all-partition array* [6], which is a sequence of distinct, non-overlapping aggregate-forming regions that completely and exactly partition a matrix of pitch classes called a *projection*. Construction of a *six-part* all-partition array results in an irregular projection of six rows and 696 pitch classes that can be partitioned into 58 aggregate regions. Figure 2 shows the beginning of the projection of Babbitt's *Sheer Pluck*. A projection is not the musical surface but, rather, a framework upon which the surface is based. Figure 3 illustrates the process of defining the first three aggregate regions in this projection.

Note in Fig. 3(b) that the first region contains an aggregate represented as a collection of row segments (from top to bottom) of length $3, 2, 1, 3, 1$, and 2. We define an *integer partition*, denoted by $\text{IntPart}(s_1, s_2, ..., s_k)$, to be a representation of an integer $n = \sum_{i=1}^{k} s_i$, as an *unordered* sum of k positive integers. For example, if $n = 12$ and $k = 6$, then one possible integer partition is $\text{IntPart}(3, 3, 2, 2, 1, 1)$. We define an *integer composition*, denoted by $\text{IntComp}(s_1, s_2, ..., s_k)$, to be a representation of an integer $n = \sum_{i=1}^{k} s_i$, as

```
11 4 3 5 9 10 10 1        11 4 3 5 9 10 10 1        11 4 3 5 9 10 10 1        11 4 3 5 9 10 10 1
6 7 7 0 2                 6 7 7 0 2                 6 7 7 0 2                 6 7 7 0 2
5 6 11 1                  5 6 11 1                  5 6 11 1                  5 6 11 1
2 9 10 8 4 3              2 9 10 8 4 3              2 9 10 8 4 3              2 9 10 8 4 3
0 0 5 4 6                 0 0 5 4 6                 0 0 5 4 6                 0 0 5 4 6
1 8 8 9 7 3 2 11          1 8 8 9 7 3 2 11          1 8 8 9 7 3 2 11          1 8 8 9 7 3 2 11
```

(a) Area needing to be covered.

(b) 1st aggregate region.

(c) 2nd aggregate region (dashed lines).

(d) 3rd aggregate region (dashed lines). Area covered completely.

Fig. 3. Process of forming a sequence of three aggregate regions in an excerpt from the projection shown in Fig. 2.

an *ordered* sum of k positive integers. For example, if $n = 12$ and $k = 6$, then IntComp(3, 3, 2, 2, 1, 1) $\neq$ IntComp(3, 2, 1, 3, 2, 1). We define a *weak integer composition*, WIntComp($s_1, s_2, ..., s_k$), to be a representation of an integer, $n = \sum_{i=1}^{k} s_i$, as an ordered sum of k *non-negative integers*. For example, if $n = 12$ and $k = 6$, then WIntComp(6, 6, 0, 0, 0, 0) is a weak integer composition. Thus, the first aggregate region in *Sheer Pluck*, shown in Fig. 3(b), represents the integer partition, IntPart(3, 3, 2, 2, 1, 1), and the integer composition, IntComp(3, 2, 1, 3, 1, 2). We further define two relations, *partitionally equivalent* and *partitionally distinct*. Two integer compositions, c and d, are *partitionally equivalent* if and only if $[c] = [d]$, where $[c]$ and $[d]$ denote the partitions containing the compositions. Two integer compositions, c and d, are *partitionally distinct* if and only if $[c] \neq [d]$.

Using our definitions above, the problem we pose with respect to the all-partition array as an exact cover asks, "Given a universe of integer compositions (when $n = 12$ and $k = 6$), denoted by S, and a projection of 696 pitch classes in six rows, denoted by A, does there exist a sequence of 58 partitionally distinct, and aggregate-forming integer compositions, s, that exactly covers A?" We call this the *projection cover problem*. Our efforts to answer this question continue work started by Bazelow and Brickle [2, pp. 282–283], that asked a similar question of all-partition arrays in four parts. However, where their research sought to construct a projection, this paper begins with a completed projection and, as a method for musical analysis, seeks to efficiently reveal its all-partition array structure by discovering how (or if) it can be partitioned.

3   Solving the Projection Cover Problem

Our proposed solution to the projection cover problem posed above, is the backtracking algorithm, BACKTRACKINGBABBITT, shown in Fig. 4. This algorithm takes a projection as input and returns a list of 58 partitionally distinct compositions chosen as partial solutions.

The algorithm begins in line 1 by computing a $6, 188 \times 6$ list of compositions (in six parts), denoted by **compositions**. Lines 2–4 initialize **cList**[cnt] to be an empty list of 58 lists, **position** to be a 1×6 vector of indices (one for each row in **projection**), and cnt to be 1 (using 1-based indexing). Line 5 begins a **while** loop where cnt is less than or equal to the number of required compositions, 58. First, it checks to see whether **cList**[cnt] has been computed (line 6). **cList** contains candidate compositions at each cnt. Candidate compositions are those compositions that satisfy the constraints of a partial solution (i.e., are partitionally distinct and form a region containing an aggregate).

If **cList**[cnt] is empty (line 6), it has not been previously computed and so it calls PARSEPROJECTION, which returns **cList** and **currentComp** (line 7). **currentComp** is initialized by PARSEPROJECTION to be the first composition in **cList**[cnt] if **cList**[cnt] is not returned empty. If, after PARSEPROJECTION, **cList**[cnt] remains empty, there are no candidate compositions at this cnt. It must then backtrack, removing the previous composition from **partialSolutions**

```
BACKTRACKINGBABBITT(projection)
1    compositions ← COMPUTECOMPOSITIONS(12, 6)
2    cList ← ⊕⁵⁸ᵢ₌₁⟨⟨⟩⟩
3    position ← ⊕⁶ᵢ₌₁⟨1⟩
4    cnt ← 1        ▶ Index into cList
5    while cnt <= 58 and cnt > 0       ▶ Number of compositions
6      if |cList[cnt]| == ⟨⟩
7        PARSEPROJECTION(projection, compositions, partialSolutions, cnt, position)
               ▶ Returns cList and currentComp
8        if |cList[cnt]| == ⟨⟩       ▶ Backtrack
9          cnt ← cnt − 1
10         position ← position − currentComp
11         partialSolutions[cnt] ← ⟨⟩
12       else       ▶ Success
13         partialSolutions[cnt] ← currentComp
14         position ← position + currentComp
15         cnt ← cnt + 1
16     else
17       currentComp ← cList[cnt][currentComp.index + 1]       ▶ Select next composition
18       if cList[cnt][currentComp.index] > |cList[cnt]|       ▶ Backtrack
19         cnt ← cnt − 1
20         position ← position − currentComp
21         partialSolutions[cnt] ← ⟨⟩
22       else       ▶ Success
23         partialSolutions[cnt] ← currentComp
24         position ← position + currentComp
25         cnt ← cnt + 1
26   return partialSolutions
```

Fig. 4. Pseudocode for implementation of BACKTRACKINGBABBITT.

(lines 8–11). **partialSolutions** is a 58×6 list of candidate compositions at each cnt selected by the algorithm to be a partial solution. If **cList**$[cnt]$ is not empty (line 12), then the algorithm has found at least one candidate composition at this cnt. The **currentComp** is stored in **partialSolutions**$[cnt]$ and both **position** and cnt are incremented (lines 13–15). **position** is equivalent to counting from 1 a distance equal to the summation of like parts from each composition in **partialSolutions** from 1 to $cnt - 1$. **position** is incremented at each cnt by **currentComp**. In Fig. 3(d), **partialSolutions** currently holds $\langle 3, 2, 1, 3, 1, 2\rangle$ and $\langle 3, 3, 3, 3, 0, 0\rangle$ and so **position** would be $\langle 7, 6, 5, 7, 2, 3\rangle$.

If the first check for whether **cList** is empty (line 6) returns false, **cList**$[cnt]$ has already been computed. This means the algorithm has backtracked at some point (line 16). It then attempts to select the next composition in **cList**$[cnt]$ by incrementing the index of **currentComp** (line 17). If there is not a next composition here because this index exceeds the size of **cList**$[cnt]$ (line 18), it must backtrack (lines 19–21). However, if there is another composition, it can proceed (lines 23–25). It continues this until it returns a complete **partialSolutions** or fails (line 26).

Figure 5 shows pseudocode for the PARSEPROJECTION function called in line 7 of BACKTRACKINGBABBITT. PARSEPROJECTION begins by creating a copy of **compositions** called **comps** (line 1). Next, it removes from **comps** compositions partitionally equivalent to those already selected as partial solutions (line 2). It then loops through the rows and columns of **comps** (lines 4–6) and initializes $aggregate$ to be an empty set (line 5). Next, it finds jth row segments in **projection** parsed by **comps**$[i][j]$ using the distance measured from **position**$[j]$ to the sum of **position**$[j]$ and **comps**$[i][j] - 1$. It stores these

```
PARSEPROJECTION(projection, compositions, partialSolutions, cList, cnt, position)
1     comps ← compositions
2     comps ← REMOVEUSEDCOMPOSITIONS(partialSolutions, comps)
3     k ← 1
4     for i ← 1 to |comps|        ▶ By row.
5         aggregate ← ∅
6         for j ← 1 to |comps|      ▶ By column.
7             if comps[i][j] ≠ 0
8                 aggregate ← projection[j][position[j]..position[j] + comps[i][j] − 1]
                      ▶ Add jth row segments to form region
9         aggregate ← REMOVEDUPLICATES(aggregate)
10        if |aggregate| == 12
11            cList[cnt][k] ← comps[i][1..6]
12            k ← k + 1
13    currentComp ← cList[cnt][1]
14    return ⟨cList, currentComp⟩
```

Fig. 5. The PARSEPROJECTION function.

segments in *aggregate* (lines 7–8). After removing any duplicate integers from *aggregate* (line 10), if *aggregate* is complete, it has found a candidate composition and saves this composition in **cList**[*cnt*][*k*] (lines 10–12). The algorithm then assigns **currentComp** to be the first composition in **cList**[*cnt*] and returns **cList** and **currentComp** (lines 13–14).

We conclude this section by providing the results of analyzing one of Babbitt's projections from both BACKTRACKINGBABBITT and those found by a human analyst. Figure 6(a) first shows the sequence of compositions found by a human analyst to partition the projection shown in Fig. 2. For comparison, Fig. 6(b) shows one of several sequences returned by BACKTRACKINGBABBITT.

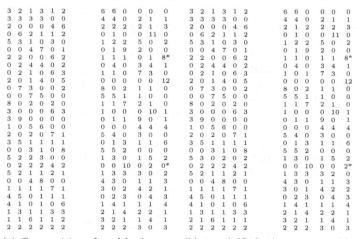

(a) Compositions found by human analyst.

(b) **partialSolutions** returned by BACKTRACKINGBABBITT.

Fig. 6. Distinct sequences of compositions that partition the projection shown in Fig. 2 as found by a human analyst in (a) and returned by BACKTRACKINGBABBITT in (b). Note asterisks (*) indicate where the sequences differ.

4   Conclusion

In this paper we suggest that analyzing Milton Babbitt's all-partition arrays represents a special case of a constraint satisfaction problem called an exact cover. We provide a backtracking algorithm called BACKTRACKINGBABBITT as a solution to this problem. This algorithm finds a sequence of 58 partitionally distinct and aggregate-forming integer compositions that exactly covers a given projection of 696 pitch class integers. We believe this algorithm is not only a more efficient way (when compared to a human analyst) to perform this analytical task for a work based on the all-partition array, but that it can be used to discover alternative analyses to those offered previously by theorists.

Acknowledgments. The work reported in this paper was carried out as part of the EC-funded collaborative project, "Learning to Create" (Lrn2Cre8). The Lrn2Cre8 project acknowledges the financial support of the Future and Emerging Technologies (FET) programme within the Seventh Framework Programme for Research of the European Commission, under FET grant number 610859.

References

1. Babbitt, M.: Set structure as a compositional determinant. J. Music Theor. **5**, 72–94 (1987)
2. Bazelow, A.R., Brickle, F.: A partition problem posed by Milton Babbitt. Perspect. New Music **14**(2), 280–293 (1976)
3. Bemman, B., Meredith, D.: From analysis to surface: generating the surface of Milton Babbitt's Sheer Pluck from a parsimonious encoding of an analysis of its pitch-class structure. In: The Music Encoding Conference, Charlottesville, VA, 20–23 May 2014
4. Eger, S.: Restricted weighted integer compositions and extended binomial coefficients. J. Integer Seq. **16**(13.1.3), 1–25 (1997)
5. Donald, K.: Dancing links. 22 February 2000. http://www-cs-faculty.stanford.edu/uno/musings.html
6. Mead, A.: An Introduction to the Music of Milton Babbitt. Princeton University Press, Princeton (1994)
7. Scott, D.S.: Programming a combinatorial puzzle. Technical report No. 1, Princeton University Department of Electrical Engineering, Princeton, NJ, 10 June 1958

Geometric Approaches

Constructing Geometrical Spaces
from Acoustical Representations

Özgür İzmirli[✉]

Department of Computer Science, Connecticut College, New London, CT, USA
oizm@conncoll.edu

Abstract. This paper presents several models for constructing geometrical spaces from acoustical representations. Through specific spectral representations and associated distance measures each model is designed to highlight or ignore certain types of relationships within the given pitch sets. Dimensionality reduction is employed to obtain low dimensional embeddings from spectral representations. The viability of these models is demonstrated for the resulting low dimensional embeddings with respect to a number of group actions including octave shifts, permutation, transposition and inversion.

Keywords: Geometrical spaces · Chord spaces · Spectral representations · Spectral distance functions

1 Introduction

Much work has been done in music theory to understand how listeners hear vertical and horizontal pitch relationships. A central question for most of this research is to understand how certain pitch combinations are heard in relation to others. Music analysis takes many forms and concerns itself with a wide range of styles, anywhere from traditional music analysis of common practice music to analysis of serial compositions. However, a divide exists between music theory and acoustical representations, making it hard to conduct music analysis directly on sound input. Electroacoustic music aside, most music analysis methods employ symbolic representations of pitches and this divide makes it hard to relate the symbolic and sonic domains. While symbolic modeling of music has received ample attention, the links between acoustical (signal based) representations and symbolic approaches have not been explored extensively. The difficulty in reliable audio-to-score transcription is possibly at the core of this gap. Most of the models that operate on acoustical input employ some form of approximation due to this difficulty. We remain cognizant of the interdisciplinary challenges in this area as outlined in [35] and our purpose in this work is to explore new ways of linking the two worlds that have accumulated sizable knowledge in their respective domains.

Low dimensional geometrical models provide intuitive representations of distance relationships between objects. With a given number of dimensions, a space

© Springer International Publishing Switzerland 2015
T. Collins et al. (Eds.): MCM 2015, LNAI 9110, pp. 245–256, 2015.
DOI: 10.1007/978-3-319-20603-5_26

is the extent in which objects can be positioned. The types of space range from the intuitive three dimensional physical space ($\mathbb{R}^3$) to less intuitive abstract spaces. A metric space, for example, needs to be accompanied by a measure of distance between the objects. Projections onto low dimensional spaces can aid in visualizing structures that may capture interesting relationships between objects. These structures could be in the form of clusters or manifolds which, in turn, enable approximate representations in a reduced number of dimensions. In this work, we are interested in exploring the existence of structures in low dimensional spaces that capture meaningful relationships between measurable quantities in a much higher dimensional feature space. While these structures may not be exact due to being projections onto the low dimensional spaces, they nevertheless aim to preserve the distance relationships between objects and their utility lies in visualization and exploration. In the remainder of the paper we describe distances based on several spectral representations and show that some well-known chord constellations in the literature can emerge using the proposed method. These representations are primarily applicable to the Western twelve-tone equal-tempered system.

The term construction of geometrical spaces refers to the process of choosing spectral representations, defining distances for these representations and visualizing the relationships between objects through a mapping onto low dimensional spaces. Otherwise, we do not formally construct these spaces. In the proposed system, the space construction is data-driven and the bounds of the space of interest will be adaptively scaled to the inputs. Moreover, the space bounds and the projection orientations will be dictated by the contributing objects and the optimization criteria. This enables us to experiment with different acoustic data sets and explore the data at the appropriate scales and localities. The way objects are interrelated affect the structure of how they will be positioned in a space. This is why suitable features and distance measures are important in constructing these spaces. Dimensionality reduction operates directly on data or pairwise distances between observed objects. We are interested in studying the nature of distances and the emergent structures of scales and chords in these spaces using dimensionality reduction.

2   Related Work

Geometrical models of music have received considerable attention throughout history. Dating back to the 1700 s, we are aware of planar geometric models of Heinichen, Kellner, Weber and Euler [23,26]. Originally proposed by Euler, the Tonnetz was revived in the nineteenth century by Oettingen and Riemann. Many models have since been inspired by the Tonnetz and it has been at the core of neo-Riemannian work (e.g. [12]).

Shepard [30] proposed multiple helical structures for modeling musical distances. Chew's three-dimensional model called the Spiral Array [8,9] is based on the Tonnetz and employs ascending pitches displaced by fifths along a helix. Together with accompanying geometric algorithms it combines pitches, chords and keys in a single model. Chuan and Chew have applied this model to audio input in the context of key finding [10].

In his book Tonal Pitch Space, Lerdahl quantified distances in chordal space and regional space separately [23]. Starting from a basic space of interval relationships, he proposed the chordal space for chord distances within a key and regional distance for key distances. Along the same lines, Gatzsche et al. proposed a geometric approach to tonality analysis from audio in which the within key and inter-key relationships were modeled separately [16].

Humphrey et al. performed chord recognition from audio by learning a function that projects audio onto Tonnetz space [19]. Gómez and Bonada described a system for the visualization of tonal music using key correlations on the surface of a torus [17]. Toiviainen's model used short-term memory and a self-organizing map for the visualization of tonal content in the symbolic as well as the audio domain [31].

Purwins et al. [26] give a history of geometric models and point out the agreement between music theory and experimental psychology in the derivation of models that embody circular relations between major and minor chords. In their seminal paper, Krumhansl and Kessler showed a 4-dimensional space housing major and minor chords based on their probe tone experiment findings [21]. The visualization is given in two plots with a separate pair of dimensions in each plot. This is an important result since it connects psychoacoustic research with geometrical models and machine listening.

Izmirli used spectral prototypes constructed from diatonic collections of instrument sounds and demonstrated the emergence of cyclic patterns of the major as well as minor keys [20]. Purwins used constant Q profiles derived from audio in applications such as key finding, style and composer classification [25]. He also demonstrated the emergence of the circle of fifths where the constant Q profiles were learned from recordings of Bach's Well-Tempered Clavier (Book II). More recently, Chacón et al. [7] used a Restricted Boltzmann Machine to learn a circle of fifths topology from musical data.

Tymoczko discussed three types of musical distance [33], namely, those based on voice leading, acoustic (frequency ratios) and total interval content. In the paper he explored the degree of interrelatedness of the seemingly different notions of musical distance and discussed the appropriate conditions under which each could be used. Sethares and Budney discuss metrics on musical data in relation to topology where they recover the circularity of octave-reduced musical scales and the circle of fifths [29]. In his book, Tymoczko showed several types of chord spaces and discussed their utility [34] (also see [32]). Callender, Quinn and Tymoczko [5] showed a number of geometrical spaces under a unified categorization of equivalence relations. Quinn proposed a method based on the Discrete Fourier Transform (DFT) to quantify the intervallic nature of any chord [27]. More recently, using this method, Amiot studied the phase behavior of the complex Fourier coefficients of pc-sets [1].

3   Linking Music Theory and Acoustical Representations

First, we cover some preliminaries that will be useful for defining the connections between the acoustical domain and music theory. While the commonly used spectral representations eagerly encode changes in timbre and pitch shift, in music theory, pitches have largely been regarded as single frequencies. In this section we first review equivalence classes and group actions. We then define the signal model for our spectral representation, the direct spectral distance and dimensionality reduction to achieve a representation in low dimensional space.

3.1   Equivalence Classes, Quotient Spaces and Group Actions

Music listening experience is shaped by similarities and dissimilarities of sound events. These similarities are complex and usually context dependent but a large number of generalizations can be made. From a music theory perspective, group actions or equivalence operations create equivalence classes. Equivalence relations include octave equivalence (O), permutation (P), transposition (T), inversion (I), and cardinality (C). Through use of the symmetries resulting from these operations, Callender, Quinn and Tymoczko [5] proposed a unified geometrical framework for understanding quotient spaces generated by combinations of these equivalence relations. Hall described contextual transformations as groups of affine linear transformations acting on these spaces [18]. We will adopt this terminology and use it in the paper to describe the nature of the spaces being explored.

A chord can be characterized by n single notes where each note is an element of a pitch class. According to this convention, an equivalence relation would connect together all chords with any arrangement of voicings over multiple octaves and with any doublings of notes. In order to do this, information regarding octave, order (voice assignments) and multiplicity need to be ignored hence creating an OPC equivalence relation. On the other hand, instead of looking for a single chord, if we are interested in differentiating chord types, that is, in obtaining classes which contain elements of the same chord type, then, we would need to employ transposition in addition to OPC. In this case, all chords with the same interval relationship between their notes would be sought. In short, an OPTC equivalence relation will define chord type. Additionally, by including inversion equivalence we can apply OPTIC operations to define set classes.

3.2   Signal Model for Synthesis

The multiset of points drawn from a pitch set constitutes a chord, denoted v. We would like to retain the ability to encode octave information so that we can compare the effects of octave equivalence on our output. Multiplicities will be ignored. The time signal for each note is obtained by summing harmonics, $k = 1..K$, weighted by their amplitudes, a_k, with randomized phases, θ_k. This is shown in the inner summation below in (1). Consequently, the composite signal

$x_i(t)$ for a particular chord v_i is constructed by a summation of the constituent notes of that chord (outer summation).

$$x_i(t) = \sum_{n \in v_i} \sum_{k=1}^{K} a_k sin(2\pi k f(n)t + \theta_k) \qquad (1)$$

where $f(n)$ is a function mapping the integer representing the note's MIDI number to its corresponding frequency assuming 12-TET and reference A4 = 440 Hz. In our experiments we have used between 10–15 harmonics (K) and assigned the amplitudes a decaying sequence $a_k = 1/k$. Insofar as the summation over the notes, order is ignored in this convention leading to P equivalence.

3.3   Direct Spectral Distance

To be able to explore the relationship between various pitch structures we need a suitable representation and some distance measure between the features representing the sounds of interest. A first choice for the spectral distance would be any one of the Euclidean, correlation, cosine, city block or Minkowski distances applied to the spectral vectors. However, we would like our distance measure to reflect pitch differences but ideally not be sensitive to attributes such as inharmonicity, tuning and spectral shape. We chose to use a correlational distance between smoothed spectra:

$$D_{i,j} = dist\left(\Gamma_p(|X_i|), \Gamma_p(|X_j|) \right) \qquad (2)$$

where X_i is the (Hann) windowed Discrete Fourier Transform of the sampled input signal for chord v_i. $|.|$ is its magnitude, Γ_p represents convolution of its argument vector and a Gaussian window of width p and $dist$ is the correlation or cosine distance between the two smoothed spectra computed over a range of DFT bins typically chosen in the frequency range 55–3000 Hz.

3.4   Visualization and Dimensionality Reduction

As is the case for most dimensionality reduction methods, Multidimensional Scaling (MDS) [22] aims to find a mapping from a high dimensional representation to a low dimensional one such that the between-element distances are preserved as much as possible. We use Sammon's nonlinear mapping as the goodness-of-fit criterion [28]. The target dimensionality is an external parameter specified by the user. When the dimensionality of the data is unknown it can be used as a tool to better understand the underlying dimensions of the original data. Dimensionality reduction methods come in many flavors and most non-linear methods aim to reveal underlying manifolds in the data. The current study is motivated by the low dimensional nature of the geometrical spaces proposed by researchers in music theory and sets out to explore how to computationally obtain similar geometrical spaces from acoustical representations. For visualization purposes, we are interested in the relative distances between our objects and acknowledge that our spectral distances may not always align with music theoretical and psychoacoustical distances. We chose MDS after trying a number of dimensionality reduction methods and finding that MDS was the most robust for our purpose.

3.5 Diatonic Sets and Associated Spaces

We first apply the spectral distance defined above directly to the acoustical representation of keys and show the emergence of the circle of fifths. The circle of fifths represents an organization of keys on a circle in which each key has one differing accidental from each of its neighbors. In this sense it forms a cyclic group [2]. For this, we first choose scales and construct chords on 12 chromatic degrees for each scale type. We then calculate the spectra and pairwise spectral distances, $D_{i,j}$, to obtain the matrix $\mathbf{D}$. Finally, we perform MDS on $\mathbf{D}$. Each key is represented by a single vector composed of the notes that belong to a diatonic scale. All notes (fundamental frequencies) are confined to a single octave leaving us in OP space. Figure 1(a) shows the result in 2D space for 12 major scales. A point represents the coordinates of a key and spline curves are drawn between neighboring keys for visualization and diagnostic purposes. Part (b) shows the same keys constructed with real piano sounds (from Univ. of Iowa Musical Instrument Samples). Next we demonstrate the circle of major and minor triads totaling 24 keys by adding 12 harmonic minor keys to the major keys of part (a). We used the harmonic minor instead of the natural minor to avoid overlap with the relative major. Part (c) shows the projection in 2D. The output reveals relationships similar to those depicted in Heinichen's Circle and Krumhansl and Kessler's MDS plot of the first two dimensions [21]. The MDS solution for the 2D case is a best-effort projection and, it turns out, that two dimensions is not optimal. By adding a third dimension we might observe what is missing in the 2D plot. Part (d) shows the view from the side of the 3D plot. Note that the rotations in the outputs are arbitrary as would be expected. It should also be noted that the robustness of the circle of fifths from major scales is not surprising in light of well-formedness scale theory [6]. For comparison, the stress values are 0.006, 0.011, 0.06 and 0.027 for parts (a) through (d) respectively.

4   Octave Equivalence: Folding the Spectrum

The direct spectral distance works reasonably well when the notes are in the same octave. However, we can suspect that it exhibits dependence on the octave placement of the notes and does not implement octave equivalence very well. We would like to test this first and compare its performance to another spectral representation that makes claims of octave independence. Originally proposed for chord recognition [15] and later popularized with an application in audio thumbnailing [3], the chromagram has become a very common representation in music audio content processing and especially in chord recognition, key finding and audio alignment. Its computation is straightforward and entails accumulating the spectra for pitch-class regions over the frequency range of interest.

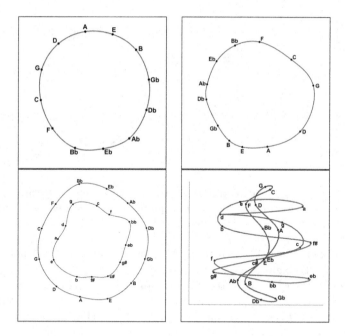

Fig. 1. Circle of Fifths: (a) synthetic, major (top left); (b) piano, major (top right); (c) synthetic, major+harmonic minor (lower left); (d) side view of (c) in 3D (lower right).

A chromagram with semitone resolution will have 12 bins per chroma vector. The chroma spectral distance is then simply calculated as $D_{i,j} = dist(C_i, C_j)$ where C_i is the chroma vector corresponding to chord vector v_i. Now we would like to compare direct and chroma spectral distances by moving the input pitch collections from OP space to P space for the circle of fifths example. This time the seven notes for the major scale are not constrained to a single octave but are randomly picked from three octaves (C4-B6) while maintaining the chroma collection of the diatonic set. The difference in the output between the two distances can be seen in Fig. 2. Both parts use the same sound input and show the superimposed output for 10 different runs. The left plot shows the MDS output for the direct spectral distance (avg. stress: 0.044) and the right one for the chroma spectral distance (avg. stress: 0.011). As expected, the chroma distance is less sensitive to octave information in this setting hence it approximates an OP output space for a P input space.

5   Chord Space: Hexachords and Cube Dance

The Tonnetz has been a central example for researchers working in geometrical music theory and neo-Riemannian theory. The note based version of the Tonnetz assigns notes to nodes in the mesh whereas its dual, the chord based version, uses chords connected with lines that represent one-step voice leading. The conversion

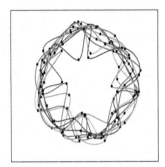

Fig. 2. Circle of fifths using (a) direct spectral distance (left); (b) chroma spectral distance (right). For each part 10 runs are superimposed with each cycle (color) representing a different run (Color figure online).

can easily be performed by identifying the chord from the three notes on the vertices of each triangle and placing that chord at the center of the triangle. The chords can then be connected to their immediate neighbors each of which will differ by one note. A particular chord based version has been named the chicken-wire torus by Douthett and Steinbach [13]. In this Tonnetz, hexachords divide the structure into four regions through LP (leading-tone exchange and parallel) transformations of the 3–11 set class (for naming in pitch-class set theory see [14]). The hexatonic collections constitute maximally smooth cycles in terms of voice leading and were shown by Cohn in his paper describing the Hyper Hexatonic System [11]. One such hexachord collection is (C, E♭, E, G, A♭, B) leading to the chord sequence (C, c, A♭, g♯, E, e). Figure 3(a) shows the chord space with chroma distance for which only these chords were input. The hexatonic cycle can be clearly seen. For this example the input chords were kept in OP space. Inputting all four hexachords results in a visual hint of the torus (although we do not draw all lines to show full toroidal connectivity – only the hexatonic cycles are shown). Part (b) shows the 3D output for the 24 chords. Douthett and Steinbach's Cube Dance structure [13] contains four additional augmented chords and is taken from their paper: part (c). The output of the MDS with the inclusion of four additional chords can be seen in (d) (stress: 0.08). As before, the rotations are arbitrary but since the structures are symmetric the resemblance can be clearly observed.

6   Fourier Balances and Intervallic Characterization of Chords

Following David Levin's 'Fourier Properties' for comparing chord quality, Quinn observed, through a concept he calls 'Fourier Balances', that the quantification of chord quality in this sense was equivalent to applying the DFT to the chord's coordinates in harmonic chord space [27]. This method essentially looks at how well the intervallic content matches even set-class prototypes. Quinn's method can also be viewed as a means of fuzzifying the interval membership if the chord

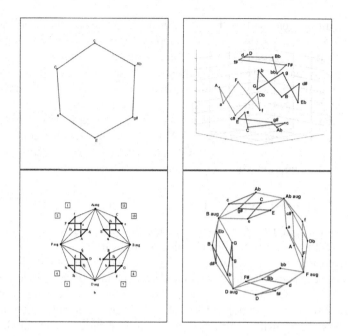

Fig. 3. (a) Single hexachord cycle (top left); (b) four hexachords in 3D (top right); (c) Douthett and Steinbach's Cube Dance – original (lower left); (d) output for same chords (lower right).

is placed in continuous space. According to Quinn's method, given a pitch-class set P for a particular chord, the harmonic character of the chord is calculated by finding the magnitude of the DFT less the zero frequency (DC) component:

$$V(k) = \left| \sum_{p \in P} e^{-2\pi jpk/12} \right|, \quad k = 1..6 \tag{3}$$

Hence producing a 6 element vector. This formula not only works for pitch-class vectors but for any pitch set as demonstrated by Callender [4] who also generalized the model for continuous harmonic spaces.

6.1   Extended Fourier Balances

The original method uses pitch numbers (MIDI pitches) to calculate the chord quality. We extend the idea and modify the equations to accommodate spectral information obtained from a semitone-spaced filter bank. Let Y represent the vector of spectral amplitudes for a range of semitones y_L to y_H of the spectrum. The chord quality can then be approximated by

$$Q(k) = \left| \sum_{p=y_L}^{y_H} Y(p)^{0.5}\, e^{-2\pi jpk/12} \right|^2, \quad k = 1..6 \tag{4}$$

again producing a 6 element vector. The intervallic distance is given by $D_{i,j} = dist(Q_i, Q_j)$ where Q_i is the vector corresponding to chord v_i. We experimentally found that compressing the filter bank output by taking the square root and squaring the magnitude of the transform worked better.

There are many frequency transforms that use logarithmic frequency spacing for obtaining Y. The variants are known as chroma pitch, semigram or the constant Q transform. For this part we use chroma pitch features described in [24]. The chroma pitches are computed from the input signal decomposed into 88 bands (A0 to C8) using a constant Q multirate filter bank. Each filter output measures the local energy (short-time mean-square power) in its subband.

6.2   Trichord Chord Type and Pitch Class Sets

Having defined a quantification for chord quality adapted to acoustical input, we show a classification of trichord chord type and pitch class sets using Forte's naming convention [14]. The 2D plot in Fig. 4(a) shows the output for 12 chord types. Only chord types with unique interval vectors have been included. For each chord type, 12 transpositionally related chords were generated using flute sounds from the Iowa database making a total of 144 chords. The chords belonging to the same class are shown within the convex hull of those points in the figure. The Forte class name pertains to all the points representing the transpositionally related chords in the convex hull. The plot shows a clear separation of classes and it would be possible to classify the chord types even from the 2D projection. Our experiments have shown that using higher number of dimensions is needed for separability in some cases with other input sounds. Next, inversionally related chords are added. The output is shown in part (b). Not much can be said about the distances between pitch class sets themselves but the inversionally related sets can be clearly seen to be clustered tightly and mostly overlapping as we would like – maybe with the exception of classes 3-2B and 3-7B which seem to experience a little bit of overlap. The two outputs reside in OPT and OPTI spaces respectively.

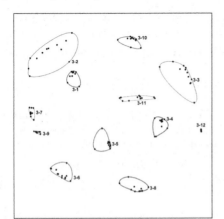 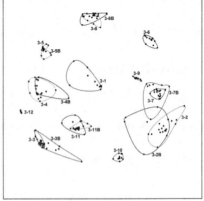

Fig. 4. (a) Chord types (left); (b) pitch class sets (right).

7 Conclusion

We have presented three models for constructing geometric spaces from acoustical representations. Both synthesized and real sounds have been used to show the emergence of well-known geometrical structures in music theory. We have also defined specific spectral representations that implement the octave, transposition and inversion equivalence relations. Visualization has been performed by dimensionality reduction on these spectral distances.

The circle of fifths is fairly robust for the major keys and the circular relationship that emerges is quite independent of the spectral characteristics of the fundamental and the overtones. However, the harmonic minor is relatively unstable due to its interval content and although the projection onto two dimensions is similar to Krumhansl's results it intrinsically seems to have higher dimensionality. We have shown that the chroma distance achieves better octave equivalence compared to the direct spectral distance. Additionally, hexatonic cycles and the Cube Dance structure have emerged quite effortlessly with the presented system. Chord type and pitch class sets show promising classification capabilities under the intervallic distance. The system has been quite robust against noise, tuning and pitch glides.

References

1. Amiot, E.: The Torii of phases. In: Yust, J., Wild, J., Burgoyne, J.A. (eds.) MCM 2013. LNCS, vol. 7937, pp. 1–18. Springer, Heidelberg (2013)
2. Balzano, G.J.: The group-theoretic description of 12-fold and microtonal pitch systems. Comput. Music J. **4**(4), 66–84 (1980)
3. Bartsch, M.A., Wakefield, G.H.: To catch a chorus: using chroma-based representations for audio thumbnailing. In: IEEE Workshop on the Applications of Signal Processing to Audio and Acoustics, pp. 15–18 (2001)
4. Callender, C.: Continuous harmonic spaces. J. Music Theor. **51**(2), 277 (2007)
5. Callender, C., Quinn, I., Tymoczko, D.: Generalized voice-leading spaces. Science **320**(5874), 346–348 (2008)
6. Carey, N., Clampitt, D.: Aspects of well-formed scales. Music Theor. Spectr. **11**(2), 187–206 (1989)
7. Chacón, C.E.C., Lattner, S., Grachten, M.: Developing tonal perception through unsupervised learning. In: Proceedings of the International Conference on Music Information Retrieval (ISMIR), Taipei, Taiwan (2014)
8. Chew, E.: Towards a mathematical model of tonality. Ph.D. thesis (2000)
9. Chew, E.: Mathematical and Computational Modeling of Tonality: Theory and Applications, vol. 204. Springer Science & Business Media, US (2013)
10. Chuan, C.H., Chew, E.: Polyphonic audio key finding using the spiral array CEG algorithm. In: International Conference on Multimedia and Expo, ICME, pp. 21–24 (2005)
11. Cohn, R.: Maximally smooth cycles, hexatonic systems, and the analysis of late-romantic triadic progressions. Music Anal. **15**, 9–40 (1996)
12. Cohn, R.: Neo-Riemannian operations, parsimonious trichords, and their "Tonnetz" representations. J. Music Theor. **41**, 1–66 (1997)

13. Douthett, J., Steinbach, P.: Parsimonious graphs: a study in parsimony, contextual transformations, and modes of limited transposition. J. Music Theor. **42**, 241–263 (1998)
14. Forte, A.: The Structure of Atonal Music. Yale University Press, New Haven (1973)
15. Fujishima, T.: Realtime chord recognition of musical sound: a system using common lisp music. In: Proceedings of the ICMC, pp. 464–467 (1999)
16. Gatzsche, G., Mehnert, M., Gatzsche, D., Brandenburg, K.: A symmetry based approach for musical tonality analysis. In: Proceedings of the International Conference on Music Information Retrieval (ISMIR), Vienna, Austria (2007)
17. Gómez, E., Bonada, J.: Tonality visualization of polyphonic audio. In: Proceedings of International Computer Music Conference, ICMC (2005)
18. Hall, R.W.: Linear contextual transformations. In: Quaderni di Matematica: Theory And. Applications of Proximity, Nearness and Uniformity (2009)
19. Humphrey, E.J., Cho, T., Bello, J.P.: Learning a robust tonnetz-space transform for automatic chord recognition. In: ICASSP, pp. 453–456. IEEE (2012)
20. Izmirli, Ö.: Cyclic-distance patterns among spectra of diatonic sets: the case of instrument sounds with major and minor scales. Tonal Theor. Digit. Age **15**, 11–23 (2008)
21. Krumhansl, C.L., Kessler, E.J.: Tracing the dynamic changes in perceived tonal organization in a spatial representation of musical keys. Psychol. Rev. **89**(4), 334–368 (1982)
22. Kruskal, J.B., Wish, M.: Multidimensional Scaling, vol. 11. Sage, California (1978)
23. Lerdahl, F.: Tonal Pitch Space. Oxford University Press, Oxford (2001)
24. Müller, M., Ewert, S.: Chroma toolbox: MATLAB implementations for extracting variants of chroma-based audio features. In: Proceedings of the International Conference on Music Information Retrieval (ISMIR), Miami, USA (2011)
25. Purwins, H.: Profiles of pitch classes circularity of relative pitch and key-experiments, models, computational music analysis, and perspectives. Ph.D. thesis, Berlin University of Technology (2005)
26. Purwins, H., Blankertz, B., Obermayer, K.: Toroidal models in tonal theory and pitch-class analysis. Tonal Theor. Digit. Age **15**, 73–98 (2008)
27. Quinn, I.: General equal-tempered harmony: parts 2 and 3. Perspect. New Music **45**(1), 4–63 (2007)
28. Sammon, J.W.: A nonlinear mapping for data structure analysis. IEEE Trans. Comput. **18**(5), 401–409 (1969)
29. Sethares, W.A., Budney, R.: Topology of musical data. J. Math. Music **8**(1), 73–92 (2014)
30. Shepard, R.N.: Geometrical approximations to the structure of musical pitch. Psychol. Rev. **89**(4), 305 (1982)
31. Toiviainen, P.: Visualization of tonal content in the symbolic and audio domains. Tonal Theor. Digit. Age **15**, 73–98 (2008)
32. Tymoczko, D.: The geometry of musical chords. Science **313**(5783), 72–74 (2006)
33. Tymoczko, D.: Three conceptions of musical distance. In: Chew, E., Childs, A., Chuan, C.-H. (eds.) MCM 2009. CCIS, vol. 38, pp. 258–272. Springer, Heidelberg (2009)
34. Tymoczko, D.: A Geometry of Music: Harmony and Counterpoint in the Extended Common Practice. Oxford University Press, Oxford (2011)
35. Volk, A., Honingh, A.: Mathematical and computational approaches to music: challenges in an interdisciplinary enterprise. J. Math. Music **6**(2), 73–81 (2012)

Geometry, Iterated Quantization and Filtered Voice-Leading Spaces

Clifton Callender$^{(\boxtimes)}$

College of Music, Florida State University, Tallahassee, FL 32306, USA
clifton.callender@fsu.edu
http://cliftoncallender.com

Abstract. A recent special issue of the *Journal of Mathematics and Music* on mathematical theories of voice leading focused on the intersections of geometrical voice-leading spaces (GVLS), filtered point-symmetry (FiPS) and iterated quantization, and signature transformations. In this paper I put forth a theoretical model that unifies all of these approaches. Beginning with the basic configuration of FiPS, allowing the n points of a filter or beacon to vary arbitrarily yields the continuous chord space of n voices (T^n/S_n). Each point in the filter space induces a quantization or *Voronoi diagram* on the beacon space. The complete space of filter and beacon is a singular fiber bundle, combining the power and generalization of GVLS with the central FiPS insight of iterated filtering by harmonic context. Additionally, any of the sixteen types of generalized voice-leading spaces described by Callender, Quinn, and Tymoczko can be used as filters/beacons to model different contexts.

Keywords: Voice-leading · Filtered point-symmetry · Geometry · Quantization · Voronoi diagram · Fiber bundle

1 Introduction

Recent mathematical theories of voice leading include geometrical approaches (Callender [3], Tymoczko [17], and Callender, Quinn, and Tymoczko [2]), filtered point-symmetry (Douthett [5] and Plotkin [9]), signature transformations (Hook [6]), and analytical applications of these approaches (Yust [20]). There is considerable interest in the potential intersections of these approaches as witnessed by a recent (2013) special issue of the *Journal of Mathematics and Music* (including contributions by Hook [7], Plotkin and Douthett [10], Tymoczko [14], and Yust [18]). In particular, one question concerns whether the highly general geometric model of voice leading encompasses the other approaches or whether the (iterated) quantization that is a feature of these other approaches stands outside of the reach of generalized voice-leading spaces.

The purpose of this paper is to briefly sketch a theoretical model that unifies all of these approaches by combining the generalizing power of the geometrical approach with the central insight of iterated filtering by harmonic context that lies

© Springer International Publishing Switzerland 2015
T. Collins et al. (Eds.): MCM 2015, LNAI 9110, pp. 257–266, 2015.
DOI: 10.1007/978-3-319-20603-5_27

at the heart of filtered point-symmetry.[1] Building on Yust's [19] response to the special issue, in particular his recasting of Plotkin and Douthett's maps of filter systems as configuration spaces that respect voice-leading distance, the approach proposed here replaces the rings of filtered point-symmetry with voice-leading spaces and adapts the filtering process to fiber bundles of induced Voronoi diagrams.

2   Configuration Spaces (I)

We begin with a relatively simple example of filtered point-symmetry—the iterated maximally-even distribution 2→5→12 shown in Fig. 1a. The innermost ring contains two "beacons" representing a strictly even two-fold division of the octave. Light emitting from each beacon passes through the *nearest* point in the middle ring.[2] This middle ring contains five "filters" representing a strictly even five-fold division of the octave. The outermost ring likewise filters the output of the middle ring, constraining the final output to the standard twelve-fold division of the octave into equal steps. (The outermost ring in a three-ring configuration usually remains fixed and can be thought of as the "tuning" ring.) The musical intuition captured by this particular configuration is that the tritone $\{D, G\sharp\}$ quantizes to the perfect fourth $\{D, G\}$ given the mediating pentatonic context $\{B\flat, C, D, F, G\}$.

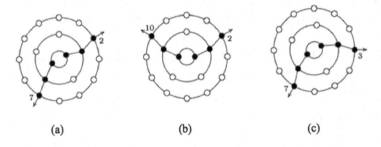

(a)   (b)   (c)

Fig. 1. Three instances of the FiPS system $2 \rightarrow 5 \rightarrow 12$.

In Fig. 1b the innermost ring is rotated clockwise by $\frac{1}{12}$ of a turn (or 30°) corresponding to transposition of the pcset by a semitone to $\{D\sharp, A\}$. Since A is closer to $B\flat$ than to G. As a result of this rotation, the dyad is now quantized to $\{D, B\flat\}$. The movement from G to $B\flat$ corresponds to a step within the mediating pentatonic context. In Fig. 1c, the innermost ring is held constant,

---

[1] This paper does not touch upon the potential intersections of Fourier-based approaches to harmony/chord quality and voice leading discussed in [1,4,11,12,16].

[2] In the usual depiction of such configurations, the output of one ring passes through the nearest point of the next ring in a *counter-clockwise* direction. Allowing outputs to pass through the nearest point in either direction makes the connection with geometrical voice-leading and, in particular, Voronoi diagrams clearer.

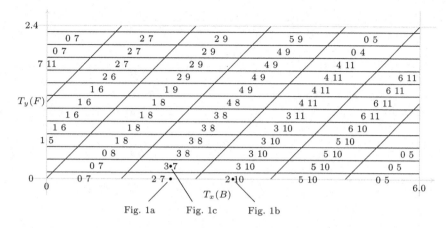

Fig. 2. The configuration space corresponding to all possible combinations of rotations of the inner and middle rings of Fig. 1. The beacon is $B = \{0,6\}$ and the filter is $F = \{12i/5\}_{i=0}^{4} = \{0, 2\frac{2}{5}, 4\frac{4}{5}, 7\frac{1}{5}, 9\frac{3}{5}\}$.

while the middle ring is rotated by $\frac{1}{60}$ corresponding to a transposition of the generic pentatonic collection by one-fifth of a semitone. The resulting transposed ring is quantized to the pentatonic collection $\{B\flat, C, E\flat, F, G\}$ (a transposition of the original context by perfect fifth). The change in mediating harmonic context causes a change in the output of the system, with the innermost ring quantized to the major third $\{E\flat, G\}$.

In order to better observe the behavior of this system we can construct a map [10] or *configuration space* [19] by allowing the two inner rings to rotate independently and arbitrarily. In the resulting configuration space, shown in Fig. 2, the horizontal axis corresponds to rotation of the beacon, while the vertical axis corresponds to rotation of the filter. The space is partitioned into regions representing combinations of beacon/filter rotations that yield the same pitch-class dyads in the *same harmonic context*. For instance, while the dyad $\{1,6\}$ occurs several times in the space, each separate occurrence is associated with a different pentatonic context, depending on the transposition of the filter. Note also that the space shown in Fig. 2 is only a portion of the complete space, which ranges through the entire octave for both the beacon and the filter.[3] We will return to configuration spaces in greater details in Sect. 5.

3   Iterated Filtering and Voronoi Diagrams

We can begin to generalize the system by allowing the two beacons of the inner-most ring to vary independently and arbitrarily through the octave, yielding

---

[3] However, because the filter and beacon are strictly even divisions of the octave, the region shown in Fig. 1 serves as a fundamental region for the entire configuration space, which can be formed by identifying the vertical and horizontal boundaries to form a torus.

```
6 6      7 7      8 8      9 9     10 10    11 11     [0 0]

   7 6      8 7      9 8     10 9     11 10     0 11

7 5      8 6      9 7     10 8     11 9      0 10     [1 11]

   8 5      9 6     10 7     11 8      0 9      1 10

8 4      9 5     10 6     11 7      0 8      1 9      [2 10]

   9 4     10 5     11 6      0 7      1 8      2 9

9 3     10 4     11 5      0 6      1 7      2 8      [3 9]

  10 3     11 4      0 5      1 6      2 7      3 8

10 2     11 3      0 4      1 5      2 6      3 7      [4 8]

  11 2      0 3      1 4      2 5      3 6      4 7

11 1      0 2      1 3      2 4      3 5      4 6      [5 7]

   0 1      1 2      2 3      3 4      4 5      5 6

0 0      1 1      2 2      3 3      4 4      5 5      [6 6]
```

Fig. 3. The "beacon space" V_2.

the voice-leading space of unordered pitch-class dyads described in [17], which is the Mobius strip, T^2/S_2. This is shown in Fig. 3, where horizontal motion corresponds to parallel voice leading (or transposition) and vertical motion corresponds to contrary voice leading. (Points corresponding to familiar twelve-tone equal-tempered dyads are labelled for orientation, but are not inherently more significant. The underlying space is continuous.) The single boundary on the top and bottom correspond to dyads with doubled pitch classes, while the left and right sides of the figure are identified by the indicated twist.

This two-note chord space, which we will refer to as V_2, now replaces the original 2-hole ring. The 5-filter can still act on this space in the following manner. Let the specific rotation of the filter be given by the set P and form the direct product $P \times P$. This product determines a set of sites, S, in V_2 (which may not be unique). Each $s \in S$ defines a cell (and equivalence class) of points in V_2 that are closer to s than any other site. (Points that are equidistant to multiple sites lie on boundaries of the cells and form their own equivalence classes.) The partition of the space into these cells yields a *Voronoi diagram*.

In Fig. 1 the middle ring is filtered by the outer ring to yield the pentatonic collection $P = \{B\flat, C, D, F, G\}$. The direct product of this collection with itself yields the sites indicated in Fig. 4a, the Voronoi diagram induced by this collection on the space T^2/S_2. The boundaries of the diagram are determined by those pitch classes that lie exactly halfway between the members of P arranged as a scale. That is, the ordered set $(0, 2, 5, 7, 10)$ gives rise to boundaries associated with the pitch classes $1, 3\frac{1}{2}, 6, 8\frac{1}{2}$, and 11. A different rotation of the filter would yield a different set of boundaries and thus a different Voronoi diagram. Figure 4b shows the Voronoi diagram induced by the rotation of the middle ring in Fig. 1c. Due to the very small rotation of the ring, there is a large amount of overlap between the regions of the two diagrams in Fig. 4. In particular, most

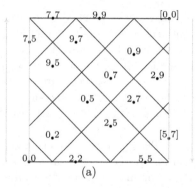 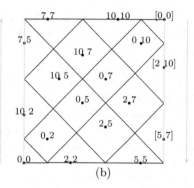

(a)   (b)

Fig. 4. The Voronoi diagram induced on V_2 by the filter $P = \{0, 2, 5, 7, 10\}$ in (a) and $P + \frac{1}{60}$ in (b). (See Tymoczko [15] Fig. 4.1.4a for a related diagram.)

of the points in the space associated with sites that include pc 9 in (a) are associated with sites that include pc 10 in (b).

4 Filter and Beacon Fiber Bundles

Continuing to generalize the system, we can allow each of the points of the filter to vary arbitrarily, yielding the space of all possible five-note chords, T^5/S_5, which we will write as V_5. Any point, P, in this filter space induces a Voronoi diagram, $\mathcal{D}_P$ on the beacon space, V_2, as described in the previous section. In general, each point induces a *unique* diagram and varying P continuously corresponds to a continuous deformation of the resulting diagram.[4] Taken together, the beacon and filter spaces form a singular fiber bundle.

Generalizing the situation for arbitrary sized beacons and filters, we begin with the beacon space, V_m, and filter space, V_n. Let the universal space, $\mathcal{U}$, be the product of the beacon and filter spaces, $\mathcal{U} = V_n \times V_m$. Each point $P \in V_n$ induces a Voronoi diagram (or set of equivalence classes) on V_m, designated $\mathcal{D}_P$. Letting P be any point in the filter space, V_n, and Q be any point in the induced Voronoi diagram, $\mathcal{D}_P$, the generalized filter system $m \to n$ is given by the singular fiber bundle $\overline{\mathcal{U}} = \overline{(P,Q)}$. These relations are summarized in the following commutative diagram:

$$
\begin{array}{ccc}
V_m & \longleftarrow \mathcal{U} = V_n \times V_m \longrightarrow & V_n \\
\downarrow & \downarrow & \nearrow \\
\mathcal{D}_P & \longleftarrow \overline{\mathcal{U}} = \overline{(P,Q)} &
\end{array}
$$

We can write $V_n \to V_m$ for $\overline{\mathcal{U}}$, which generalizes the filter $m \to n$. (The arrows are reversed since we are interested in the action of the "filter," V_n, on

---

[4] There are cases involving singularities in either the beacon or the filter space where these properties break down, but the details are not essential for present purposes.

the voice-leading space, V_m.) The voice-leading spaces can be chained arbitrarily, such as $V_z \to V_y \to \cdots \to V_n \to V_m$, where V_z induces a Voronoi diagram on V_y, the sites (equivalence classes) of which induce a Voronoi diagram on V_x, and so forth. This generalizes the multiple filter configuration $m \to n \to \cdots \to y \to z$.

While we have been assuming that beacons and filters are spaces of unordered pitch classes, this is not a necessary requirement. Any of the sixteen voice-leading spaces, depending on combinations of octave (O), permutation (P), transposition (T), and inversion (I) equivalence, described by Callender, Quinn, and Tymoczko [2] may function as a beacon or filter. That is, any arbitrary voice-leading space may induce a Voronoi diagram on any other voice-leading space, though there are some subtleties (particularly involving permutation equivalence) that are beyond the scope of the present paper.

5   Configuration Spaces (II)

The configuration space of Fig. 2 is an example of a filtered voice-leading space. It is the product of Voronoi diagram on two one-dimensional spaces: (1) the beacon space, which is $B = \{0, 6\} + x\mathbf{1}$ as x varies through the octave and (2) the filter space, which is $F + y\mathbf{1}$, where $F = \{12i/5\}_{i=0}^{5-1} \in \mathbb{T}^5/\mathcal{S}_5$ and $\mathbf{1} = (1, \ldots, 1)$. The boundaries of each Voronoi diagram are given by the intersection of horizontal lines with the vertical axis in the case of the filter space and the diagonal lines with the horizontal axis in the case of the beacon space. Further, the boundaries result from the *convolution* of each space with the intermediate points in the space that is filtering it. For example, the tuning space in this example is simply the fixed point, $\mathbb{Z}_{12}$, the intermediate points of the tuning space are $T' = \{\frac{1}{2}, 1\frac{1}{2}, \ldots, 11\frac{1}{2}\}$. The boundaries on the vertical axis are the direct sum of these intermediate points with the inverse of the filter space: $T' \oplus -F$. Similarly for the boundaries on the horizontal axis. The symmetry of Fig. 2 is a direct result of the symmetry of the beacon, filter, and tuning spaces (or inner, middle, and outer rings).[5]

It is not necessary to limit configuration spaces to the product of strictly even chords transposed through the octave. Figure 5 shows the configuration space for a beacon of stacked fifths, $\{0, 2, 7\}$, a filter of dominant ninth chords, $\{0, 2, 4, 7, 10\}$, with both beacon and filter transposed through the octave, and a tuning of $\mathbb{Z}_{12}$. (As before, the figure shows only a portion of the complete configuration space.) Again, the boundaries in the configuration space arise through convolution. The lack of symmetry in Fig. 5 is a direct result of the lack of symmetry in the beacon and filter spaces.

At this point it should be obvious that there is no need to limit configuration spaces to the products of chords rigidly transposed through the octave. Any collection of n paths through voice-leading spaces can give rise to an n-dimensional configuration space.[6] Boundaries in these spaces will occur wherever a voice in a

---

[5] I would like to thank my colleagues Eriko Hironaka and Paolo Aluffi for a number of helpful mathematical suggestions.

[6] Indeed, depending on how the filters associated with these paths are ordered, there will be $n!$ different configuration spaces.

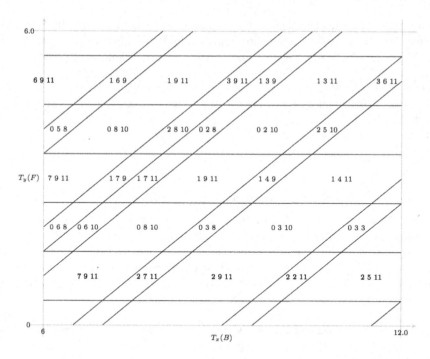

Fig. 5. The configuration space generated by the beacon space $B + x\mathbf{1}$, filter space $F + y\mathbf{1}$, and tuning space $\mathbb{Z}_{12}$. ($B = \{0, 2, 7\}$ and $F = \{0, 2, 4, 7, 10\}$.)

path crosses over the intermediate point between two adjacent voices in the filtering path. (In general, these boundaries will not run parallel as in Figs. 2 and 5.)

6   Synthesis

From the vantage point of $\bar{\mathcal{U}}$ it is easier to see that the many different approaches discussed in the introduction are manifestations of the same space. In what follows we briefly considered the specifics of each approach in terms of the broader approach posited here. (All of the approaches below assume a "highest-level filter" corresponding to 12-tone equal temperament [12tet], or $\mathbb{Z}_{12}$.)

Chord Spaces. Chapters 4 and 5 of Tymoczko [15] focus on the relation between scales/macroharmonies and the geometry of chord space. In the context of the present paper, this is similar to considering the manner in which a single harmonic context filters a given chord space:

$$\mathbb{Z}_{12} \to P \in \mathbb{T}^n/\mathcal{S}_n \to \mathbb{T}^m/\mathcal{S}_m.$$

Sequences of harmonic contexts are dealt with as discrete Voronoi diagrams rather than continuous changes of diagrams resulting from continuous paths in the filter space. Tymoczko argues that sequences of collections arise due to voice-leading proximity in the larger chord space, but the explicit connection drawn here between voice-leading spaces and iterated quantization is not present.

Filtered Point-Symmetry. As we have seen, Plotkin and Douthett [10] focus on equivalence classes generated by continuous transposition of strictly even sets in chord space:

$$\mathbb{Z}_{12} \to P + x\mathbf{1} \to Q + y\mathbf{1},$$

where

$$P = \{12i/n\}_{i=0}^{n-1} \in \mathbb{T}^n/\mathcal{S}_n,$$
$$Q = \{12j/m\}_{j=0}^{m-1} \in \mathbb{T}^m/\mathcal{S}_m.$$

Thus, FiPS is a theory of the configuration spaces arising from the product of Voronoi diagrams on transposition paths through the center of chord spaces.

Yust [19] generalizes Plotkin and Douthett above by taking Q to be any point in $\mathbb{T}^m/\mathcal{S}_m$. For example, instead of plotting configuration spaces using only augmented triads, Yust uses "generic" consonant triads, e.g., $\{0, 3.5, 7\}$, that lie exactly halfway between major and minor triads.

Signature Transformations. Hook's signature transformations [6] quantize arbitrarily large ordered and unordered pitch and pitch-class sets to a given diatonic collection. Thus it can be viewed as a discrete version of

$$\{12i/7\}_{i=0}^{6} \in \mathbb{T}^7/\mathcal{S}_7$$

filtering the associated quotient spaces: $\mathbb{R}^n$, $\mathbb{R}^n/\mathcal{S}_n$, $\mathbb{T}^n$, and $\mathbb{T}^n/\mathcal{S}_n$. Since they can act on four different types of voice-leading spaces, signature transformations are quite flexible. While the use of filters drawn exclusively from seven-note chord space might suggest a limitation of this approach, from the perspective of filtered voice-leading spaces it is easy to see how to generalize these transformations to be drawn from chord space of any cardinality.

7   Extensions

We conclude with suggestions for developing this approach to address a greater variety of musical situations and consider the ways in which these novel techniques connect with more traditional musical practice. Given the abstract nature of $\bar{\mathcal{U}}$, we can also consider situations involving arbitrary paths (not limited to transpositions) through these spaces and a greater variety of the sixteen possible generalized voice-leading spaces [2]. These types of situations arise naturally in contemporary music, especially that associated with spectral composition, where passages are governed by a smooth interpolation between melodies, chords, or rhythmic figures (among others). These interpolations are generally constrained by some fixed structure such as a tuning system or a grid of rhythmic pulses. Such processes are easy to find in the music of Magnus Lindberg [8], Kaija Saariaho [3], Gerard Grisey and Tristan Murail [13], among many others.

Example 1. Consider an interpolation between two chords in m-note pitch space. At each point in the interpolation the instantaneous chord should be quantized to the nearest possible subset of one of the twelve equal tempered acoustic collections, e.g., the twelve transpositions of $A = \{0, 2, 4, 6, 7, 9, 10\} \in \mathbb{T}^7/\mathcal{S}_7$. The corresponding $\bar{\mathcal{U}}$ is

$$\mathbb{Z}_{12} \to A + x\mathbf{1} \to \mathbb{R}^m.$$

Example 2. Consider a similar situation to Example 1 in which instantaneous chords are quantized to the nearest possible subset of one of the equal tempered members of the set class containing an arbitrary pcset P. The corresponding $\bar{\mathcal{U}}$ is

$$\mathbb{Z}_{12} \to \pm P + x\mathbf{1} \to \mathbb{R}^m.$$

Example 3. Consider a similar situation to Examples 1 and 2 in which instantaneous chords are quantized to the nearest quarter-tone approximation of a harmonic spectra of no more than n partials. The corresponding $\bar{\mathcal{U}}$ is

$$\mathbb{Z}_{24} \to P + x\mathbf{1} \to \mathbb{R}^m,$$

where $P = \{24 \times \log_2 i\}_{i=1}^n$ and $x \in \mathbb{R}$.

While there are routines for calculating these types of quantized interpolations as part of IRCAM's OpenMusic software for computer-aided composition that allow composers to experiment with these techniques, filtered voice-leading spaces suggest a means of charting the resulting spaces in a more systematic fashion.[7]

There is much to be gained from approaches that combine voice-leading spaces and iterated quantization. Many diatonic voice-leading routines result from uniformly ascending or descending voice leading that is broken into stages, or micro-transpositions in the language of [20], quantized to the underlying diatonic collection. Compelling chromatic variations can arise by systematically changing the mediating harmonic context. Contemporary composers continue to work with smooth voice leading and interpolations mediated by harmonic context, be it scalar collections, other set classes, spectra, etc. An approach that combines the insight of iterated quantization with the full generality of geometric voice-leading spaces can allow us to understand these common threads of musical practice.

References

1. Amiot, E.: The Torii of phases. In: Yust, J., Wild, J., Burgoyne, J.A. (eds.) MCM 2013. LNCS (LNAI), vol. 7937, pp. 1–18. Springer, Heidelberg (2013)
2. Callender, C., Quinn, I., Tymoczko, D.: Generalized voice-leading spaces. Science **320**, 346–348 (2008)

---

[7] OpenMusic is a visual programming environment for computer assisted composition designed and developed by the Musical Representations Team (Gerard Assayag, head) at IRCAM. Visit http://forumnet.ircam.fr/product/openmusic/.

3. Callender, C.: Continuous transformations. Music Theor. Online **10**(3) (2004)
4. Callender, C.: Continuous harmonic spaces. J. Music Theor. **51**(2), 277–332 (2007)
5. Douthett, J.: Filtered point-symmetry and dynamical voice-leading. In: Douthett, J., Hyde, M.M., Smith, C.J. (eds.) Music Theory and Mathematics: Chords, Collections, and Transformations, pp. 72–106. University of Rochester Press, NY (2008)
6. Hook, J.: Signature transformations. In: Douthett, J., Hyde, M.M., Smith, C.J. (eds.) Music Theory and Mathematics: Chords, Collections, and Transformations, p. 137160. University of Rochester Press, NY (2008)
7. Hook, J.: Contemporary methods in mathematical music theory: a comparative case study. J. Math. Music **7**(2), 89–102 (2013)
8. Martin, E.: Harmonic progression in Magnus Lindberg's *Twine*. Music Theor. Online **6**(1) (2010)
9. Plotkin, R.: Transforming Transformational Analysis: Applications of Filtered Point-Symmetry. Ph.D. dissertation, University of Chicago (2010)
10. Plotkin, R., Douthett, J.: Scalar context in musical models. J. Math. Music **7**(2), 103–125 (2013)
11. Quinn, I.: General equal-tempered harmony (introduction and part I). Perspect. New Music **44**(2), 6–50 (2006)
12. Quinn, I.: General equal-tempered harmony: parts II and III. Perspect. New Music **45**(1), 114–158 (2007)
13. Rose, F.: Introduction to the pitch organization of French spectral music. Perspect. New Music **34**(2), 6–39 (1996)
14. Tymoczko, D.: Goemetry and the quest for theoretical generality. J. Math. Music **7**(2), 127–144 (2013)
15. Tymoczko, D.: A Geometry of Music: Harmony and Counterpoint in the Extended Common Practice. Oxford University Press, USA (2011)
16. Tymoczko, D.: Set-class similarity, voice leading, and the fourier transform. J. Music Theor. **52**(2), 251–272 (2008)
17. Tymoczko, D.: The geometry of musical chords. Science **313**, 72–74 (2006)
18. Yust, J.: Tonal prisms: iterated quantization in chromatic tonality and Ravel's 'Ondine'. J. Math. Music **7**(2) (2013)
19. Yust, J.: A space for inflections: following up on JMM's special issue on mathematical theories of voice leading. J. Math. Music **7**(3), 175–193 (2013)
20. Yust, J.: Distorted continuity: chromatic harmony, uniform sequences, and quantized voice leadings. Music Theor. Spect. **37**(1), 120–143 (2015)

Using Fundamental Groups and Groupoids of Chord Spaces to Model Voice Leading

James R. Hughes[(⊠)]

Department of Mathematical and Computer Sciences,
Elizabethtown College, 1 Alpha Drive,
Elizabethtown, PA 17022, USA
hughesjr@etown.edu

Abstract. We model voice leading using tools from algebraic topology, principally the fundamental group, the orbifold fundamental group, and related groupoids. Doing so is a natural extension of modeling voice leading by continuous paths in chord spaces. The resulting algebraic precision in the representation of voice leadings and their concatenations allows for new distinctions between voice crossing cases, and enhanced connections with other approaches to voice leading.

Keywords: Chord spaces · Orbifolds · Fundamental group · Fundamental groupoid · Gestures · Voice leading · Homotopy · Groups · Groupoids

1 Introduction

Among mathematical explorations of music, few topics are as tantalizing as voice leading, which embodies the rich interplay between melody and harmony, and epitomizes the transcendant relationship between individual and ensemble. Insofar as voice leading is a specific means by which one musical situation changes to another, and describing change with precision is well-served by mathematics, voice leading is a natural candidate for abstract mathematical modeling. Voice leading has been modeled in [10] as mathematical functions between sets, as groups or groupoids of transformations on a graph or lattice in [5] and [14], as transformations between pitch-class sets in [11], and as multisets of ordered pairs drawn from specified multisets in [17]. All of the above models are discrete, focusing exclusively on the starting and ending situations, and treating the mediating process between them as an "instantaneous event" [12]. In contrast, recent geometrical and topological models for voice leading [3,4,12,17] incorporate the mediating process in the form of continuous paths between states.

To mathematicians, the use of continuous paths to study relationships among geometrical and topological objects is familiar territory: much of algebraic topology depends on careful development of the notion of a "path," which enables precise definition of important topological invariants, such as the fundamental group and groupoid of a space. Hence, the invocation of paths to study voice

© Springer International Publishing Switzerland 2015
T. Collins et al. (Eds.): MCM 2015, LNAI 9110, pp. 267–278, 2015.
DOI: 10.1007/978-3-319-20603-5_28

leading opens a wide realm of possibilities for the mathematical study of voice leading.[1] Some of these possibilities have already been explored, with significant results: in the orbifold geometry of chords in [4] and elsewhere, and in the theory of gestures developed by Mazzola and Andreatta in [12]. It is the goal of this paper to make more explicit connections between the mathematics of paths and continuous models for voice leading. One immediate payoff of this effort is greater precision in modeling concatenation of voice leadings, which enables resolution of certain voice crossing ambiguities. Of more general benefit is deeper reconciliation among geometric, transformational, and gestural approaches.

2 Precise Paths, Path Composition, and Homotopy

We now extend the notion of a "path" beyond that of a "generalized line segment" [18]. All of the mathematical material in this section and the next is standard for a first course in algebraic topology (e.g., [13]). We presuppose familiarity with topological spaces and continuous functions between them.

Definition 1. *Let X be a topological space. A* path *in X is a continuous function $p : [a, b] \to X$, where $[a, b]$ is a closed interval in the real numbers $\mathbb{R}$. The* initial point *of p is $p(a)$ and the* terminal point *of p is $p(b)$.*

Note that a path in X is an example of a *gesture* [12]. We will assume the parameter set of a path is the unit interval $I = [0, 1]$; this does not really pose any restriction since any closed interval can be shifted and scaled to coincide with I. We include the following for later reference:

Definition 2. *A topological space X is said to be* path connected *if a path from x_0 to x_1 exists for any x_0 and x_1 in X.*

Even with the parameter set restricted to I, it should be clear that there is infinite variability among paths between any pair of points. Allowing such variability has musical relevance: taking $X = \mathbb{R}$ as pitch space for a single voice, and the parameter t as time, path variability corresponds in music to the variety of ways a voice can move (continuously) from one pitch to another (or even depart from and return to the same pitch): it could slide smoothly and linearly over the entire allotted time interval (as in *Vers le blanc* [3]); it could remain on the starting pitch until a very small subinterval of time, during which the pitch slides quickly enough to the ending pitch to be perceived as an instantaneous change; it could move in quick stair-step fashion through the notes of a scale, as in a fast run played by a woodwind instrument; or it could engage in some combination of these, as in the clarinet solo opening Gershwin's *Rhapsody in Blue* (Fig. 1). Path variability— perhaps at the discretion of the performer— is

---

[1] It may seem that by invoking topological invariants, we lose important geometrical information that allows for computation of the sizes of voice leadings. However, in the cases we consider, each path class can be represented by a unique geodesic, through which the geometrical information can be recovered.

Fig. 1. Clarinet solo in opening measures of Gershwin's *Rhapsody in Blue*. Most performances convert the last fifth or so of the glissando into a smooth portamento; for voice leading purposes the glissando is a path from F3 to B♭5.

Fig. 2. Excerpt from Josquin's *Ave Maria*, showing only the alto and bass parts. On the word "Dei" the bass voice executes a written-out diminution; for voice leading purposes it is a path from F3 down to C3.

acknowledged at least as early as Fux [8], in giving "diminution" as the reason for prohibition of direct motion from an imperfect consonance into a perfect consonance (see Fig. 2).[2]

Another standard mathematical operation on paths, concatenation or composition, also has musical relevance. For example, in the first three measures of the Josquin excerpt shown in Fig. 2, we might want to equate the two-voice path from (G4, C3) to (G4, G3) followed by the path from (G4, G3) to (A4, F3) with the single path from (G4, C3) to (A4, F3). In another example from [18], pp. 81–82, paths in two-note chord space are decomposed into concatenations of pure parallel and pure contrary paths. (These examples illustrate a problem of ambiguity in modeling voice leadings using continuous paths. Specifically, in [18] and other sources, a voice leading modeled by a path that touches a mirror singularity corresponding to two voices sounding the same pitch class is assumed to contain a voice crossing; however, the alto and bass in the Josquin excerpt sound the same pitch class in the second measure, thereby touching a mirror singularity, but they do not cross. Later, we will see how the ambiguity can be resolved by means of orbifold paths.) We model concatenation of voice leadings by path composition, defined as follows.

---

[2] Admittedly, if subsequent notes in a voice are staccato or separated by a rest, modeling the voice as a continuous function of time breaks down; however, if one allows, as suggested in a footnote in [3], that the first note is retained in the listener's memory for some time beyond that of actual sound production, at least a perceptual sense of continuity can be retained. This idea was also suggested to the author by Richard Cohn in a recent conversation.

Definition 3. *If $p : I \to X$ and $q : I \to X$ are two paths such that $p(1) = q(0)$, the* composition $p * q$ *of p and q is the path given by $p(2t)$ for $0 \le t \le 1/2$ and $q(2t - 1)$ for $1/2 < t \le 1$.*

Note that, in order for the composition $p * q$ to be defined, the terminal point of p must coincide with the initial point of q, so that composition of arbitrary ordered pairs of paths is not necessarily defined, which means composition fails to be a binary operation on the set of paths in X.

Having allowed for path variability in our mathematical model, we now introduce a mathematically precise way of ignoring path differences that are unimportant from a voice leading point of view. Specifically, we define an equivalence relation on paths using the idea of continuous deformation. The motivation for invoking this particular equivalence is twofold: first, it agrees with notions of equivalence for voice leadings already in use (perhaps implicitly) in existing literature; and second, it is a standard tool in mathematics, whose invocation enables deployment of some extremely powerful ideas.

Definition 4. *Let p and q be two paths in X such that $p(0) = q(0) = x$ and $p(1) = q(1) = y$ (i.e., the initial and terminal points are x and y, respectively, for both paths). A* homotopy rel end points *from p to q is a continuous function $F : I \times I \to X$ such that $F(s, 0) = p(s)$ and $F(s, 1) = q(s)$ for $0 \le s \le 1$, and $F(0, t) = x$ and $F(1, t) = y$ for $0 \le t \le 1$. The paths p and q are called* homotopic rel end points *if such an F exists; in this case we write $p \simeq q$.*

Note that for each value of s F gives a path from x to y, and $s = 0$ gives p and $s = 1$ gives q, so F is a continuously varying family of paths from x to y between p and q. Also note that, in our setting, all homotopies will be rel end points, so until further notice we will refer simply to homotopy and assume it is rel end points. We note also that in the context of gesture theory, a homotopy is an example of a *hypergesture* [12]. For a musical example of homotopy, consider the trill at the beginning of the Gershwin excerpt (Fig. 1). There is a continuous family of variations in which the pitch variation in the trill decreases continuously from a whole tone through a semitone, to a slight vibrato, to no variation at all; hence the path of the trilled note is homotopic to that of a non-trilled note.

It is a standard exercise in algebraic topology to prove that homotopy is an equivalence relation, and that path composition is well-defined for homotopy classes of paths. That is to say, if p and q are composable paths and we denote their homotopy classes by $[p]$ and $[q]$, then for $p' \simeq p$ and $q' \simeq q$ we have $[p * q] = [p' * q']$, so we can define $[p] * [q]$ unambiguously as $[p * q]$. Another standard exercise is to prove that composition of homotopy classes of paths is associative; that is, $([p] * [q]) * [r] = [p] * ([q] * [r])$ for all p, q, and r for which the indicated compositions are defined. Furthermore, if for $x \in X$ we denote by 1_x the constant path at x, then we have $[1_x] * [p] = [p]$ for any path p with initial point x, and $[p] * [1_y] = [p]$ for any path p with terminal point y. Finally, if p is a path from x to y, and we define p^{-1} by $p^{-1}(t) = p(1-t)$ for $0 \le t \le 1$ (so p^{-1} is the path from y to x obtained by traversing p backwards), then $[p] * [p^{-1}] = [1_x]$ and $[p^{-1}] * [p] = [1_y]$. We can now give our model for voice leading.

Definition 5. *Let C be a path connected chord space, and let c_1 and c_2 be chords in C. A voice leading in C from c_1 to c_2 is a homotopy class of paths in C with initial point c_1 and terminal point c_2.*

We note that, under this definition, if we were to replace the bass part in the third measure of the Josquin excerpt shown in Fig. 2 with the single note F3 held for the entire measure (i.e., remove the diminution), the voice leading from the start of the third measure to the start of the fourth measure would be unchanged. Equating the two versions is very much in the spirit of [8]; in using one version to justify prohibition of the other, it is clear that Fux intended to consider the two versions identical in some sense, at least from a voice leading point of view. The equivalence relation of homotopy makes the identification precise.

3   The Fundamental Group and Covering Spaces

By now it should be clear that, but for the fact that arbitrary ordered pairs of paths need not be composable, homotopy classes of paths under path composition satisfy all the defining axioms for a group. The usual expedient at this juncture is to choose a specific member x_0 of X, called a *base point*, and consider only those paths whose initial and terminal points are x_0 (loops at x_0). Such paths are all composable with one another, and the set of their homotopy classes does indeed form a group, called the *fundamental group of the pair* (X, x_0), and denoted $\pi_1(X, x_0)$. If X is path connected, then it can be shown that $\pi_1(X, x_0)$ and $\pi_1(X, x_1)$ are isomorphic, so it is common to dispense with explicit reference to the base point, and write simply $\pi_1(X)$. The fundamental group is *functorial* in the sense that a given continuous mapping between topological spaces induces a specific homomorphism between their fundamental groups. Because it allows one to unleash the computational power of algebra in the study of topology, the mathematical importance of the fundamental group is hard to overestimate.

For our first example of a fundamental group, consider n-dimensional Euclidean space $\mathbb{R}^n$. Choose an arbitrary base point $x_0 \in \mathbb{R}^n$, and an arbitrary loop p at x_0, so that $p : I \to \mathbb{R}^n$, $p(0) = p(1) = x_0$, and p is continuous. The mapping $H : I \times I \to \mathbb{R}^n$ given by $H(t, s) = x_0 + (1 - s)(p(t) - x_0)$ gives a homotopy from p to the constant path at x_0, so there is only one homotopy class of loops at x_0; hence $\pi_1(\mathbb{R}^n)$ is the trivial group consisting of only one element. A path-connected space with trivial fundamental group is said to be *simply connected*. It is a straightforward exercise to prove that, in a simply connected space, any two paths having the same initial and terminal points are (path) homotopic.

Let us pause to consider the musical relevance of the preceding paragraph. In [18], Tymoczko uses Euclidean n-space $\mathbb{R}^n$ to model n-voice music (ordered pitch space). He defines a *voice leading in pitch space* to be an equivalence class of pairs of elements of (n-voice) ordered pitch space under the uniform operation of permutation. A representative of such a voice leading is simply an ordered pair of points in $\mathbb{R}^n$. There is no need to specify a path, because the voice leading is completely determined by the starting and ending points. In agreement with this is the fact that in $\mathbb{R}^n$ all paths from one specified point to another are homotopic to each other. If a path is needed, any path between the specified endpoints will do; for example, Tymoczko in [18] chooses the straight line segment for its geometrical properties.

We now introduce *covering spaces*. These are of high importance both mathematically and musically, because they mediate in a very precise, direct way between continuous and discrete models. Recall that every topological space is axiomatically equipped with a collection of "open sets" that enable precise definition of continuity, and a "homeomorphism" is a bijective, bicontinuous function. In Definitions 6 and 7 below, E and B are topological spaces, "map" is synonymous with "function," and "surjective" means every element of B is the image of at least one element of E under p.

Definition 6. *Let $p : E \to B$ be a continuous surjective map. The open set U of B is said to be* evenly covered *by p if the inverse image $p^{-1}(U)$ can be written as the union of disjoint open sets V_α in E such that for each α, the restriction of p to V_α is a homeomorphism of V_α onto U.*

The usual visual image associated with the above definition is a disjoint (possibly infinite) collection of copies of U floating above U; p maps all of the copies of U in E down to U itself in B. In the next definition, a "neighborhood" of a point is an open set containing that point.

Definition 7. *Let $p : E \to B$ be continuous and surjective. If every point b of B has a neighborhood U that is evenly covered by p, then p is called a* covering map, *and E is said to be a* covering space *of B.*

The classic first (nontrivial) example of a covering space one sees is the function p from the real line $\mathbb{R}$ to the unit circle S^1 given by $p(x) = (\cos 2\pi x, \sin 2\pi x)$. This function also models an important musical example; namely, that of octave equivalence. If we take the unit of length in $\mathbb{R}$ to be one octave, then p is the precise function that maps pitch space to pitch-class space. If we choose the base point in pitch-class space S^1 to be the pitch class C, then its inverse image under p (called the *fiber* over C) will consist of all the particular pitches whose pitch class is C; it is an infinite, discrete subspace of pitch space $\mathbb{R}$. The situation is illustrated in Fig. 3.

In topological parlance, if $p : E \to B$ and $f : X \to B$ are (continuous) functions, and $\tilde{f} : X \to E$ is such that $p \circ \tilde{f} = f$, then $\tilde{f}$ is called a *lifting* of f. One of the most important facts about covering spaces concerns the existence and uniqueness of liftings of paths and path homotopies:

Lemma 1. *If $p : E \to B$ is a covering map, and $p(e_0) = b_0$, then any path in B with initial point b_0 has a unique lifting to a path in E with initial point e_0, and any path homotopy of a path in B with initial point b_0 has a unique lifting to a path homotopy in E of a path with initial point e_0.*

The significance of the above lemma is that it can be used to define a correspondence between homotopy classes of loops at b_0 in B (that is, elements of $\pi_1(B, b_0)$) and elements of the fiber $p^{-1}(b_0)$. In fact, this correspondence can be

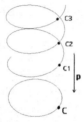

Fig. 3. The octave equivalence covering map from pitch space to pitch-class space, with pitch class C and part of the fiber over it shown.

viewed as a group action of $\pi_1(B, b_0)$ on the fiber, which can be extended to the entire covering space E, so that we obtain a homomorphism of $\pi_1(B, b_0)$ into the group of *covering transformations* (i.e., fiber-preserving homeomorphisms) of E. If E is simply connected, then this homomorphism is an isomorphism and E is called the *universal cover* of B; in the general case (assuming E is path connected) the group of covering transformations is a quotient of $\pi_1(B, b_0)$.

For topologists, one of the immediate payoffs of the situation described above is that the problem of finding $\pi_1(B, b_0)$ is converted to finding the group of covering transformations of the universal cover of B (if it exists). In the case of the covering $\phi : \mathbb{R} \to S^1$, $\mathbb{R}$ is known to be simply connected, and the group of covering transformations is readily seen to be the group of translations by integer amounts, so $\pi_1(S^1)$ is the additive group $\mathbb{Z}$ of integers. Elements of $\pi_1(S^1)$ correspond to the (signed) number of times a path winds around the circle S^1 before returning to the base point. Musically, if we model voice pitch-class voice leadings as homotopy classes of paths in pitch-class space, this result means that the pitch-class voice leadings of a one-note chord to itself are in one-to-one correspondence with the integers, interpreted as leaps up or down by some (whole) number of octaves. This set agrees with (one-voice) pitch-class voice leadings from a pitch class to itself as defined in [18]. In particular, as in [18] ("the specific path matters!") our model distinguishes between paths that travel up or down, and between paths that travel different numbers of octaves. Note, however, that we have obtained more than just an enumerative correspondence: we have made an explicit transition (via path liftings) from a continuous model for voice leadings (paths) to a discrete one (integers), and the discrete model comes equipped with a group structure. Moreover, the group is none other than the group whose action on pitch space $\mathbb{R}$ gave us octave equivalence (and hence pitch-class space S^1) in the first place. The ability to identify the set of voice leadings of a chord to itself in some chord space with the set of symmetry operations by which the chord space is defined can be extended beyond the particular example of one-note chords under octave equivalence, but to do so we will need to introduce additional mathematical machinery.

4   The Orbifold Fundamental Group

If B is an orbifold, its ordinary fundamental group $\pi_1(B)$ does not account for the behavior of paths at orbifold singularities. For example, the fundamental group of the Möbius band (two-note chord space in [18]) is the additive group of integers,[3] but this misses the orbifold structure by which the boundary acts as a mirror for paths that are projections of straight line segments in $\mathbb{R}^2$, a feature that is essential for the chord space voice leading model [18]. To capture the additional structure algebraically, we need to make use of the *orbifold fundamental group*, an extension of the ordinary fundamental group. Note that, unlike the ordinary fundamental group and ordinary covering spaces, the material on orbifolds in this section is not typically covered in a first course on algebraic topology. The classic reference is [16]; other useful references are [6] and [7].

Before presenting a definition of the orbifold fundamental group, we provide some mathematical background on manifolds, group actions, and orbifolds. A *manifold* is a space that looks locally like $\mathbb{R}^n$. The precise definition uses the language of local models (charts and atlases, see [15] or [9]); we omit it for brevity. A group Γ *acts* on a space X if there is a function $\Gamma \times X \to X$, with the image of an ordered pair (α, x) denoted by $\alpha \cdot x$, which is compatible with the group structure of Γ in the following sense: First, for all $\alpha, \beta \in \Gamma$ and $x \in X$, we have $(\alpha\beta) \cdot x = \alpha \cdot (\beta \cdot x)$, and second, if 1 denotes the identity element of Γ, we have $1 \cdot x = x$. Two examples of group actions that are of particular importance to us are $\mathbb{Z} \times \mathbb{R} \to \mathbb{R}$ by $k \cdot x = x + k$ (translation in $\mathbb{R}$ by integer amounts, musically interpretable as transposition by some fixed amount in pitch space), and $\Sigma_n \times \mathbb{R}^n \to \mathbb{R}^n$ by $\sigma \cdot (x_1, x_2, \ldots, x_n) = (x_{\sigma(1)}, x_{\sigma(2)}, \ldots, x_{\sigma(n)})$, where Σ_n is the group of permutations of n items and $\sigma \in \Sigma_n$ (musically interpretable as the permutation operation applied to ordered sequences of pitches). The *orbit* of a point $x \in X$ under a given action of Γ on X is the set $\{\alpha \cdot x | \alpha \in \Gamma\}$; the quotient space obtained by collapsing each orbit to a point is denoted X/Γ. Note, for example, that the orbit of $0 \in \mathbb{R}$ under the translation action of $\mathbb{Z}$ is $\mathbb{Z}$ itself, and that the orbit of a point $(x_1, x_2, \ldots, x_n) \in \mathbb{R}^n$ under the permutation action described above consists of $n!$ points if the coordinates of the point are all distinct, but fewer if some of the coordinates are the same; in the extreme case the orbit has just one point if all the coordinates are the same. If $x \in X$, the *isotropy group* of x is the subgroup Γ_x of Γ defined by $\Gamma_x = \{\alpha \in \Gamma | \alpha \cdot x = x\}$. Under the translation action of $\mathbb{Z}$ on $\mathbb{R}$, every point of $\mathbb{R}$ has trivial isotropy group; on the other hand, under the permutation action a point $(x, x, \ldots, x)$ on the diagonal of $\mathbb{R}^n$ has the whole group Σ_n as its isotropy group.

An *orbifold* is a generalization of a manifold in which quotient spaces $\mathbb{R}^n/\Gamma$ replace $\mathbb{R}^n$ as the local model, where Γ is either finite or acts properly (i.e., with finite isotropy). The precise, general definition is complicated [16]. Fortunately, all of the orbifolds that concern us belong to the tractable family of *developable* orbifolds: global quotients of manifolds by discrete groups acting properly.

---

[3] Topologists see this immediately by noticing that S^1 is a deformation retract of the Möbius band.

The definition we will give of the orbifold fundamental group is due to William Thurston [6]. The main idea of the definition is to generalize covering space theory to orbifolds. First, the definition of a covering map is modified such that the mapping from each sheet to an evenly covered open set is allowed to be a quotient map corresponding to a group action, rather than a homeomorphism, yielding the definition of an *orbifold covering projection*. Next, the definitions of universal cover and covering transformations are modified accordingly. Finally, rather than identifying the group of covering transformations of the universal cover with the fundamental group via a theorem as before, the identification is used to *define* the orbifold fundamental group:

Definition 8. *Let Q be a connected orbifold. The* orbifold fundamental group *of Q, denoted $\pi_1^{orb}(Q)$, is the group of covering transformations of the universal orbifold cover $p : \tilde{Q} \to Q$.*

For the developable orbifolds of interest to us musically, the universal orbifold cover is $\mathbb{R}^n$, and the orbifold fundamental group coincides with the group Γ whose action on $\mathbb{R}^n$ defines the orbifold. In the case of developable orbifolds, the relationship between the orbifold fundamental group and the ordinary fundamental group is captured neatly in a theorem of Armstrong ([1,7]):

Theorem 1. *Let Γ act properly by homeomorphisms on a connected, simply connected, locally compact metric space X, and let Γ' be the normal subgroup of Γ generated by the elements which have fixed points in X. Then the fundamental group of the orbit space X/Γ is isomorphic to the factor group Γ/Γ'.*

For example, in the case of the two-note chord space $C^2 = \mathbb{R}^2/\Gamma$, where the action of Γ combines both octave equivalence and permutational equivalence, Γ is isomorphic to the semidirect product $(\mathbb{Z} \times \mathbb{Z}) \rtimes \Sigma_2$. In this case the subgroup Γ' is the normal subgroup of Γ containing elements of the form $(k, -k) \cdot \sigma$ with $k \in \mathbb{Z}$ and $\sigma \in \Sigma_2$; this is isomorphic to the semidirect product $\mathbb{Z} \rtimes \Sigma_2$. The quotient Γ/Γ' is isomorphic to $\mathbb{Z}$ and is the fundamental group of the Möbius band, which is the underlying topological space of $\mathbb{R}^2/\Gamma$ [18]. Note that when $\sigma = \tau$ is nontrivial (swap coordinates), the fixed point set of $(k, -k) \cdot \tau$ in $\mathbb{R}^2$ is the line $y = x + k$; note also that the orbifold quotient $\mathbb{R}^2/\Gamma'$ is the strip between two consecutive such lines and has underlying space homeomorphic to $[0, 1] \times \mathbb{R}$.

As in the case of ordinary covering spaces, if $Q = \mathbb{R}^n/\Gamma$ is a developable orbifold, and we choose a basepoint x_0 of Q that has trivial isotropy group $\Gamma_{x_0} = 1$, then the elements of $\pi_1^{orb}(Q, x_0)$ are in one-to-one correspondence with homotopy classes of (ordinary) paths from a specified point $\tilde{x}_0$ in the fiber of the basepoint x_0 of Q to any point $\tilde{x}$ in that fiber (including $\tilde{x}_0$ itself), and since $\mathbb{R}^n$ is simply connected, there is only one such homotopy class for a given pair $(\tilde{x}_0, \tilde{x})$. Hence the elements of $\pi_1^{orb}(Q, x_0)$ are in one-to-one correspondence with line segments in $\mathbb{R}^n$ from $\tilde{x}_0$ to points of the fiber (including the constant line segment at $\tilde{x}_0$). If Q is a chord space, the projections of such line segments are the generalized line segments in Q from x_0 to itself, which are the voice leadings from x_0 to itself [4]. If, however, the basepoint x_0 has non-trivial isotropy group Γ_{x_0} (as would happen in a chord space if x_0 were on the boundary, corresponding to a chord with one or more pitch classes occurring more than once), then there is not a one-to-one correspondence between elements of $\pi_1^{orb}(Q, x_0)$ and line segments in $\mathbb{R}^n$ from $\tilde{x}_0$ to points of the fiber of x_0. Rather, for every such line segment there is a distinct member of $\pi_1^{orb}(Q, x_0)$ for each member of Γ_{x_0}.

We now examine what all this means musically. Denote the orbifold of n-note chord space as defined in [18] by C^n. Recall $C^n = \mathbb{R}^n/\Gamma$, where Γ is the group of transformations corresponding to octave equivalence and permutational equivalence. In this case, Γ is the wreath product of $\mathbb{Z}$ with the symmetric group Σ_n, or equivalently the semidirect product $\mathbb{Z}^n \rtimes \Sigma_n$. From Thurston's definition we obtain that $\pi_1^{orb}(C^n) = \Gamma$. Hence, if we are modeling voice leadings as homotopy classes of (orbifold) paths in C^n, the set of voice leadings from a given n-voice chord to itself can be identified with Γ. For $n = 2$, we can make the identification explicit as follows. Since in this case $\Gamma = (\mathbb{Z} \times \mathbb{Z}) \rtimes \Sigma_2$, any element of Γ can be written uniquely as a product $(m, n) \cdot \sigma$ where m and n are integers, and σ is either the nontrivial element τ (swap coordinates) or trivial element 1 (do not swap) of Σ_2. Musically, this means that any voice leading from a two note chord to itself can be represented by transposing the first voice by m octaves (up or down, depending on whether m is positive or negative), the second voice by n octaves, and then possibly swapping the two voices (σ).

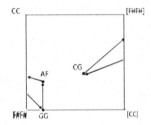

Fig. 4. A sequence of paths in chord space representing the Josquin excerpt in Fig. 2.

For a particular musical example, consider the Josquin excerpt in Fig. 2 again. The four-measure passage shown begins and ends on the same chord, so it represents an element of $\pi_1^{orb}(C^2) = \Gamma = (\mathbb{Z} \times \mathbb{Z}) \rtimes \Sigma_2$. Moreover, neither voice changes octave, and the voices do not swap, so the four-measure sequence represents the identity element $(0, 0) \cdot 1$ of $\pi_1^{orb}(C^2)$. A sequence of directed line segments in C^2 representing the passage is shown in Fig. 4 (the segments wrap around between the right and left sides as described in [18]). It is important to note that, unlike in the case of ordinary (non-orbifold) coverings, for a given loop in C^2 and specified lifting of the base point, there can be more than one lifting to $\mathbb{R}^2$. Specifically, in the Josquin excerpt, suppose the bass and alto switch parts (remaining in their respective octaves) in the second measure, when both parts are on G (i.e., at the point where the loop in C^2 contacts the boundary). In this case the loop traced in C^2 is unchanged, but the element of $\pi_1^{orb}(C^2)$ represented is now $(-1, 1) \cdot \tau \neq (0, 0) \cdot 1$, so the two voice leadings differ.

This example illustrates the need for $\pi_1^{orb}(C^2)$ as opposed to $\pi_1(C^2)$; the latter is not sensitive to the difference between the two voice leadings. In general, an orbifold path carries more data than the (ordinary) path to which it projects in the underlying quotient space; the additional data distinguishes between multiple liftings. Details on defining orbifold paths can be found in [7].

5   The Fundamental Groupoid

Most voice leadings do not begin and end at the same chord, just as most paths in a space are not loops. The usual generalization of a group to use in such a situation[4] is a *groupoid*. A groupoid is typically defined using category theory [2]. A *category* $\mathcal{C}$ consists of a class $\text{ob}(\mathcal{C})$ of *objects* and, for each x, y in $\text{ob}(\mathcal{C})$, a set $\mathcal{C}(x, y)$ of *morphisms in $\mathcal{C}$ from x to y*. For each triple (x, y, z) of objects, there is an associative composition function $* : \mathcal{C}(x, y) \times \mathcal{C}(y, z) \to \mathcal{C}(x, z)$, and for each object x there is an identity morphism $1_x \in \mathcal{C}(x, x)$ such that for $g \in \mathcal{C}(w, x)$ we have $g * 1_x = g$ and for $f \in \mathcal{C}(x, y)$ we have $1_x * f = f$.[5] A morphism $f \in \mathcal{C}(x, y)$ is called an *isomorphism* if there exists a morphism $f^{-1} \in \mathcal{C}(y, x)$ such that $f * f^{-1} = 1_x$ and $f^{-1} * f = 1_y$. If $\text{ob}(\mathcal{C})$ is a set (as opposed to a proper class), then $\mathcal{C}$ is called a *small* category.

Definition 9. *A* groupoid *is a small category in which every morphism is an isomorphism.*

Let X be a topological space. The category πX whose objects are the points of X, and for which $\pi X(x, y)$ consists of the homotopy classes of paths from x to y, forms a groupoid, called the *fundamental groupoid of X*. If C is a chord space, then voice leadings in C coincide with the morphisms of πC.

As in the case of the fundamental group, if C is an orbifold, we need to make a distinction between the ordinary fundamental groupoid πC, for which the morphisms are homotopy classes of ordinary paths, and the orbifold fundamental groupoid $\pi^{orb} C$, for which the morphisms of will be homotopy classes of orbifold paths. Defining classes of orbifold paths in the general case is complicated, but for a developable orbifold $Q = \mathbb{R}^n / \Gamma$, and a given pair of points x, y in Q with trivial isotropy groups, the members of $\pi^{orb}(x, y)$ are in one-to-one correspondence with homotopy classes of (ordinary) paths in $\mathbb{R}^n$ from a particular member $\tilde{x}_0 \in \mathbb{R}^n$ of the fiber over x to members of the fiber over y. Again, since $\mathbb{R}^n$ is simply connected, there is only one such path class for a given pair $(\tilde{x}_0, \tilde{y})$ of points in $\mathbb{R}^n$. If Q is a chord space, and we take the line segment in $\mathbb{R}^n$ from $\tilde{x}_0$ to $\tilde{y}$ as a representative path, its projection in Q is a generalized line segment. If x or y has non-trivial isotropy group, then there are multiple orbifold path classes corresponding to each generalized line segment. If both isotropy groups are non-trivial, then accounting for the multiplicity is complicated, but if just one (say Γ_x) is non-trivial, then the orbifold path classes corresponding to a generalized line segment themselves correspond to members of Γ_x.

---

[4] i.e., where not all ordered pairs can be combined using an operation that would otherwise be a group operation.

[5] This reverses the usual order for functional composition, for compatibility with path composition.

Returning to the Josquin example, note that there are two orbifold path classes corresponding to the generalized line segment from GG to AF, since the isotropy group of GG is Σ_2. The multiplicity is necessary for composition of voice leadings to be well defined; clearly we must distinguish the voice leadings represented by $(G, C) \xrightarrow{(2,5)} (A, F)$ and $(G, C) \xrightarrow{(-2,9)} (F, A)$, but to do so we must distinguish those represented by $(G, G) \xrightarrow{(2,-2)} (A, F)$ and $(G, G) \xrightarrow{(-2,2)} (F, A)$.

Acknowledgments. I am especially grateful to George Dragomir for explaining orbifold paths and related matters. I am also grateful for communications with Hans Boden, Richard Cohn, Michael Davis, Guerino Mazzola, and Dmitri Tymoczko. I thank the anonymous reviewers for their many helpful suggestions.

References

1. Armstrong, M.A.: The fundamental group of the orbit space of a discontinuous group. Proc. Camb. Philos. Soc. **64**, 299–301 (1968)
2. Brown, R.: Topology and Groupoids. Booksurge, Charleston (2006)
3. Callender, C.: Continuous transformations. Music Theor. Online **10**(3) (2004)
4. Callender, C., Quinn, I., Tymoczko, D.: Generalied voice-leading spaces. Science **320**, 346–348 (2008)
5. Cohn, R.: Maximally smooth cycles, hexatonic systems, and the analysis of late-Romantic triadic progressions. Music Anal. **15**, 9–40 (1996)
6. Davis, M.: Lectures on orbifolds and reflection groups. https://people.math.osu.edu/davis.12/papers/lectures. Accessed September 2014
7. Dragomir, G.: Closed geodesics on orbifolds. Ph.D. thesis, McMaster University (2011)
8. Fux, J.: Gradus ad parnassum (1725). In: Mann, A. (ed.) The Study of Counter-Point. Norton, New York (1965)
9. Hirsch, M.: Differential Topology. Springer, New York (1976)
10. Lewin, D.: Some ideas about voice-leading between Pcsets. J. Music Theor. **42**, 15–72 (1998)
11. Lundberg, J.: A theory of voice-leading sets for post-tonal music. Ph.D. thesis, Eastman School of Music (2012)
12. Mazzola, G., Andreatta, M.: Diagrams, gestures, and formulae in music. J. Math. Music **1**, 23–46 (2007)
13. Munkres, J.: Topology. Prentice Hall, Upper Saddle River (2000)
14. Popoff, A.: Generalized inversions and the construction of musical group and groupoid actions. http://arxiv.org/pdf/1402.1455.pdf. Accessed November 2014
15. Spivak, M.: A Comprehensive Introduction to Differential Geometry, 3rd edn. Publish or Perish, Houston (1999)
16. Thurston, W.: The Geometry and Topology of 3-Manifolds. Princeton University Press, Princeton (1997)
17. Tymoczko, D.: Scale theory, serial theory, and voice leading. Music Anal. **27**, 1–49 (2008)
18. Tymoczko, D.: A Geometry of Music: Harmony and Counterpoint in the Extended Common Practice. Oxford University Press, Oxford (2011)

All-Interval Structures

Robert W. Peck[✉]

Louisiana State University, Baton Rouge, LA, USA
rpeck@lsu.edu

Abstract. All-interval structures are subsets of musical spaces that incorporate one and only one interval from every interval class within the space. This study examines the construction and properties of all-interval structures, using mathematical tools and concepts from geometrical and transformational music theories. Further, we investigate conditions under which certain all-interval structures are Z (or GISZ) related to one another. Finally, we make connections between the orbits of all-interval structures under certain interval-groups and the sets of lines and points in finite projective planes. In particular, we conjecture a correspondence that relates to the co-existence of such structures.

Keywords: All-interval chords · Geometrical music theory · Transformational music theory · Generalized interval systems · Z relation · Projective planes

1 Introduction

Figure 1 presents the opening of the E-major fugue from Book II of *The Well-Tempered Clavier* by J.S. Bach. The reduction in Fig. 2 shows three trichords whose unordered pitch-class contents are related to one another by diatonic transposition. One pitch class, E, is invariant to all three of these trichords (as indicated in the figure with open noteheads). Moreover, each of the remaining six scale degrees in E major is represented once and only once in the reduction. As we will see below, these properties allow us to label the seven points of a Fano plane (the smallest finite projective plane) with the seven degrees of an E-major scale, such that the set of point labels on each of the plane's seven lines represents a diatonic transposition of the above trichords (Fig. 3) (see [1] for a discussion of related graphic representations).

It is possible to label the points of a Fano plane in this manner with only one other class of trichords in E major, the diatonic transpositions of {E, G♯, A}. Not only are these seven trichords diatonic inversions of the above, but they share with them another important property: namely, the members of these two diatonic transposition classes are all-interval structures. That is, any such trichord contains one and only one interval in each of the three diatonic interval classes: seconds/sevenths, thirds/sixths, and fourths/fifths. All-interval structures of varying sizes occur in other contexts, as well. In atonal theory, the two set-classes of all-interval tetrachords–i.e., orbits of the pitch-class sets {0, 1, 4, 6}

© Springer International Publishing Switzerland 2015
T. Collins et al. (Eds.): MCM 2015, LNAI 9110, pp. 279–290, 2015.
DOI: 10.1007/978-3-319-20603-5_29

Fig. 1. Bach, *The Well-Tempered Clavier*, Book II, Fugue in E, opening

Fig. 2. Reduction of Fig. 1

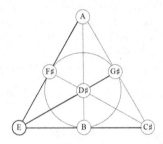

Fig. 3. Fano plane with points labeled in E major, lines as transpositions of {E,F♯,A}; lines in bold represent the trichords from Fig. 2

and {0, 1, 3, 7} under the action of the order-24 dihedral transposition and inversion group–are particularly significant. Allen Forte begins *The Structure of Atonal Music* [2] with the example of the chord {3, 5, 8, 9}, which appears at the end of the first song in Schoenberg's *George Lieder*, Op. 15 (Fig. 4). He writes, "This pitch combination, which is reducible to one form of the all-interval tetrachord, has a very special place in atonal music."

Important distinctions exist between these tetrachords and the above trichords, beyond their obvious difference of cardinality. For instance, we cannot label the points of a projective plane with twelve chromatic pitch classes in a manner similar to the diatonic example above, as no such relevant incidence structure exists with twelve points. What are the implications of this difference? Is there a projective plane with some number of points other than twelve, all

Fig. 4. Schoenberg, *George Lieder*, Op.15, No. 1 (ending)

of whose lines we can label with the members of transposed all-interval tetra-chords? If so, do projective planes exist on which to model all-interval structures of any cardinality in this way? All-interval structures are of considerable music-analytical and compositional interest. They present a maximum amount of intervallic information in as minimal a space as possible. Using tools from geometrical and transformational music theory, as well as from mathematical graph theory, group theory, and projective geometry, this article explores such structures in terms of traditional interval systems, as in the examples above, and in terms of generalized abelian and non-abelian interval systems.

2 All-Interval Structures, Intervals, and Generalized Interval Systems

We begin with some preliminary definitions.

Definition 1. A *space* P is a set of musical objects. Spaces must allow for the computation of intervals among their members (see below). For the purposes of this study, we consider all spaces to be finite, and not to include any redundancies of their elements. For example, the set of twelve chromatic pitch-classes commonly used in atonal music theory is a space.

Definition 2. An *interval* i is a particular relation between two distinct members of a space P (i.e., we do not consider prime intervals). That is, $i = (x, y) \in P \times P$, where $x \neq y$. We construe intervals according to two prominent models in the music-theoretic literature: first, as equivalence classes of directed, non-zero distances, in the manner of [3]. Second, following [4], we consider intervals as non-identity group elements in generalized interval systems, including cyclic, abelian, and non-abelian examples.

Definition 3. An *interval class* $[i]$ is an equivalence class of intervals. In terms of the metric definition of intervals above, an interval class contains all intervals that have the same distance (undirected intervals). In the generalized-interval-system model, an interval class contains a group element and its inverse (the interval class for an involution contains only that element).

Definition 4. A *structure* S is a subset of a space P.

Definition 5. A structure S in a space P is *all-interval* if (a) every undirected interval in P is represented by some interval in S; and (b) one and only one undirected interval in S exists for each undirected interval in P.

Definition 6. A structure S in a space P is *all-directed-interval* if (a) every directed interval in P is represented by some directed interval in S; and (b) one and only one directed interval in S exists for each directed interval in P.

2.1   Intervals as Equivalence Classes of Distances

Given a space P of size n, we find a triangular number, $\binom{n}{2}$, of distinct unordered two-member subsets in P, and $2\binom{n}{2}$ ordered two-member subsets. Accordingly, there exist $\binom{n}{2}$ undirected intervals among the members of P, and twice that number of directed intervals. A tally of the former conforms to a generalization of Forte's [2] interval-class vector, and of the latter to the non-trivial entries in Lewin's interval vector, based on the interval function from P to itself [4]. We note a one-to-one correspondence between the undirected intervals and the $n(n-1)/2$ edges in a complete graph with n vertices, and between the directed intervals and the $n(n-1)$ arcs (directed edges) of a complete directed graph (digraph) with n vertices. We may therefore equip such graphs and their subgraphs with appropriate vertex and edge (arc) labels to represent intervals in these spaces.

The E-major diatonic space P for the example from the Bach fugue above is cyclic of order 7. We may therefore map its elements to $\mathbb{Z}_7$ with E $\mapsto$ 0, F$\sharp$ $\mapsto$ 1, etc. Figure 5 presents a complete graph that illustrates this space. An undirected interval between any two of these pitch-classes is given by the lesser of the two differences mod 7: $x - y$ or $y - x$. The intervals partition into three equivalence classes: 1, 2, and 3. We observe that the subgraph for the trichord $\{0,1,3\}$ (shown in bold) is itself a complete graph on three vertices. It contains one and only one of each of these intervals. Accordingly, an interval-class vector for this structure reckoned in the manner of Forte reads $\langle 1, 1, 1 \rangle$; it is an all-interval structure for this space.

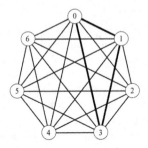

Fig. 5. Complete graph on seven vertices, labeled with $\mathbb{Z}_7$; subgraph for $\{0,1,3\}$ in bold

As a directed graph, each one of the three edges in the $\{0, 1, 3\}$ subgraph of Fig. 5 is replaced by two arcs with alternate heads and tails. Any one of these six arcs corresponds to a directed interval in P. Between any two vertices x and y there exist two arcs of opposite direction. Whereas these arcs have the same distance, they agree with distinct ordered subsets, (x, y) and (y, x), of the modular diatonic space. For this trichord, each one of the directed intervals represents a unique equivalence class within the space, based on its distance and direction. As the subgraph also presents one and only one interval in each directed interval class–as indicated by the non-trivial entries in its interval vector [3111111]–we may say that, in addition to being an all-interval structure for this space, $\{0, 1, 3\}$ is also an all-directed-interval structure.

Our example above from the *George Lieder* presents a different situation, as illustrated in Fig. 6. Here, the complete subgraph on the vertex set $\{3, 5, 8, 9\}$ indicates an all-interval tetrachord for the space of the twelve chromatic pitch-classes. It contains one and only one interval in each of the six interval classes in $\mathbb{Z}_{12}$. Accordingly, its interval-class vector reads $\langle 1, 1, 1, 1, 1, 1 \rangle$. However, $\{3, 5, 8, 9\}$ is not also all-directed-interval. The arc from vertex 3 to 9 and that from 9 to 3 represent the same equivalence class of directed intervals: each can be expressed as either $+6$ or -6. Indeed, any cyclic group of even order contains such an element, the involution that divides the order of the group in half. Consequently, the equivalence class for this interval contains more than one representative in the tetrachord, as indicated by variance in the non-trivial entries of interval vector [411111211111].

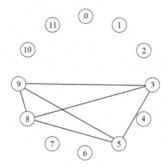

Fig. 6. All-interval subgraph for $\{3, 5, 8, 9\}$ in $\mathbb{Z}_{12}$

2.2 Intervals in Generalized Interval Systems

Intervals in [4] appear as elements of groups that have simply transitive (regular) actions on the members of a space. Therefore, we assume that type of action for all interval groups described below; further, we assume all groups to be finite. If G is a permutation group that acts on a set $S = \{a, b, c, d\}$, and if the circular permutation $g = (a, b, c, d)$ is an element of G, then we say that a relates to b by the directed interval g. Similarly, b relates to c by g, etc.; but a different directed interval, g^{-1}, relates b to a, c to b, etc. Yet another directed interval, g^2, relates a to c, c to a, and so on. Therefore, the ordering within such dyadic subsets of S, with reference to cycles in G, is of the essence. The theory of generalized interval systems does not possess a corresponding notion of interval class (undirected interval), hence our proposal in Definition 3. In this way, the members of unordered dyadic subsets of S, such as those determined by adjacencies in the cycles of g and g^{-1} above are related by intervals within the same interval class, $g, g^{-1} \in [g]$. In contrast, $\{a, c\}$ and $\{b, d\}$ relate by an interval in a different interval class, $[g^2]$, which consists exclusively of the involution g^2.

Let V be the subset of involutions in a group G, and let $G^{\#}$ be the subset of all non-identity elements in G. We determine the number n of interval classes in G according to the following formula:

$$n = \left\lfloor \frac{|V| - |G^{\#}|}{2} \right\rfloor \tag{1}$$

For a structure to be all-interval in relation to some group G, the total number n of interval classes in G needs to be a triangular number.[1] In terms of the cyclic interval groups that underlie our earlier examples, $\mathbb{Z}_n$ has no involutions when n is odd, and it has one involution, $n/2$, when n is even. Therefore, $\mathbb{Z}_7$ has three interval classes. $\mathbb{Z}_{12}$ has six, including one that consists of an involution. The group of intervals, however, need not be cyclic to allow for all-interval structures; indeed, it need not even be abelian. For instance, G could be the order-8 dihedral subgroup of the usual, order-24 transposition and inversion group, which has a regular action on the space of major and minor triads in the octatonic pitch-class collection $\{0, 1, 3, 4, 6, 7, 9, 10\}$. The group has five involutions: T_6 and the four inversions, I_1, I_4, I_7, and I_{10}. As a result, this non-abelian subgroup has six interval classes. The four-member subset of triads that contains C major, E-flat major, C minor, and F-sharp minor–which relate to one another by intervals in each of those interval classes–serves an example of an all-interval structure for this subgroup and space (Fig. 7).

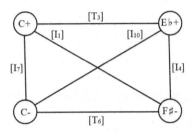

Fig. 7. A non-abelian all-interval structure

The orders of interval groups for all-interval structures of a given size may vary accordingly from smaller (groups that contain a maximal number of involutions) to larger (groups that contain a minimal number of involutions). Consequently, information regarding the number of involutions in a group is useful [5]. Let S be an all-interval structure of size k. Then, S has intervals in $t = \binom{k}{2}$ interval classes. The smaller groups that may potentially accommodate this number of interval classes are among the generalized dihedral groups–i.e., those which are a semidirect product of an abelian group and $\mathbb{Z}_2$ (see [6]). Half or more than half of the elements in these groups are involutions.[2] For instance, five of the

---

[1] N.B.: The condition of triangularity is necessary, but not sufficient. For instance, 21 is a triangular number, but there are no all-interval heptachords in $\mathbb{Z}_{43}$, as 6 is not a power of a prime (see below).

[2] Among the 46 groups of order 32 that are not generalized dihedral, only 2 have half or more of their elements that are not involutions, but the number of interval classes for both these groups is 25, which is not a triangular number. Similarly, of the 257 groups of order 64 that are not generalized dihedral, half or more than half the elements of three of these groups are involutions, but none of them have a triangular number of interval classes: 49 for one such group, and 51 for the other two.

six interval classes in D_8 are involutions. It is the smallest group that allows for all-interval tetrachords (i.e., there are no groups of order 7 with six elements as involutions, and no other groups of order 8 with five involutions).

Of particular interest, all the non-identity elements of $\mathbb{Z}_2^n$ are involutions. For instance, $\mathbb{Z}_2^2$ is the smallest group structure that allows for all-interval trichords; it has three interval classes. Similarly, the smallest group structure in which we find all-interval hexachords is $\mathbb{Z}_2^4$, with its fifteen interval classes. However, $2^n - 1$ is generally not a triangular number. The only other example of practical size (i.e., less than 10000) is $2^{12} - 1 = \binom{91}{2} = 4095$. This case has an interesting application in pitch-class set theory. Let S be the power set for $\mathbb{Z}_{12}$, i.e., the 4096 pitch-class sets. Construct a hypercube graph whose vertices are labeled with the members of S, such that an edge connects any two vertices whose labels differ by the inclusion or exclusion of a single pitch class. As such, any such edge represents a unit of Hamming distance ([7] discusses this structure in greater detail). We conjecture the existence of a class of all-interval 91-chords for this subgroup and space.

Whereas no single category of smallest groups exists for all-interval structures of any size, one class of the largest groups is always available: $\mathbb{Z}_{2t+1}$, which contains a triangular number of interval classes and no involutions. For instance, the largest group that allows for all-interval tetrachords is $\mathbb{Z}_{13}$; any larger involution-free group would have the wrong number of interval classes. Nevertheless, some sizes of all-interval structures allow for more than one isomorphism class of largest groups. In cases where $2t + 1$ is not prime, we may find other group structures. In particular, Theorem 18.68 in [8] states that for any planar cyclic difference set with $\lambda = 1$ and $k \equiv 2 \pmod 3$, there exists a non-abelian difference set with the same parameters. For example, the largest groups for $t = 10$ are of order 21, for which there are two isomorphism classes. Here we find all-interval pentachords ($k = 5$) not only in $\mathbb{Z}_{21}$, but also in $\mathbb{Z}_7 \rtimes \mathbb{Z}_3$, which is non-abelian and contains no involutions.

These highest-order groups also have a connection to two-dimensional projective geometry. A classical result in that field states that a finite projective plane of order q exists for any q that is a power of a prime p, and that such a plane has $q^2 + q + 1$ points [9, p. 501]. We note here preliminarily that for any all-interval structure of size k, where $k = q + 1$, we may label the points of a projective plane of order q in the manner of our introductory examples, based on following equivalence.

$$q^2 + q = 2\binom{k}{2}, \quad \text{for any } q = p^n \tag{2}$$

Such structures are also known in combinatorics as *planar difference sets* [10].

3   The Z Relation

The Z relation, systematized by Forte [2], is a classical topic in pitch-class set theory. It relates pitch-class sets with the same interval content, particularly those that are otherwise not related by transposition and/or inversion. Among other significant, more recent studies, Lewin [11] expands this notion to generalized interval systems, including those that incorporate non-abelian groups; Mandereau et al. [12] extend it to multisets and beyond, using the Patterson function of crystallography (as a more general cognate to Lewin's interval function).

In Sect. 1, we noted that both {E, F♯, A} and {E, G♯, A} are all-interval trichords within the E-major diatonic pitch-class space. As such, they have the same interval-class vector, $\langle 1, 1, 1 \rangle$; hence, they are Z related to one another. We say, however, that this Z relation is trivial, as all sets that are related by transposition or inversion have the same interval content, and these two sets are related by diatonic inversion. Nevertheless, we note a further correspondence: these trichords are members of the same orbit of pitch-class sets under the action of the group of affine transformations on that space. The full set of fourteen all-interval trichords that are available in E major comprises that orbit entirely.

In reference to our geometric model of all-interval structures in Sect. 2.1, affine transformations do not preserve distances, but they do preserve ratios of distances. Consequently, because of the unique composition of all-interval structures in modular spaces (i.e., their possessing one and only one of each and every interval within the modular space), such transformations have the effect of permuting the set of intervals, while also preserving them as a set.[3] For the cyclic groups $\mathbb{Z}_n$, where $n = 2t$ or $n = 2t + 1$, such affine groups consist of the maps $x \mapsto ax + b$, where $b \in \mathbb{Z}_n$ and a is an invertible element, which yields $n \times \Phi(n)$ affine transformations. However, the orbit of all-interval structures under $\mathrm{Aff}(\mathbb{Z}_n)$ is not necessarily of the same size. For instance, $|\mathrm{Aff}(\mathbb{Z}_{12})| = 12 \cdot 4 = 48$, which is the same number as all-interval tetrachords in that space (two T/I set-classes of size 24 each). In contrast, $|\mathrm{Aff}(\mathbb{Z}_{13})| = 13 \cdot 12 = 156$, but its orbit of all-interval tetrachords is of size 52 (two T/I set classes of size 26 each, where each tetrachord is stabilized as a set by three members of the affine group).

In the theory of generalized interval systems, we may find structures with the same intervallic content that are related by neither GIS transposition nor GIS inversion[4] (nor, in the non-abelian case, by members of the group of interval-preserving transformations, the centralizer of the interval group in the symmetric group on the space): Lewin's GISZ-related sets [11]. However, reference to affine

---

[3] This situation does not imply that all Z relations are the result of affine transformations for sets with flat interval distributions other than 1. For instance, $\mathbb{Z}_{31}$ contains several different (31,15,7) difference sets; hence, they are Z-related. However, they are not all related to one another by multiplication (from personal correspondence with Jonathan Wild). See [8, p. 420] for more information in this connection.

[4] "In an abstract GIS, the operation of GIS-transposition by interval i is well-defined by the formula int(s, Ti(s)) = i. That is, for any object s, the Ti-transform of s is that unique object Ti(s) which lies the interval i from s." [11, p. 42]. "In an abstract GIS, given any objects y and v (where v may be the same object as y), the operation I of y/v GIS-inversion is well-defined by the formula int(v, I(s)) = int(s, y). The formula expresses a pertinent intuition: given any object s, its inverted transform I(s) lies intervallically in relation to v, exactly as y lies intervallically in relation to s." [11, p. 43].

transformations in certain generalized interval systems is not appropriate (e.g., those in which intervals are not comparable to vectors; see [13]). Nevertheless, we find a suitable analog: the affine group for $\mathbb{Z}_n$ is also precisely the normalizer of $\mathbb{Z}_n$ in the symmetric group on $\mathbb{Z}_n$, noting that the centralizer (Lewin's group of interval-preserving transformations) is a subgroup of the normalizer.

$$\mathrm{Aff}(\mathbb{Z}_n) = \mathrm{N}_{\mathrm{Sym}(\mathbb{Z}_n)}\mathbb{Z}_n \tag{3}$$

As a result, we observe that, for a generic group G with an action on a space S, the orbit of any all-interval structure under the action of the normalizer of G in the symmetric group on S, $\mathrm{N}_{\mathrm{Sym}(S)}G$, comprises the set of all-interval structures available in the space.

For example, let G be the non-abelian group of order 57, isomorphic to $\mathbb{Z}_{19} \rtimes \mathbb{Z}_3$, with a regular action on a space P. This group has twenty-eight interval classes, with none that contain involutions. We are able to find 6498 all-interval octachords in P. The normalizer N of G in the symmetric group on P is of order 19494. The orbit of an all-interval octachord S under N is of size 6498. Yet, there exist only $6498/3 = 2166$ all-interval octachords in P that are related to S by GIS transposition/inversion and/or some interval-preserving operation. Consequently, we find non-trivially GISZ-related triples in the orbit of S under N. (This fact can be verified as a consequence of the existence of more traditionally Z-related tetrachordal triples in $\mathbb{Z}_{19}$.)

3.1 All-Interval Structures and Projective Planes

As we indicated above, a significant relationship exists between all-interval structures and finite projective planes. We explore here in more depth their general connection, and we consider some particular examples of small-order (i.e., $q \leq 11$) structures. A finite projective plane consists of a set of points and a coincident set of lines. Essentially, the following three axioms summarize its structure [14]:

A1. Any two distinct points are contained in one and only one line.
A2. Any two distinct lines contain one and only one point in common.
A3. There exist four points, no three of which are contained in one line.

A standard result in projective geometry states that a Desarguesian projective plane of order q exists for any q that is a power of a prime: $\mathrm{PG}(2, q)$. Such a structure contains $q^2 + q + 1$ points, and the same number of lines. Each line is incident with $q + 1$ points, and each point is incident with $q + 1$ lines.

For our purposes, let G be an interval group that has a regular action on a space P of size $q^2 + q + 1$. Let S be an all-*directed*-interval structure of size $q + 1$ in P, and call S^G the orbit of S under G. There is a structural affinity between S^G and $\mathrm{PG}(2, q)$ that permits us to label the plane's points in the manner of Fig. 3. Lewin's (directed) interval vector provides a tally of invariants that result under transformations by the members of an interval group. For any all-directed-interval structure S, all entries in the interval vector for S contain 1 s.

Therefore, we observe that one and only element in S will be held invariant under transformation by any $g \in G$. This fact agrees with the second axiom, A2, above: any and all pairs of transformations of S under the members of G have only one element in common. Moreover, A1 is a consequence of the lack of involutions in the regular action of G. For any set $S^g \in S^G$, the dyadic subset of S^g between whose members the interval $h \in G$ obtains is stabilized only by the identity element of the group; therefore, these elements can appear together in only one set. Finally, A3 prohibits the so-called degenerate planes. It also follows from the regular action of the interval group. In short, we are able to axiomatize S^G in a manner that corresponds precisely to A1–3 above.

The smallest projective plane is PG(2, 2), the Fano plane, which we considered above in Fig. 3. We may label the points and lines of this plane with all-interval trichords, where the intervals are defined in terms of a group of order 7. As there is only one isomorphism class of groups of this order, $\mathbb{Z}_7$, all such trichords must be defined in terms of an appropriate cyclic group. Indeed, interval groups of size 7 are discussed frequently in the literature, particularly because of the traditional example of diatonic pitch-class space. As we noted previously, there are two orbits of all-interval trichords under the action of the affine group on such a space. For $\mathbb{Z}_7$, these orbits are transposition classes of $[0, 1, 3]$ and $[0, 2, 3]$, which are inversionally related. Hence, they are only trivially Z-related.

The second smallest projective plane is PG(2, 3). It contains $3^2 + 3 + 1 = 13$ points (as well as lines). Its points and lines can be labeled with G-orbits all-interval tetrachords. Again, we find only one isomorphism class of interval groups for its prime order: $\mathbb{Z}_{13}$. Unlike $\mathbb{Z}_7$, however, the orbit of an all-interval tetrachord under Aff($\mathbb{Z}_{13}$) yields four transposition classes, not merely two. These four transposition classes partition into two pairs whose pitch-class set members are also related to each other by inversion; hence, we note a Z-relation between mod-13 T/I set classes $[0, 2, 3, 7]$ and $[0, 1, 4, 6]$. Lewin [15] discusses some musically relevant spaces of order 13 in connection with this plane, although he does not make a connection to interval groups or all-interval structures. We suggest here the additional possibility of a 13-member "white-key space," embedded in 24-ET, similar to the maximally even embedding of the usual 7-note white-key space in 12-ET.

PG(2, 4) is the first projective plane for which the number of points, $4^2 + 4 + 1 = 21$, is not prime. As a result, we find more than one relevant isomorphism classes of interval groups: $\mathbb{Z}_{21}$, which is abelian; and $\mathbb{Z}_7 \rtimes \mathbb{Z}_3$, which is non-abelian. $\mathbb{Z}_{21}$ allows for all-interval pentachords, but presents no non-trivial Z relations among them. The orbit of any one such pentachord (e.g., $\{0, 2, 7, 8, 11\}$) under the action of $N_{\mathrm{Sym}(\mathbb{Z}_{21})}\mathbb{Z}_{21}$ is of size $21 \cdot 2 = 42$ (i.e., it is merely the mod-21 T/I set class). Similarly, in the non-abelian case, we find no GISZ relations among the 294 all-interval pentachords in the orbit of an all-interval structure under the action of $N_{\mathrm{Sym}(\mathbb{Z}_{21})}(\mathbb{Z}_7 \rtimes \mathbb{Z}_3)$. In [15], Lewin also discusses some musical representations of PG(2, 4), but, again, not in the context of interval groups or all-interval structures. Nevertheless, one of his representations is of special interest in the non-abelian case: its points are labeled with members of the set of twenty-one diatonic unordered dyads. Here, the three diatonic set classes each contain seven dyads, which can be modeled well by diatonic transposition under $\mathbb{Z}_7$; then, $\mathbb{Z}_3$ has an action that permutes the set of three set classes.

In terms of all-interval structures, PG(2, 5) is similar to the other planes above that have a prime number of points. The single associated interval group is cyclic of order 31. Such an interval group is appropriate to the study of 31-ET, which relates to a significant music-historical tradition dating back to the 17th century [16]. Each of the six lines in PG(2, 5) has six points, which can be labeled with the members of G-orbits of all-interval hexachords. In $\mathbb{Z}_{31}$, we find ten transposition classes of all-interval hexachords, which partition into five Z-related mod-31 T/I set classes. Likewise, PG(2, 8) has a prime number of points. The number of points for PG(2, 9) and PG(2, 11) are not prime, but each presents only one cyclic isomorphism class of interval groups (in both cases, any semi-direct product is also a direct product).

PG(2, 7) is structurally similar to PG(2, 4): it contains a non-prime number of lines and points, 57, and has one associated class of abelian interval groups and one non-abelian. As we discussed in Sect. 3, we may label these lines and points with all-interval octachords, whose intervals derive from either $\mathbb{Z}_{57}$, or from $\mathbb{Z}_{19} \rtimes \mathbb{Z}_3$. In the abelian case, these octachords comprise six T/I Z-related set classes. In the non-abelian case, we find GISZ-related triples.

The general question of the existence of projective planes of orders that are not prime powers is still open. The nonexistence of a projective plane of order 6 has been proven [17], based on Euler's famous "Graeco-Latin-square" conjecture [18], and our computer testing indicates that no all-interval heptachords exist in $\mathbb{Z}_{43}$. Likewise, the non-existence of a projective plane of order 10 has been proven by massive computer calculation [19]. Not surprisingly, we have been unable to locate any all-interval 11-chords in $\mathbb{Z}_{111}$.

4 Conclusions

Several open questions exist for future work. A more general study of all-interval structures that incorporates tools from geometrical music theory would be of interest. The spaces investigated here that employ such techniques are comparatively rudimentary. Do all-interval structures exist in other, more complex musical spaces (e.g., non-modular, non-Cartesian, etc.)? If so, what do they look (or, sound) like? Are there sets of all-interval structures that can be modeled with non-Desarguesian (e.g., Hall or Hughes) planes? What can be said about all-interval 13-chords? Do any exist (and are there any associated projective planes of order 12)? There are six isomorphism classes of groups of order 156 that have 78 interval classes (with no involutions). These may all be tested for all-interval structures by computer, but is there also a more abstract, logical answer to such existence questions?

All-interval structures are of considerable music-compositional and analytical significance, particularly in the post-tonal repertoire. The frequent incorporation of the well-known all-interval tetrachords in $\mathbb{Z}_{12}$ by composers such as Berg,

Babbitt, and Carter, and many others, is a testament to the usefulness of their intervallic efficiency. It is our hope that the generalizations and specific examples that appear in this study will shed new light on further uses of these types of structures.

References

1. Gamer, C., Wilson, R.: Microtones and projective planes. In: Fauvel, J., Flood, R., Wilson, R. (eds.) Music and Mathematics: From Pythagoras to Fractals. Oxford University Press, Oxford (2003)
2. Forte, A.: The Structure of Atonal Music. Yale University Press, New Haven (1977)
3. Tymoczko, D.: A Geometry of Music: Harmony and Counterpoint in the Extended Common Practice. Oxford University Press, Oxford (2011)
4. Lewin, D.: Generalized Musical Intervals and Transformations. Yale University Press, New Haven (1987)
5. Aschbacher, M., Meierfrankenfeld, U., Stellmacher, B.: Counting involutions. Ill. J. Math. **45**, 1051–1060 (2001)
6. Popoff, A.: Building generalized Neo-Riemannian groups of musical transformations as extensions. J. Math. Music **7**, 55–72 (2013)
7. Peck, R.W.: A hypercube-graph model for n-tone rows and relations. In: Yust, J., Wild, J., Burgoyne, J.A. (eds.) MCM 2013. LNCS, vol. 7937, pp. 177–188. Springer, Heidelberg (2013)
8. Jungnickel, D., Pott, A., Smith, K.: Difference sets. In: Rosen, K. (ed.) Handbook of Combinatorial Designs, 2nd edn, pp. 419–435. Chapman & Hall/CRC, Boca Raton (2007)
9. Lidl, R., Niederreiter, H.: Finite Fields, 2nd edn. Cambridge University Press, Cambridge (1997)
10. Bruck, R.H.: Difference sets in a finite group. Trans. Am. Math. Soc. **78**, 464–481 (1955)
11. Lewin, D.: Conditions under which, in a commutative GIS, two 3-Element sets can span the same assortment of GIS-Intervals; notes on the non-commutative GIS in this connection. Intégral **11**, 37–66 (1997)
12. Mandereau, J., Ghisi, D., Amiot, E., Andreatta, M., Agon, C.: Z-relation and homometry in musical distributions. J. Math. Music **5**, 83–98 (2011)
13. Tymoczko, D.: Generalizing musical intervals. J. Music Theor. **53**, 227–254 (2009)
14. Hall Jr., M.: Projective planes. Trans. Am. Math. Soc. **54**, 229–277 (1943)
15. Lewin, D.: Some compositional uses of projective geometry. Perspect. New Music **42**, 12–63 (2004)
16. Rossi, L.: Sistema musico, overo, Musica speculativa: dove si spiegano i più celebri sistemi di tutti i tre generi. Nella stampa episcopale, per Angelo Laurenzi (1666)
17. Bose, R.C.: On the application of the properties of Galois fields to the problem of construction of Hyper-Graeco-Latin squares. Sankhy **3**, 323–339 (1938)
18. Euler, L.: Recherches sur une nouvelle espèce de quarrés magiques. Vehr. Zeeuwsch. Genootsch. Wetensch. Vlissengen **9**, 85–239 (1782)
19. Lam, C.W.H.: The search for a finite projective plane of order 10. Amer. Math. Mon. **98**, 305–318 (1991)

Unifying Tone System Definitions: Ordering Chromas

Tobias Schlemmer[(✉)]

TU Dresden, 01066 Dresden, Germany
Tobias.Schlemmer@tu-dresden.de

Abstract. This article suggests a tone system definition that uses abstract ordered groups as interval groups. The common operation of octave identification is replaced by a folding operation for ordered sets that is based on the factorisation of groups by normal subgroups. The resulting structure is used to provide a chroma system definition that can preserve a certain amount of the order relation. Additionally, a path is shown that allows the integration of the theory of Generalized Interval Systems into the extensional language that has been proposed by Rudolf Wille and Wilfried Neumaier.

Keywords: Order relation · Directed graph · Tone system · Chroma system · Binary relation orbifold · Generalized interval system (GIS)

1 Introduction

The pitch-order of tones plays an important role in everyday music perception and performance. This linear order mainly describes melodic properties of the underlying tone system. In the discussion about consonance, dissonance or harmonic properties of a tone system other description schemes like the Tonnetz have been used. While the pitch-order has been used in the discussion about these constructions, the linear order is often not fully compatible with the underlying structure. For example the Tonnetz of just octaves, just fifths, and just thirds give rise to a tree-dimensional order relation that is linked to the harmonic distance between certain tones.

Order relations cannot be preserved under octave identification. However, the perception of Shepard tones suggests that the notions "higher" and "lower" for pitches are still convenient to a certain extent for such chroma systems.

Musical language is full of ambiguities. Some of them arise from the usual laziness, that makes the scientific life easier like the different meanings of a "fifth" as scale step, interval, chordal or harmonic form (cf. [15,16]), some have been imported from other scientific fields as the language of American set theory [4]. The latter is also an example where a keen development of the mathematical language can improve the readability. Such a suggestion has been made in the 80's by Rudolf Wille for the German language. The current article tries to extend this language to aspects of mathematical models that have been developed independently.

© Springer International Publishing Switzerland 2015
T. Collins et al. (Eds.): MCM 2015, LNAI 9110, pp. 291–302, 2015.
DOI: 10.1007/978-3-319-20603-5_30

The development of an extensional language for music theory by Rudolf Wille and Wilfried Neumaier has been necessary by the need for a musical programming language to drive the famous MUTABOR instrument [5,9,17]. Thus, its design goals include extensionality, simplicity and intuition. So it reuses as many musical notions as possible while it dissolves ambiguities in order to provide a language that can be understood by humans and used as a programming language. It turns out that many of the different mathematical concepts already have their counterparts in the musical language. Some of them have been used mainly in a more psychological context.

This reference to existing musical language makes it a useful and accessible tool for algebraic and set theoretic considerations, as it encourages the reader to use also the intuition that comes from the other fields. So he could do some calculations using his intuition and verify the individual steps afterwards. Considering mathematical music theory, this is a big advantage over computational models like the denotator theory [7]. The latter has been proven to be useful for expressing conceptual structures in computational environments [8]. However, it has the mathematical disadvantage that the reader has to memorise a large system of new axioms and rules in order to use it. Many of these axioms and rules do not directly lead to the solution of their problem.

The present article will provide a suggestion for a mathemusical extensional language, that is based mainly on the ideas of Rudolf Wille [11,13,16] and David Lewin [6]. We follow the presentation of Wille's theory as it is the only one that is guided by the intermediate philosophical interpretation, which is necessary to develop a mathematical model that is close to a natural phenomenon.

After some preliminary notes in Sect. 3 the main mathematical results are mentioned that are needed to generalise Wille's language to arbitrary ordered groups. It turns out that the common interpretation of chroma systems as cyclic ordered structures cannot be directly generalised to the multi-dimensional case without losing its relational character. Instead, the group theoretic language of binary relation orbifolds, as discussed in [1,2,14,19,20], has been reformulated using small categories in order to provide a better interpretation in an order-like fashion.

In Sect. 4 this theory is then applied to the basic notions of Wille's extensional language in order to derive a generalised chroma system definition that is equipped with a structure that is as near to a partial order relation as possible.

The final section touches on further questions that are related to the present work.

2   Preliminaries

A *directed graph* G is a tuple $G = (V, E, \sigma, \tau)$, such that V is a set, whose elements are called *vertices*, E is another set whose elements are called *edges* or *arrows*, $\sigma : E \to V$ and $\tau : E \to V$ are mappings. The set of arrows from a node $v \in V$ to a node $w \in V$ is denoted by $\mathfrak{Mor}_G(v, w)$. In the same way the operators $\mathfrak{Mor}_G := E$ and $\mathfrak{Ob}\, G := V$ refer to set of the edges and the set of the vertices of the graph G. A directed graph together with an associative partial binary operation $*$ is called *small category* iff for all edges $\mathfrak{a}, \mathfrak{b} \in E$ with $\tau\mathfrak{a} = \sigma\mathfrak{b}$ the product $\mathfrak{a} * \mathfrak{b}$ exists and for each node $v \in V$ the set $\mathfrak{Mor}_G(v, v)$ together with the operation $*$ is a monoid, whose neutral element is called *identity* of v, which is denoted by id_v.

A *preordered group* $(\mathbb{G}, \leq)$ with underlying set G is a group $\mathbb{G}$ that is equipped with a preorder relation $\leq$ that fulfils the implication $a \leq b \Rightarrow xay \leq xby$ for all elements $a, b, x, y \in G$. If $\leq$ is an order relation it is called *(partially) ordered group* or *po-group*. An element $a \geq 1$ is called *infinitesimal* with respect to another element $b \geq 1$ iff its cyclic group lays below b: $\forall x \in \langle a \rangle_\mathbb{G} : x \leq b$.

3 Folding Ordered Groups

This section shortly describes the mathematical background of the chroma definition from the following section. It is closely related to the notion of a binary relation orbifold that has been used by Daniel Borchmann and Bernhard Ganter [1,2], while the general idea has been described by Monika Zickwolff [19,20].

Let $(\mathbb{G}, \leq)$ be an ordered group. Then, the group $\mathbb{G}$ acts semiregularly on the ordered set $(G, \leq)$ and for each element $a \in G$ the left multiplication with a is *translative*, as defined in the following way: any element $x \in G$ satisfies $x \leq ax$ iff all elements $y \in G$ also satisfy $y \leq ay$. The same is true for right multiplication by a. We call a permutation group $\mathbb{H}$ *translative* on a set with a binary relation (e.g. a simple directed graph) iff all elements $h \in \mathbb{H}$ are translative.

For every graph (V, E) and every permutation group $\mathbb{P}$ that acts on the set of vertices V, we can define an equivalence relation $\approx$ in the following way:

1. For all vertices $a, b \in V$, we define $a \approx b$ iff there exists an element $p \in \mathbb{P}$ such that $a^p = b$, and
2. For all edges $(a, b), (c, d) \in \leq$, we define $(a, b) \approx (c, d)$ iff there exists an element $p \in \mathbb{P}$ such that $(a^p, b^p) = (c, d)$.

This equivalence relation allows us to form the quotient graph and we define $(V, E) /\!/ \mathbb{P} := (G/_{\approx}, E/_{\approx})$, which we call *graph orbifold* corresponding to binary relation orbifolds as defined in [2].

It is well-known that the order relation $\leq$ gives rise to a small category $\mathfrak{K}$. Throughout this section let N be a fixed normal subgroup of $\mathbb{G}$. Then we call the graph orbifold $(G, \leq) /\!/ N$ an *order orbifold*. Let T be a transversal of the vertex partition that is generated by the equivalence $\approx$, i.e. the equivalence classes of all vertices $a \in \mathbb{G}$ fulfil the equation $|[a]_\approx \cap T| = 1$. Then a *canonical annotation* A for each edge $\mathfrak{a} \in \mathfrak{Mor}_\mathfrak{K}$ exists such that

$$A([\mathfrak{a}]) = hg^{-1}, \text{ where } g, h \in N \text{ and } \exists s, t \in T : s^g = \sigma\mathfrak{a} \text{ and } t^h = \tau\mathfrak{a}.$$

On the other hand each group $\mathbb{G}'$ can be considered as a category with one vertex whose arrows are the group elements. When $\mathfrak{K}'$ is a small category, $\mathbb{G}'$ is a group and $A : \mathfrak{K}' \to \mathbb{G}'$ is a contravariant functor, then the tuple $(\mathfrak{K}', A, \mathbb{G}')$ is called a *representation*.

Every representation can be unfolded into a category $\mathfrak{K}' \circ_A \mathbb{G}'$ in the following way: The vertices of $\mathfrak{K}' \circ_A \mathbb{G}'$ are given by the set $\mathfrak{Ob}\,\mathfrak{K}' \times \mathfrak{Mor}_{\mathbb{G}'}$, and the arrows between two nodes (v, g) and (w, h) are given by the set

$$\mathfrak{Mor}_{\mathfrak{K}' \times \mathfrak{Mor}_{\mathbb{G}'}} \big((v, g), (w, h)\big) := \{\mathfrak{a} \in \mathfrak{Mor}_{\mathfrak{K}'}(v, w) \mid A(\mathfrak{a}) = hg^{-1}\}$$

together with the composition from the category $\mathfrak{K}'$.

For every order orbifold a canonical annotation gives rise to a canonical representation. It can be unfolded into a category that is isomorphic to the original order relation. In the case of a po-group the unfolded po-group representation is also isomorphic to the original po-group, in the sense of po-groups.

For every po-group the orbifold inherits the property that every group element $g \in \mathbb{G}$ induces an automorphism of the category via right multiplication with g. In a similar way we can define infinitesimal automorphisms: an automorphism $\varphi \in \mathfrak{Aut}\,\mathfrak{K}'$ is called *infinitesimal* with respect to another automorphism $\psi \in \mathfrak{Aut}\,\mathfrak{K}'$ iff there exists a vertex $v \in \mathfrak{Ob}\,\mathfrak{K}'$ such that for all automorphisms $\xi \in \langle \varphi \rangle$ every arrow $\mathfrak{a} \in \mathfrak{Mor}_{\mathfrak{K}'}\big(v, \psi(v)\big)$ can be written as a composite $\mathfrak{a} = \mathfrak{b} * \mathfrak{c}$ where $\mathfrak{b} \in \mathfrak{Mor}_{\mathfrak{K}'}\big(v, \xi(v)\big)$ and $\mathfrak{c} \in \mathfrak{Mor}_{\mathfrak{K}'}(\xi(v), \psi(v))$ are two arrows that have $\xi(v)$ as a common vertex.

Let (V, E, σ, τ) be a directed graph, $\mathbb{G}$ a group and $A' : E \to \mathbb{G}$ a mapping. Then A' can be extended to a contravariant functor A'' from the path category $\mathfrak{P}$ of (V, E, σ, τ) to the group $\mathbb{G}$ considered as category. Furthermor, let $\sim$ be the equivalence relation that is defined for any two arrows $\mathfrak{a}, \mathfrak{b} \in \mathfrak{Mor}_{\mathfrak{P}}$ by

$$\mathfrak{a} \sim \mathfrak{b} :\Leftrightarrow \sigma\mathfrak{a} = \sigma\mathfrak{b} \text{ and } \tau\mathfrak{a} = \tau\mathfrak{b} \text{ and } A''(\mathfrak{a}) = A''(\mathfrak{b})$$

Then a representation $\big(\mathfrak{K}', B_H, \mathbb{H}\big)$ *is generated by* $\big((V, E, \sigma, \tau), A'', \mathbb{G}\big)$ iff: the groups $\mathbb{G}$ and $\mathbb{H}$ are isomorphic, the category $\mathfrak{K}'$ is isomorphic to the quotient $\mathfrak{P}/_{\sim}$ of the path category of the graph (V, E, σ, τ) by the relation $\sim$, and the following diagram commutes:

$$
\begin{array}{ccc}
\mathfrak{K}' & \longleftrightarrow & \mathfrak{P}/_{\sim} \\
\downarrow{\scriptstyle B_H} & & \downarrow{\scriptstyle A} \\
\mathbb{H} & \longleftrightarrow & \mathbb{G}
\end{array}
$$

In this case the tuple $\big((V, E, \sigma, \tau), A'', \mathbb{G}\big)$ is called a *generator* of the representation $(\mathfrak{K}', A, \mathbb{G})$.

Let $e = \big((V, E, \sigma, \tau), A'', \mathbb{G}\big)$ be a generator of the representation $(\mathfrak{K}', A, \mathbb{H})$. Then, e is called

Simple iff the graph (V, E, σ, τ) is a simple graph,

Antisymmetric iff all edges of the the graph (V, E, σ, τ) between two vertices $v, w \in V$ have the same direction (i.e. $\mathfrak{Mor}_{(V,E,\sigma,\tau)}(v, w) = \emptyset$ or $\mathfrak{Mor}_{(V,E,\sigma,\tau)}(w, v) = \emptyset$).

$\mathbb{H}$-automorphic, where $\mathbb{H} \leq \mathfrak{Aut}\,\mathfrak{K}'$ iff $\mathbb{H}$ is a subgroup of the automorphism group of the category $\mathfrak{K}'$ and induces a subgroup of $\mathfrak{Aut}(V, E, \sigma, \tau)$,

Complete iff every arrow $\mathfrak{a} \in \mathfrak{Mor}_{\mathfrak{K}'}$ has its counterpart in E whenever there exist two arrows $\mathfrak{b}, \mathfrak{c}$ such that $\mathfrak{c}$ is generated by a simple edge in E and one of the equations $\mathfrak{c} = \mathfrak{a} * \mathfrak{b}$ or $\mathfrak{c} = \mathfrak{b} * \mathfrak{a}$ hold.

Optimal iff it is antisymmetric, $\mathfrak{Aut}\,\mathfrak{K}'$-automorphic, complete and the graph (V, E, σ, τ) is maximal in the set of all graphs that fulfil these conditions with respect to graph embeddings.

Not every ordered group does permit an antisymmetric and complete generator. For example the additive group of the rational numbers $(\mathbb{Q}, +, \leq)$ can be folded by the subgroup of the powers of three $\langle\{3^z \mid z \in \mathbb{Z}\}\rangle$, but the corresponding representation does not have an antisymmetric and complete generator.

If the automorphism group of the category $\mathfrak{K}'$ has infinitesimal automorphisms that are somehow related to arrows of the same category, there is no complete and antisymmetric generator. Nevertheless, in many cases there exists a generator that is somehow complete up to infinitesimality. So far no characterisation has been provided. It is an open question to resolve this problem.

4   Tone Structures

Tones and intervals are the fundamental bricks in music theory. Tones are often considered as elements of ordered sets, where the ordering may be achieved by the analysis of the pitches of the tones, their metric position, their duration, or other parameters. This section provides the basic notions of a language that takes these aspects into account. Some of the definitions are illustrated by musicological remarks. However, these examples cannot be discussed in full detail here.

The basic musical structure in this paper is the notion of a tone structure. It can be used to express a huge variety of structures, including most western scores or even those compositions as performative structures that include additional data. Another application is the description of temperaments.

Definition 1. *A* tone structure *is a tuple* $\mathbb{T} = (T, \delta, \mathbb{I}, P)$ *where* T *is a set whose elements are called* tones, $\mathbb{I}$ *is a preordered group whose elements are called* intervals *and whose neutral interval* $1_{\mathbb{I}}$ *is called* prime, P *is a set of mappings with domain* T *called intrinsic tone parameters of* $\mathbb{T}$, $\delta : T \times T \to \mathbb{I}$ *is a mapping which satisfies the following equations for all elements* $t_1, t_2, t_3 \in T$:

$$\delta(t_2, t_3)\delta(t_1, t_2) = \delta(t_1, t_3) \tag{1}$$

A tone $t_1 \in T$ *is called* higher *than another tone* $t_2 \in T$, *denoted by* $t_1 > t_2$, *iff the interval* $\delta(t_2, t_1) > 1_{\mathbb{I}}$ *i.e. strictly larger then the prime. In that case* t_2 *is called* lower *than* t_1. *If the interval between them is equivalent to the prime, then both are considered to be* equally high, *and this is denoted by* $t_1 \sim t_2$.

As the equivalence of the tone preorder is derived from the equivalence of the interval preorder, we use the same symbol $\sim$ for both equivalence relations if the correct meaning can be derived from the context.

Proposition 1. *For each tone structure the relation $<$ is a strict order relation. Furthermore $\sim$ is an equivalence relation, while $\leq := <\cup\sim$ is a preorder relation.*

Proposition 2. *Let $t_1, t_2, t_3, t_4 \in T$ be four tones with the equivalences $t_1 \sim t_3$ and $t_2 \sim t_4$. Then the intervals $\delta(t_1, t_2)$ and $\delta(t_3, t_4)$ are also equivalent, i.e. the relation $\delta(t_1, t_2) \sim \delta(t_3, t_4)$ holds.*

Tone structures were first described by Wilfried Neumaier and Rudolf Wille [13]. In contrast to this definition they use the ordered additive Group $\mathbb{R}\mathcal{O}$ as the fixed interval group ("Größenbereich") that is not part of the signature of the tone system.

Tone parameters can be anything that seems to be useful. Some common parameters include pitch, loudness, onset, duration, end and different kinds of articulations. Having this in mind, we can define a *composition* as a tone structure that has at least two out of the three parameters onset, duration and end as tone parameters.

When it comes to transpositions, we must distinguish between transposition-ally invariant parameters and transposition-dependent parameters. For example in Rudolf Wille's definition of a composition the metric parameters onset, duration and end are transpositionally invariant, while a pitch parameter is transposition-dependent. However, we explicitly also want to consider such tone structures where metric information is included in the interval group. David Lewin discusses such tone structures in more detail [6].

Obviously, for each tone structure and every parameter, we can find a super-tone-structure or a sub-tone-structure, such that this parameter is transpositionally invariant or transposition-dependent. Many of these structures have a useful musical interpretation. Thus, the differentiation between transposition-dependent and transpositionally invariant parameters must be provided by the application and cannot be defined in the general mathematical model.

Definition 2. *Let $\mathbb{T} = (T, \delta, \mathbb{I}, P)$ be a tone structure, P' a subset of its parameters and let $i \in I$ be an interval. A P'-invariant transposition by i is a partial injective mapping $T_i : T \to T$ such that for each tone $t \in \operatorname{Dom} T_i$ and each parameter $p \in P'$ the equations $\delta\big(t, T_i(t)\big) = i$ and $p(t) = p\big(T_i(t)\big)$ hold.*

In many cases transpositions can be used in a simplified way. If the set of transpositionally invariant tone parameters is fixed and either clear from the context or the given property is defined for every subset of the tone parameters, we omit the set of tone parameters. If not otherwise stated every transposition of a set of transpositions refers to the same set of transpositionally invariant parameters.

One might prefer to consider each transposition to be defined for all tones, however this does not describe all musical situations. For example, both in medieval music as well as in music performances the tone space is limited and limits the opportunities for transpositions.

In general a transposition by an interval $i \in I$ is not necessarily unique. In order to allow a minimum of algebraic structure, we characterise certain sets of transpositions as being consistent with respect to the composition of mappings in the following way:

The transpositions of a tone structure form a monoid with a partial left associative monoid action on the tone set. The prime transposition is the neutral element and the transposition with empty domain is an absorbing element. As this condition is not preserved by the subset formation, we must enforce it.

Definition 3. *Let* $\mathbb{T} = (T, \delta, \mathbb{I}, P)$ *be a tone structure, a set S of transpositions is called* (transpositional) consistent *iff there exists a partial mapping $\vartheta : I \to S$, that is both injective and surjective such that for all intervals $i \in I$ and $j \in I$ and all tones $t \in T$ the following equation holds whenever ϑi and ϑj are defined and $t \in \mathrm{Dom}(\vartheta i)$ and $\vartheta i(t) \in \mathrm{Dom}(\vartheta j)$:*

$$(\vartheta i \circ \vartheta j)(t) := \vartheta j\big(\vartheta i(t)\big) = \vartheta(ij)(t). \tag{2}$$

This allows us to ensure a limited algebraic structure for transposable tone structures. David Lewin [6] uses tone systems (see below). So for each interval i there exists at most one transposition by i. Rudolf Wille's first definition [16] is similar. Wilfried Neumaier [10] circumvents the problem, as he forces all tone parameters except pitch to be invariant under transpositions. This clashes with the general intuition of a transposition in musical analysis, as certain parameters like articulations or voice assignments may be different between transpositions. Later in their joint work [12] the definition was a little bit fuzzy, so Jan Thomas Winkler applied Wille's original definition to tone structures [18]. The resulting relational definition restricts the domain of a transposition to the subset of tones where it is uniquely defined by an interval. The latter condition is violated in many tone structures. So we need another suggestion for the definition.

Definition 4. *Let* $\mathbb{T} = (T, \delta, \mathbb{I}, P)$ *be a tone structure and J a set of intervals. The tone structure $\mathbb{T}$ is called J-*transposable *iff there exists a transpositionally consistent set of transpositions S such that for every interval $i \in J$, a transposition $T_i \in S$ by i exists, which is defined for all tones $t \in T$. The tone structure $\mathbb{T}$ is called* transposable *iff it is $\delta[T \times T]$-transposable.*

For tone systems (see below) another type of functions plays a similar role as the transpositions.

Definition 5. *Let* $\mathbb{T} = (T, \delta, \mathbb{I}, P)$ *be a tone structure. A (partial) mapping μ is called* interval preserving *iff the following equation holds for all tones $t_1, t_2 \in T$:*

$$\delta(t_1, t_2) = \delta\big(\mu(t_1), \mu(t_2)\big) \tag{3}$$

In the same way as the transpositions, the interval preserving mappings can be considered as a monoid with a partial right associative monoid action on the tone set.

Another aspect of tone structures is their ability to provide the pool from which tones are chosen for a certain application. Such a supply of tones is called a tone system:

Definition 6. *A tone structure* $\mathbb{T} = (T, \delta, \mathbb{I}, P)$ *is called a* tone system *iff it fulfils the following conditions:*

1. *The interval group* I *is a po-group, and*
2. *For any two tones* $t_1, t_2 \in T$ *the interval between them is the prime (i.e.* $\delta(t_1, t_2) = 1_{\mathbb{I}}$*) iff* $t_1 = t_2$.

Proposition 3. *For a tone system the preorder of the tones is an order relation.*

If we fix in a tone system $(T, \delta, \mathbb{I}, P)$, a tone $t_0 \in T$ and define the annotation function

$$\beta : T \to I : t \mapsto \delta(t_0, t), \tag{4}$$

then the transpositions can be described as left multiplication by an interval, while the right multiplication corresponds to interval preserving mappings. Consequently a transposition T_i is also interval preserving if its interval i commutes with all intervals from $\delta[T \times T]$ and an interval preserving (partial) map is a transposition if it commutes with all intervals from the same set.

Now, we are in the position to integrate David Lewin's definition of a GIS. As every generalisation is always also a specialisation in a certain sense, we omit the attribute "generalised" to allow further generalisations of this structure without confusion. In order to distinguish the structures from other areas it is more useful to add the scope, if necessary, e.g. "algebraic interval system" or "formal tone system".

Definition 7. *A tone structure* $(T, \delta, \mathbb{I}, P)$ *is called* interval structure, *if it is transposable and its interval mapping* δ *is surjective. Furthermore, if it is a tone system, it is called an* interval system.

In order to justify the different notions of tone systems and interval systems, we should have a look at the origin of both definitions. David Lewin uses interval structures in order to analyse the relationships between different parts of the same musical composition. Rudolf Wille and his team used tone systems to define a tuning language. Thus, there must be a theory that can be used to describe retunings. This difference will become clearer, as we define the corresponding morphisms.

When we want to characterise a retuning as a morphism, it is a transformation by a small interval that retains the relation $\sim$.

Definition 8. *Let* $\mathbb{T}_1 = (T_1, \delta_1, \mathbb{I}_1, P_1)$ *and* $\mathbb{T}_2 = (T_2, \delta_2, \mathbb{I}_2, P_2)$ *be two tone structures. A mapping* $\varphi : T_1 \to T_2$ *is called* tone structure morphism *if it fulfils the following condition: for all tones* $t_1, t_2 \in T_1$ *such that* $\delta_1(t_1, t_2) = 1_{\mathbb{I}_1}$ *follows* $\delta_2(\varphi(t_1), \varphi(t_2)) = 1_{\mathbb{I}_2}$.

Obviously, such a morphism preserves equality between intervals between certain pairs of tones, but it does not necessarily preserve the order relation or the interval group. Both properties of such morphisms can be found in music theory. This will be illustrated in the following sentences.

Consider, for example, a Tonnetz of just thirds, fifths and octaves with interval group $\mathbb{Z}^3$ preordered by the frequency of the tones. When we factor out both the syntonic and the diatonic commas, the resulting structure is a chroma structure (see Definition 10) with a chroma interval group that is isomorphic to $\mathbb{Z}$. But the preorder relation gets transformed into something that is not an order relation. However, it is a common and quite successful practice to consider the resulting structure as an (equally tempered) tone system. This shows that tone system morphisms do not necessarily have to be order preserving. Exploring further musicological implications of such morphisms would exceed the scope of this article.

Tuning a composition from an equal tempered tuning to something else, e.g. some well-tempered tuning or mean-tone, will destroy the group structure in most cases. Most of these tunings have different intervals between scale steps that have the same interval in the equal tempered tuning. This topic has been touched in [15].

Interval structure morphisms have been introduced by Fiore et al. [3].

Definition 9 (Fiore et al.). *Let* $\mathbb{T}_1 = (T_1, \delta_1, \mathbb{I}_1, P_1)$ *and* $\mathbb{T}_2 = (T_2, \delta_2, \mathbb{I}_2, P_2)$ *be two tone structures. A pair* (φ, ψ) *of mappings* $\varphi : T_1 \to T_2$ *and* $\psi : \mathbb{I}_1 \to \mathbb{I}_2$ *is called an* interval structure morphism *if* φ *is a tone structure morphism,* ψ *is a group morphism and the following diagram commutes:*

$$
\begin{array}{ccc}
T_1 \times T_2 & \xrightarrow{\varphi \times \varphi} & T_2 \times T_2 \\
\downarrow{\scriptstyle \delta_1} & & \downarrow{\scriptstyle \delta_2} \\
\mathbb{I}_1 & \xrightarrow{\psi} & \mathbb{I}_2
\end{array}
\tag{5}
$$

If both tone structures are tone systems the pair (φ, ψ) *is called* interval system morphism.

Sub-tone-systems are considered as sub-structures that are tone systems. Using the usual definition of automorphisms and partial automorphisms, we can state that every interval preserving map exists as the tone mapping of a partial interval structure automorphism whose interval group homomorphism is the identity.

Most of the literature about music theory is not based on tones, but on classes of tones that are formed by grouping tones according to similar behaviour. Again, they can be described by the coincidence of certain tone parameters. But that's not the way they are used. The usual definition is based on certain subgroups of the interval group.

Definition 10. *Let* $\mathbb{T} = (T, \delta, \mathbb{I}, P)$ *be a tone structure and* $N \lhd \mathbb{I}$ *a normal subgroup of the interval group. Furthermore let* T_N *be a transpositionally consistent set of transpositions and* $\vartheta : N \to T_N$ *the corresponding mapping. The tuple* (C, δ_N, G, P_N) *is called* chroma structure *of* $\mathbb{T}$ *iff it fulfils the following conditions:*

1. *The structure* $G = (V, E)$ *is a directed graph,* $A : \mathfrak{Mor}_{\mathbb{I} /\!/ N} \to N$ *is a canonical annotation, and* $\mu : E \to \mathfrak{Mor}_{\mathbb{I} /\!/ N}$ *such that the tuple* $(G, \mu \circ A, N)$ *is a* $\mathbb{I}_{/\!/ N}$-*automorphic generator of the representation* $(\mathbb{I} /\!/ N, A, N)$, *where we identify the set of nodes* V *with* $\mathbb{I}_{/\!/ N}$ *via the isomorphism* ι.

2. *C is a set, there exists a bijective mapping* ν : $T/\!/N \to C$, *and* δ_N : $C \times C \to V$ *is a mapping such that the following diagram commutes:*

$$
\begin{array}{ccc}
T \times T & \xrightarrow{\ \ \ \ \ \ \ \ \ \ \delta \ \ \ \ \ \ \ \ \ \ } & \mathbb{I} \\
\downarrow{\scriptstyle \cdot/\!/N} & & \downarrow{\scriptstyle \cdot/N} \\
T/\!/N \times T/\!/N & \xrightarrow{\ \nu\ } C \times C \xrightarrow{\ \delta_N\ } V \xrightarrow{\ \iota\ } & I/N
\end{array}
\qquad (6)
$$

3. P_N *is the set of tone parameters which are constant on the equivalence classes* t^N *for each tone* $t \in T$.

In this case the elements of C *are called* chromas *and the vertices of the graph* G *are called* chroma intervals. *For* $c, d \in C$, *we call* c *"lower or equally high" to* d *iff* G *contains an edge from* $\delta_N(c, c)$ *to* $\delta_N(c, d)$.

   Furthermore, the tuple (C, δ_N, G, P_N) *is called* chroma system (chroma interval structure; chroma interval system) *if* T *is a tone system (interval structure; interval system).*

   The chroma system (C, δ_N, G, P_N) *is called* simple, $\mathbb{H}$-automorphic, antisymmetric, *or* A-complete *if the generator* $(G, \mu \circ A, N)$ *has this property.*

Note that a chroma structure does not uniquely define the underlying tone structure. Given a chroma structure, the tone structure depends on the annotation, which is not always unique with respect to the chroma system.

   As an example, consider the Tonnetz of just octaves, just fifths and just thirds as abstract tone system $(\mathbb{Z}^3, -^T, (\mathbb{Z}, \leq)^3, \emptyset)$ with the distance mapping $a -^T b := b - a$. The octave, the Pythagorean and the syntonic commas generate a subgroup $N := \langle \{(1, 0, 0), (0, 12, 0), (0, -4, 1)\} \rangle$. Let us consider the graph $G := (\mathbb{Z}_{12}, \mathbb{Z}_{12} \times \{1, 2, 3\}, \sigma, \tau)$ with $\sigma(i, j) := i$, $\tau(i, 1) := i + 1$, $\tau(i, 2) := i + 4$, and $\tau(i, 3) = i + 5$, and the annotation $A : \mathbb{Z}_{12} \times \{1, 2, 3\} \to \mathbb{Z}^2$ that is given by the formulae $A(i, 1) := (0, 4, 0)$ for $i \in \{3, 7, 11\}$, $A(i, 2) := (0, 0, 3)$ for $i \in \{8, 9, 10, 11\}$, $A(i, 3) := (0, 4, 0)$ for $i \in \{3, 7\}$, $A(i, 3) := (0, 0, 3)$ for $i \in \{8, 9, 10\}$, $A(11, 3) := (0, 4, 3)$, and $A(i, j) = (0, 0, 0)$ otherwise. Then the tuple (G, A, N) is a simple, antisymmetric and complete generator for the orbifold $(\mathbb{Z}, \leq)^3 /\!/N$ that is not optimal, and $(\mathbb{Z}_{12}, -^T, G, \emptyset)$ is a chroma system of $(\mathbb{Z}^3, -^T, (\mathbb{Z}, \leq)^3, \emptyset)$ with the same properties.

   An optimal chroma structure is very close to the description of a chroma system that fulfils all the requirements that are commonly used in the literature. But there are some limitations which will be discussed in the following section.

5   Conclusion and Further Questions

In this article, an extensional standard language for music has been extended to cover certain aspects of interval structures and order relations.

   Relational aspects as studied in American set theory [4] are not in the scope of the present article. However, they also have their handy counterparts in the theory. The original definitions can be directly applied to the tone system definition that is given in this paper. A short English description of the corresponding definitions has been used in [15].

The mathematical structure that has been used throughout this text allows us to consider order relational aspects also in folded structures like chroma systems. The definition of a chroma structure in this article is based on the broad usage of chroma systems in the literature. While many articles implicitly refer to octave identification, others properly describe the abstraction on which their mathematical structure is based on. Only in very few cases the usual mathematical questions of uniqueness and the relationships between different levels of abstraction are discussed.

In the relationship between tone structures and chroma structures two main aspects are important. Firstly, the chroma system cannot distinguish between tones of the same chroma. Secondly the expansion of a chroma system into a tone system is not unique. It depends on an annotation function that can provide different kinds of group extensions. Thus, the unfolding of a chroma system is not uniquely defined.

Optimal chroma systems are very handy as long as there are no infinitesimal elements in the interval group. Such systems are called Archimedian. This implies that all aspects are equally important with respect to the order relation. However, one can consider an interval group where both the pitch order and the temporal order of chromas are important, while one gets precedence over the other. In such cases one would model the corresponding tone structure using an interval group with infinitesimal elements. If the structure is then folded according to a subgroup of the infinitesimal elements, the above definition of an optimal chroma system fails even if there exists an optimal chroma system in the equally-important case.

Nevertheless in such cases we can use the latter as the best chroma system and try to construct a chroma system with similar properties from the original tone system. Such a chroma system we would call a chroma system that is optimal up to infinitesimality. As such systems exist in many cases, it would be useful to have a characterisation of the corresponding generators available. This question is linked to a description of sufficiently order-convex sets in the interval order, that may be used in the discussion of scales. Both topics offer open problems and are subject of further research activities.

References

1. Borchmann, D.: Context orbifolds. Diplomarbeit, Technische Universität Dresden (2009)
2. Borchmann, D., Ganter, B.: Concept lattice orbifolds – first steps. In: Ferré, S., Rudolph, S. (eds.) ICFCA 2009. LNCS, vol. 5548, pp. 22–37. Springer, Heidelberg (2009)
3. Fiore, T.M., Noll, T., Satyendra, R.: Morphisms of generalized interval systems and PR-groups. J. Math. Music 7(1), 3–27 (2013). http://dx.doi.org/10.1080/17459737.2013.785724

4. Forte, A.: The Structure of Atonal Music, 3rd edn. Yale University Press, New Haven (1979)
5. Ganter, B., Henkel, H., Wille, R.: MUTABOR - ein rechnergesteuertes Musikinstrument zur Untersuchung von Stimmungen. Acustica **57**(4–5), 284–289 (1985)
6. Lewin, D.: Generalized Musical Intervals and Transformations. Oxford University Press, New York (2007)
7. Mazzola, G., Göller, S., Müller, S.: The Topos of Music: Geometric Logic of Concepts, Theory, and Performance. Birkhäuser, Basel (2002)
8. Milmeister, G.: The Rubato Composer Music Software. Computational Music Science. Springer, Heidelberg (2009)
9. Mutabor team: Mutabor - the dynamic tempered piano. Website (2012). http://www.math.tu-dresden.de/~mutabor/, (Archived by WebCite® at http://www.webcitation.org/6ASACEUxB). Accessed 5 September 2012
10. Neumaier, W.: Was ist ein Tonsystem?: Eine historisch-systematische Theorie der abendländischen Tonsysteme, gegründet auf die antiken Theoretiker Aristoxenos, Eukleides und Ptolemaios, dargestellt mit Mitteln der modernen Algebra. In: Quellen und Studien zur Musikgeschichte von der Antike bis in die Gegenwart, vol. 9. Lang, Frankfurt am Main (1986)
11. Neumaier, W.: Eine exakte Sprache der Musiktheorie. Musiktheorie **4**(1), 15–25 (1989)
12. Neumaier, W., Wille, R.: Extensionale Standardsprache der Musiktheorie: Eine Schnittstelle zwischen Musik und Informatik. In: Informatik Spektrum (1989)
13. Neumaier, W., Wille, R.: Extensionale Standardsprache der Musiktheorie - eine Schnittstelle zwischen Musik und Informatik. In: Hesse, H.P. (ed.) Mikrotöne III: Bericht über das 3. internationale Symposium Internationale Symposium "Mikrotonforschung, Musik mit Mikrotönen, Ekmelische Musik" Veröffentlichungen der Gesellschaft für Ekmelische Musik, vol. 6, pp. 149–167. Helbling, Innsbruck (1990)
14. Schlemmer, T.: Annotating lattice orbifolds with minimal acting automorphisms. In: Szathmary, L., Priss, U. (eds.) CLA 2012, pp. 57–68, October 2012. http://cla.inf.upol.cz/papers.html
15. Schlemmer, T., Andreatta, M.: Using formal concept analysisto represent chroma systems. In: Yust, J., Wild, J., Burgoyne, J.A. (eds.) MCM 2013. LNCS, vol. 7937, pp. 189–200. Springer, Heidelberg (2013)
16. Wille, R.: Mathematische Sprache in der Musiktheorie. In: Fuchssteiner, B., Kulisch, U., Laugwitz, D., Liedl, R. (eds.) Jahrbuch Überblicke Mathematik 1980, pp. 167–184. Bibliographisches Institut, Mannheim (1980)
17. Wille, R.: Eulers Speculum Musicum und das Instrument MUTABOR. DMV-Mitteilungen **4**, 9–12 (2000)
18. Winkler, J.T.: Algebraische Modellierung von Tonsystemen. Beiträge zur be-grifflichen Wissensverarbeitung. Verlag Allgemeine Wissenschaft – HRW e.K., Mühltal (2009)
19. Zickwolff, M.: Darstellung symmetrischer Strukturen durch Transversale. Contrib. Gen. Algebra **7**, 391–403 (1991)
20. Zickwolff, M.: Rule exploration: first order logic in formal concept analysis. Dissertation, Technische Hochschule Darmstadt (1991)

A Categorical Generalization
of Klumpenhouwer Networks

Alexandre Popoff[1], Moreno Andreatta[2]([✉]), and Andrée Ehresmann[3]

[1] IRCAM, 119 Rue de Montreuil, 75011 Paris, France
al.popoff@free.fr
[2] IRCAM/CNRS/UPMC, Paris, France
Moreno.Andreatta@ircam.fr
[3] LAMFA, Université de Picardie, Amiens, France
andree.ehresmann@u-picardie.fr

Abstract. This article proposes a functorial framework for generalizing some constructions of transformational theory. We focus on Klumpenhouwer Networks for which we propose a categorical generalization via the concept of set-valued poly-K-nets (henceforth PK-nets). After explaining why K-nets are special cases of these category-based transformational networks, we provide several examples of the musical relevance of PK-nets as well as morphisms between them. We also show how to construct new PK-nets by using some topos-theoretical constructions.

Keywords: Transformational theory · K-nets · PK-nets · Category theory

1 Introduction

Since the publication of pioneering work by David Lewin [1,2] and Guerino Mazzola [3,4] respectively in the American and European formalized music-theoretical tradition, transformational approaches have established themselves as an autonomous field of study in music analysis. Surprisingly, although group action-based theoretical constructions, such as Lewin's "Generalized Interval Systems" (GIS), are naturally described in terms of categories and functors, the categorical approach to transformational theory remains relatively marginal with respect to the major trend in math-music community [5–7].

Within the transformational framework, which progressively shifted the music-theoretical and analytical focus from the "object-oriented" musical content to the operational musical process, Klumpenhouwer Networks (henceforth K-nets), as observed by many scholars, have stressed the deep synergy between set-theoretical and transformational approaches thanks to their anchoring in both group and graph theory [10]. Following David Lewin's [11] and Henry Klumpenhouwer's [12] original group-theoretical description, theoretical studies have mostly focused until now on the underlying algebra dealing with the automorphisms of the T/I group or of the more general T/M affine group [11,13].

T. Collins et al. (Eds.): MCM 2015, LNAI 9110, pp. 303–314, 2015.
DOI: 10.1007/978-3-319-20603-5_31

This enables one to define the main notions of positive and negative isographies, a notion which can easily be extended to more isographic relations by taking into account the group generated by affine transformations, together with high-order isographies via the recursion principle. This surely provides a computational framework for building networks of networks (and so on, in a recursive way) but it somehow misses the interplay between algebra and graph theory, which is well captured by category theory and the functorial approach. Since a prominent feature of K-nets is their ability of instantiating an in-depth multi-level model of musical structure, category theory seems nowadays the most suitable mathematical framework to capture this recursive potentiality of the graph-theoretical construction.

Following the very first attempt at formalizing K-nets as limits of diagrams within the framework of denotators [14], we propose in this article a categorical construction, called poly-K-nets and taking values in **Sets** (henceforth PK-nets). This construction generalizes the notion of K-nets in various ways. K-nets theory usually distinguishes two main types of isography: positive and negative isography. Positive isography corresponds to two directed graphs having the same transpositions, but having inversion operators that differ by a constant value. Negative isography corresponds to two directed graphs in which the subscripts of the transpositional and inversional operators sum to constant values, respectively equal to 0 and different from 0. Figure 1 shows four K-nets, the first two of which (K-nets (a) and (b)) are positively isographic, whereas the other two (K-nets (c) and (d)) have the same node-content as the K-nets (a) and (b), but their arrows are labeled in such a way that they are not isographic.

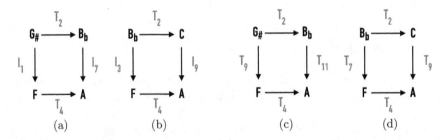

Fig. 1. Four K-nets, the first two of which ((a) and (b)) are (positively) isographic since the transpositions in K-net (b) are the same as those in (a), while the respective inversions in (b) have subscripts 2 more than those of (a). The second two of the four K-nets ((c) and (d)) do not have any isographic relation, although their node content is the same as K-nets (a) and (b)

This clearly suggests that the concept of isography is highly dependent on the selection of specific transformations and asks for more general settings in which isographic networks remain isographic when the nodes are preserved and the family of transformations between the nodes is changed. This has been one of the main motivations for introducing PK-nets as natural extensions of K-nets (Sect. 2). Moreover, PK-nets enable one to compare in a categorical framework digraphs with different cardinalities, a fact that occurs very often in many musically interesting analytical situations. They also realize Lewin's intuition that transformational networks do not necessarily have groups as support spaces, since one can define PK-nets in any category. In this article we focus on the category $\mathbf{PKN_R}$ of PK-nets of constant form R whose (co)limit structure is described in Theorem 1. Morphisms of $\mathbf{PKN_R}$ correspond to a change of a musical context, as in the case of transformations between elements of a cyclic group. Morphisms of PK-nets clearly show the structural role of natural transformations by which one can generalize the case of isographic K-nets (Sect. 3). In particular they enable us to define K-nets which remain isographic for any choice of transformations between the original pitch-classes (or pitch-class sets). The problem of determining the morphisms between PK-nets naturally leads to the topos-theoretic formalization of the main construction we have introduced in this paper, as it will be detailed in the final section.

2   From K-Nets to PK-nets

We begin this section by giving the definition of a PK-net. In the rest of this paper, all functors are covariant.

2.1   Definition of PK-nets

Let $\mathbf{C}$ be a category, and S a functor from $\mathbf{C}$ to the category **Sets** with non empty values. Such a functor corresponds to an action of the category $\mathbf{C}$ on the disjoint union of the sets $S(c)$ for the objects $c \in \mathbf{C}$ [15].

Definition 1. *Let Δ be a small category and R a functor from Δ to* **Sets** *with non empty values. A PK-net of form R and with support S is a 4-tuple (R, S, F, ϕ), in which F is a functor from Δ to $\mathbf{C}$, together with a natural transformation $\phi\colon R \to SF$.*

The definition of a PK-net is summed up by the following diagram:

The usual K-nets are a particular case of PK-nets in which

1. $\mathbf{C}$ is the group T/I of transpositions and inversions, considered as a single-object category and the functor $S\colon T/I \to$ **Sets** defines the usual action of T/I on the set $\mathbb{Z}_{12}$ of the twelve pitch classes,
2. Δ is the graph describing the K-nets, the functor R associates the singleton $\{X\}$ to each object X of Δ, and the natural transformation ϕ reduces to a map from the objects of Δ to the image of SF.

Within the framework of denotators, Guerino Mazzola and Moreno Andreatta have proposed in [14] a generalized definition of a K-net as an element of the limit of a diagram R of sets (or modules). We can compare the notion of PK-net with this notion of K-net as follows: if (R, S, F, ϕ) is a PK-net, the functor $\lim \colon \mathbf{Sets}^{\Delta} \to \mathbf{Sets}$ maps the natural transformation $\phi \colon R \to SF$ to the map $\lim_{\phi} \colon \lim R \to \lim SF$ from the set of K-nets of form R to the set of K-nets of form SF. Thus a PK-net does not represent a unique K-net, but the set of K-nets associated to SF and a way to 'name' them (via $\lim_{\phi}$) by the K-sets of R. The PK-net reduces to a K-net if $R(X)$ is a singleton for each object X of Δ.

Remark 1. In Definition 1 and in the sequel, the category **Sets** can be replaced by any category H to obtain the notion of a P(oly-)K-net in H (developed in a paper in preparation). Let us note two interesting cases:

i. H is a category of presheaves: the networks considered in [14] correspond to PK-nets in $\mathbf{Mod}_{\mathbf{Z}}^{@}$ of the form (R, S, F, ϕ) where $\mathbf{C} = T/I$ (p. 104), and to PK-nets in a category of presheaves, with F an identity, which they call "network of networks" (pp. 106–107) and they show how to define iterated networks using powerset constructions.

ii. H is the category $Diag(\mathbf{C})$ of diagrams in a category $\mathbf{C}$; in particular, if $\mathbf{C}$ is the category $\mathbf{PKN_R}$ of morphisms of PK-nets, a PK-net in $Diag(\mathbf{PKN_R})$ gives a notion of "PK-net of PK-nets", and by iteration of the $Diag$ construction we can define a hierarchy of PK-nets of increasing orders (without recourse to powerset constructions as in [14]).

In the more general case, the category $\mathbf{C}$ provides the musically relevant transformations, whose action on some musical objects (pitch classes, chords, etc.) is given by the functor S. The form of a PK-net, given by the functor R, provides a diagram of elements which are identified to the musical objects by the functor F and the natural transformation ϕ. The definition of PK-nets provides advantages over K-nets, some of which are detailed in the examples below.

Example 1. The functor R allows one to consider sets $R(X)$, $X \in \Delta$, whose cardinality $|R(X)|$ is greater than 1.

For example, let $\mathbf{C}$ be the group T/I, considered as a single-object category, and consider its natural action on the set $\mathbb{Z}_{12}$ of the twelve pitch classes, which defines a functor $S \colon T/I \to \mathbf{Sets}$. Let Δ be the interval category, i.e. the category with two objects X and Y and precisely one non-trivial morphism $f \colon X \to Y$, and consider the functor $F \colon \Delta \to T/I$ which sends f to T_4.

Consider now a functor $R \colon \Delta \to \mathbf{Sets}$ such that $R(X) = \{x_1, x_2, x_3\}$ and $R(Y) = \{y_1, y_2, y_3, y_4\}$, and such that $R(f)(x_i) = y_i$, for $1 \leq i \leq 3$. Consider the natural transformation ϕ such that $\phi(x_1) = 0$, $\phi(x_2) = 4$, $\phi(x_3) = 7$, and $\phi(y_1) = 4$, $\phi(y_2) = 8$, $\phi(y_3) = 11$, and $\phi(y_4) = 2$. Then (R, S, F, ϕ) is a PK-net of form R and support S which describes the transposition of the C major triad to the E major triad subset of the dominant seventh E^7 chord. This functorial construction is shown in Fig. 2.

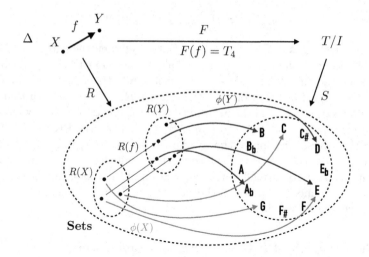

Fig. 2. Diagram showing the functorial construction underlying the definition of PK-nets as applied to the Example 1

Example 2. The definition of PK-nets allows one to consider networks of greater generality than the usual K-nets.

Consider the category $\mathbf{C} = T/I$ and the functor $S \colon T/I \to \mathbf{Sets}$ as in the previous example, and consider the category Δ with one single-object X and one non-trivial morphism $f \colon X \to X$ such that $f^2 = id_X$. Consider now the functor $F \colon \Delta \to T/I$ which sends f to $I_1 \in T/I$.

If we restrict ourselves to functors $R \colon \Delta \to \mathbf{Sets}$ such that $R(X)$ is a single-ton, then there exists no natural transformation $\phi \colon R \to SF$, since the equation $\phi(x) = 1 - \phi(x)$ has no solution in $\mathbb{Z}_{12}$. However, it is possible to consider a functor R such that $R(X) = \{x_1, x_2\}$, with $R(f)(x_1) = x_2$ and *vice-versa*, and a natural transformation ϕ which sends x_1 to 0 and x_2 to 1. Then (R, S, F, ϕ) is a valid PK-net of form R and support S.

Example 3. In addition to groups, the definition of PK-nets allows one the use of any category $\mathbf{C}$. Thus, PK-nets can describe networks of musical objects being transformed by the image morphisms of $\mathbf{C}$ through S.

Consider for example the monoid $M = \{(u, 2^v) \mid u \in \mathbb{Z}[\frac{1}{2}], u \geq 0, v \in \mathbb{Z}\}$ whose multiplication law is given by the following equation:

$$(u_1, 2^{v_1}) * (u_2, 2^{v_2}) = (u_2 + u_1 \cdot 2^{v_2}, 2^{v_1 + v_2}). \tag{1}$$

The monoid M is generated by the elements $a = (1, 1)$ and $b = (0, 1/2)$ and has presentation $M = \langle a, b \mid a^2 b = ba \rangle$. It can be considered as a discrete monoid version of Lewin's continuous group of time-span transformations.

Recall that a time-span, in the sense of Lewin [2], is a pair (t, d), where $t \in \mathbb{R}$ is called the *onset* of the time-span, and $d \in \mathbb{R}, d > 0$, is called its *duration*. Consider the set $T = \{(t, 2^\delta) \mid t \in \mathbb{Z}[\frac{1}{2}], \delta \in \mathbb{Z}\}$ of *dyadic time-spans*, equipped with the action $M \times T \to T$ given by the following equation:

$$(u, 2^v) \cdot (t, 2^\delta) = (t + 2^\delta \cdot u, 2^{\delta+v}) \tag{2}$$

This action defines a functor $S\colon M \to$ **Sets**. Let then the category **C** be the monoid M, and S be the functor as defined above. Let Δ be the interval category, and consider the functor $F\colon \Delta \to M$ which sends the non-trivial morphism $f\colon X \to Y$ to $(2, 1/2) \in M$. Consider the functor $R\colon \Delta \to$ **Sets** such that $R(X)$ and $R(Y)$ are singletons, and the natural transformation $\phi\colon R \to SF$ which sends $R(X)$ to $\{(1,1)\} \subset T$ and $R(Y)$ to $\{(3, 1/2)\} \subset T$. Then the PK-net (R, S, F, ϕ) describes the transformation of the dyadic time-span $(1, 1)$ into $(3, 1/2)$ by the element $(2, 1/2)$ of the monoid M. Observe that, contrary to the group of Lewin, no element of M can describe the transformation of the time-span $(3, 1/2)$ to the time-span $(1, 1)$, since the action of the elements of M only translates time-spans by a positive amount of time.

2.2   The Category of PK-nets with Constant Form

Let Δ be a small category and R a functor from Δ to **Sets** with non empty values. One can form a category **PKN$_R$** of PK-nets of constant form R, according to the following definition.

Definition 2. *The category* **PKN$_R$** *has*

- *objects which are PK-nets (R, S, F, ϕ) of form $R\colon \Delta \to$* **Sets***, and*
- *morphisms between PK-nets (R, S, F, ϕ) and (R, S', F', ϕ') which are pairs (L, λ), where L is a functor from* **C** *to* **C′***, and λ is a natural transformation from S to $S'L$ such that $\phi' = (\lambda F) \circ \phi$.*

The following theorem describes part of the structure of **PKN$_R$**. We omit here the proof, which is rather technical.

Theorem 1. *The category* **PKN$_R$** *is complete, and has all connected colimits.*

Musically speaking, a morphism (L, λ) of PK-nets of constant form R can be interpreted as a change of musical context, through the change of functor from S to S'. We give some examples of such morphisms below.

Example 4. Let the category **C** be the cyclic group $G = \mathbb{Z}_{12}$, generated by an element t of order 12. Consider the action of t on the set $\mathbb{Z}_{12}$ of the twelve pitch classes given by $t \cdot x = x + 1, \forall x \in \mathbb{Z}_{12}$. This defines a functor $S\colon G \to$ **Sets**, which corresponds to the traditional action of $\mathbb{Z}_{12}$ by transpositions by semitones. Consider now the action of t on the set $\mathbb{Z}_{12}$ of the twelve pitch classes given by $t \cdot x = x + 5, \forall x \in \mathbb{Z}_{12}$. This defines another functor $S'\colon G \to$ **Sets**, which corresponds to the action of $\mathbb{Z}_{12}$ by transpositions by fourths. Let L be the automorphism of G which sends $t^p \in G$ to t^{5p} in G, $\forall p \in 1, \ldots, 12$, and let λ be the identity function on the set $\mathbb{Z}_{12}$. It is easily checked that λ is a natural transformation from S to $S'L$.

Let (R, S, F, ϕ) be the PK-net wherein Δ is the interval category, F is the functor from Δ to G which sends the non-trivial morphism $f\colon X \to Y$ of Δ

to t^{10} in G, R is the functor from Δ to **Sets** which sends the objects of Δ to singletons, and ϕ is the natural transformation which sends $R(X)$ to $\{0\} \subset \mathbb{Z}_{12}$ and $R(Y)$ to $\{10\} \subset \mathbb{Z}_{12}$. This PK-net describes the transformation of C to $B\flat$ by a transposition of ten semitones.

By the morphism of PK-nets (L, λ) introduced above, one obtains a new PK-net (R, S', F', ϕ'), wherein the functor $F' = LF$ sends $f \in \Delta$ to $t^2 \in G$, and the natural transformation $\phi' = (\lambda F) \circ \phi$ sends $R(X)$ to $\{0\} \subset \mathbb{Z}_{12}$ and $R(Y)$ to $\{10\} \subset \mathbb{Z}_{12}$. This new PK-net describes the transformation C to $B\flat$ by a transposition of two fourths.

Example 5. We give here an example of a morphism between a PK-net of beats and a PK-net of pitches. Figure 3 shows a passage from the final movement of Chopin's Piano Sonata Nr. 3, op. 58 in B minor, wherein the initial six-notes motive is raised by two semitones every half-bar.

Fig. 3. A passage from the final movement of Chopin's Piano Sonata Nr. 3 op. 58

Let the category **C** be the infinite cyclic group $G = \mathbb{Z}$, generated by an element t. Let $\mathbb{Z}$ be the set of equidistant beats of a given duration and consider the action of t on this set given by $t \cdot x = x + 1$, $\forall x \in \mathbb{Z}$. This action defines a functor $S : G \to$ **Sets**.

Let the category **C'** be the cyclic group $G' = \mathbb{Z}_{12}$, generated by an element t' of order 12. Consider the set $U = \{u_i \mid i \in \mathbb{Z}_{12}\}$ of the twelve successive transpositions of the pitch class set $u_0 = \{10, 11, 0, 3, 4\}$, and consider the action of t' on U given by $t' \cdot u_i = u_{i+1 \pmod{12}}$, $\forall i \in \mathbb{Z}_{12}$. This defines a functor $S' : G' \to$ **Sets**.

Let (R, S, F, ϕ) be the PK-net wherein

- Δ defines the order of the ordinal number 4 (whose objects are labelled X_i),
- F is the functor from Δ to G which sends the non-trivial morphisms $f_{i,i+1} : X_i \to X_{i+1}$ of Δ to t in G,
- R is the functor from Δ to **Sets** which sends the objects X_i of Δ to singletons $\{x_i\}$, and
- ϕ is the natural transformation which sends $R(X_i)$ to $\{i\} \subset \mathbb{Z}$.

This PK-net describes the successive transformations of the initial set by translation of one half-bar in time. Let (R, S', F', ϕ') be the PK-net wherein

- F' is the functor from Δ to G' sending the non-trivial morphisms $f_{i,i+1} \colon X_i \to X_{i+1}$ of Δ to t'^2 in G',
- ϕ' is the natural transformation which sends $R(X_i)$ to $\{u_{2i}\} \subset U$.

This PK-net describes the successive transformations of the initial set $\{10, 11, 0, 3, 4\}$ by transpositions of two semitones.

Consider the functor $L \colon G \to G'$ which sends t to t'^2, together with the natural transformation $\lambda \colon \mathbb{Z} \to U$ given by $\lambda(x) = u_{2x \pmod{12}}$. As compared to a traditional K-net approach which would typically focus on the pitch-class set transformation, the morphism of PK-nets (L, λ) allows us to describe the relation between the translation in time and the transposition in pitch.

Given the knowledge of two functors $S \colon \mathbf{C} \to \mathbf{Sets}$ and $S' \colon \mathbf{C}' \to \mathbf{Sets}$, and a functor $L \colon \mathbf{C} \to \mathbf{C}'$, there may not always exist a natural transformation $\lambda \colon S \to S'L$. The following theorem gives a sufficient condition on S for the existence of the natural transformation $\lambda \colon S \to S'L$.

Theorem 2. *Let $S \colon \mathbf{C} \to \mathbf{Sets}$, $S' \colon \mathbf{C}' \to \mathbf{Sets}$, and $L \colon \mathbf{C} \to \mathbf{C}'$ be three functors, where S' has non empty values. If S is a representable functor, then there exists at least one natural transformation $\lambda \colon S \to S'L$.*

Proof. If S is a representable functor, then there exists a natural isomorphism $\mu \colon S \to \mathrm{Hom}(c, -)$ for some object c of $\mathbf{C}$. We therefore need to prove that there exists at least one natural transformation $\lambda' \colon \mathrm{Hom}(c, -) \to S'L$, as the composition $\lambda' \circ \mu$ will give the desired natural transformation λ. By Yoneda Lemma, the natural transformations from $\mathrm{Hom}(c, -)$ to $S'L$ are in bijection with the elements of $S'L(c)$, thus there exists at least one λ' since $S'L(c)$ is supposed to be non empty. $\qquad\square$

An immediate corollary of this result is that, given a PK-net (R, S, F, ϕ) where S is a Generalized Interval System (GIS), a functor $S' \colon \mathbf{C}' \to \mathbf{Sets}$, and a functor $L \colon \mathbf{C} \to \mathbf{C}'$, one can always form a new PK-net $(R, S', F' = LF, \phi')$. Indeed, from a result of Vuza [16] and Kolman [8], a GIS is known to be equivalent to a simply transitive group action on a set, which is in turn equivalent to a representable functor from the group (as a single-object category) to **Sets**. The previous theorem can then be used to form the new PK-net.

Note that given a functor $S \colon \mathbf{C} \to \mathbf{Sets}$ and a functor $L \colon \mathbf{C} \to \mathbf{C}'$, it is known that there always exists a functor $S_K \colon \mathbf{C}' \to \mathbf{Sets}$ and a natural transformation $\kappa \colon S \to S_K L$ obtained by the Kan extension [17] of S along L. Any other natural transformation $\lambda \colon S \to S'L$, where S' is a functor from $\mathbf{C}'$ to **Sets**, factors through it.

3 Application of PK-net Morphisms to Isographic Networks

We have seen previously that, given two functors $S \colon \mathbf{C} \to \mathbf{Sets}$ and $S' \colon \mathbf{C}' \to \mathbf{Sets}$, a morphism of PK-nets is a pair (L, λ) where L is a functor from $\mathbf{C}$ to $\mathbf{C}'$, and λ is a natural transformation from S to $S'L$ such that

$\phi' = (\lambda F) \circ \phi$. One notable feature which is directly derived from the definition of the natural transformation λ in a PK-Net morphism is that, given two objects X and Y in $\mathbf{C}$, and two elements $x \in S(X)$, $y \in S(Y)$ such that $y = S(f)(x)$ for some morphism $f \in \mathbf{C}$, we have

$$\lambda(y) = \lambda(S(f)(x)) = S'L(f)(\lambda(x)), \tag{3}$$

In other words, whatever the transformation f in $\mathbf{C}$ which relates the elements x and y, their images by λ are related by the image transformation $L(f)$.

This property is all the more interesting in the case $S = S'$, which covers the case of isomorphic networks (see below). However, in the general case, the problem of determining the existence of a natural transformation $\lambda: S \to SL$ given the functors L and S has no obvious solution. It can nevertheless be solved for some particular cases: we consider here the case when $\mathbf{C}$ is a "generalized" group of transpositions and inversions, with an application to isographic networks.

Let $\mathbf{C}$ be the dihedral group D_{2n} of order $2n$ whose presentation is given by $\langle T, I \mid T^n = I^2 = ITIT^{-1} = 1 \rangle$. By analogy with the T/I group, the elements of D_{2n} are designated by $T_n = T^n$, and by $I_n = T^n I$. Consider the set $\mathbb{Z}_n$ of pitch classes in n-equal temperament (n-TET), equipped with the action of D_{2n} given by $T \cdot x = x + 1$, and $I \cdot x = -x$, for all $x \in \mathbb{Z}_n$. This defines a functor $S: \mathbf{C} \to \mathbf{Sets}$. The following theorem establishes the existence of natural transformations $\lambda: S \to SL$, for a given automorphism L of D_{2n}.

Theorem 3. *Let L be an automorphism of $\mathbf{C} = D_{2n}$. Then:*

- *if n is even, there exists either 0 or 2 natural transformations $\lambda: S \to SL$.*
- *if n is odd, there exists exactly one natural transformation $\lambda: S \to SL$.*

Proof. From known results about dihedral groups, an automorphism L of D_{2n} sends, for all $p \in \mathbb{Z}_n$, the elements $T_p \in D_{2n}$ to T_{kp}, and the elements $I_p \in D_{2n}$ to I_{kp+l}, where $k, l \in \mathbb{Z}_n$, with $\gcd(k, n) = 1$. Assume that there exists a natural transformation $\lambda: S \to SL$, which defines (by an abuse of notation) a function $\lambda: \mathbb{Z}_n \to \mathbb{Z}_n$. Given any element x of the set $\mathbb{Z}_n$, the definition of a natural transformation imposes $T_{kp} \cdot \lambda(x) = \lambda(T_p \cdot x)$ for all $p \in \mathbb{Z}_n$, which leads to the equation $kp + \lambda(x) = \lambda(p + x)$. By setting $x = 0$, we have that $\lambda(p) = kp + \lambda(0)$, for all $p \in \mathbb{Z}_n$. Similarly, given any element x of the set $\mathbb{Z}_n$, the definition of λ imposes the equation $I_{kp+l} \cdot \lambda(x) = \lambda(I_p \cdot x)$, which leads to $kp + l - \lambda(x) = \lambda(p - x)$, for all $p \in \mathbb{Z}_n$. We therefore obtain a condition on $\lambda(0)$ given by the equation $kp + l - \lambda(0) = kp + \lambda(0)$, which reduces to $l - \lambda(0) = \lambda(0)$. If n is odd, this equation has exactly one solution, given by $\lambda(0) = l/2$ if l is even, and by $\lambda(0) = (n + l)/2$ if l is odd. If n is even and l is odd, then the equation has no solution. Finally, if n and l are even, this equation has two solutions $\lambda(0) = l/2$ and $\lambda(0) = (n + l)/2$. $\square$

We now give an application to isographic networks. We have previously introduced two isographic K-nets (see the Fig. 1, networks (a) and (b)). These can be considered as PK-nets (R, S, F, ϕ) and $(R, S, F' = LF, \phi')$, wherein the category $\mathbf{C}$ corresponds to the usual T/I group, the functor $S: \mathbf{C} \to \mathbf{Sets}$ corresponds to the usual action of T/I on the set $\mathbb{Z}_n$ of pitch classes, and the functor

$L: T/I \rightarrow T/I$ is the automorphism which sends $T_p \in T/I$ to T_p, and $I_p \in T/I$ to I_{p+2}. Figure 1 presents one PK-net (R, S, F'', ϕ'') with the same pitch classes, wherein the functor F'' labels the arrows with transpositions (Fig. 1, networks (c)). Figure 1 also shows the PK-net (b) where arrows are labelled with transpositions: the two PK-nets (c) and (d) of Fig. 1 are not isographic, which could be deduced from the fact that, in this particular case, we have $F'' = LF''$.

By the previous theorem, there exists two natural transformations from S to SL, given by the functions $\lambda_1(x) = x + 1$ and $\lambda_2(x) = x + 7$, for $x \in \mathbb{Z}_{12}$. By the PK-net morphisms (L, λ_1) and (L, λ_2) applied to (R, S, F, ϕ), two new PK-nets are obtained, which are represented in Fig. 4. The reader can verify that, for any other choice of transformations between the original pitch-classes, these PK-nets remain isographic to the initial one.

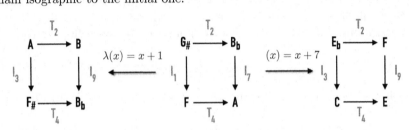

Fig. 4. Isographic PK-nets

4   PK-nets and Topoi

It is a well-known result that for any small category $\mathbf{C}$, the category of functors $\mathbf{Sets}^{\mathbf{C}}$ is a topos. The category $\mathbf{Sets}^{\mathbf{C}}$ therefore has a subobject classifier Ω, and for any subobject $A \in \mathbf{Sets}^{\mathbf{C}}$ of an object $B \in \mathbf{Sets}^{\mathbf{C}}$, there exists a *characteristic map* $\chi_A : B \rightarrow \Omega$. Topoi have found applications in music theory, for example in the work of Mazzola [5] and more recently in the work of Noll, Fiore and Satyendra [6,9].

In the context of PK-nets, the characteristic map can be considered as a morphism of PK-nets. Let (R, S, F, ϕ) be a PK-net of form R and of support $S \in \mathbf{Sets}^{\mathbf{C}}$. Let A be a subobject of S: this defines a characteristic map $\chi_A : S \rightarrow \Omega$ which is equivalent to a morphism of PK-nets $(\mathrm{id}_{\mathbf{C}}, \chi_A)$. This morphism thus defines a new PK-net (R, Ω, F, ϕ'). We detail below a concrete example based on the monoid introduced in Example 3.

Example 6. Consider the monoid $M = \{(u, 2^v) \mid u \in \mathbb{Z}[\frac{1}{2}], u \geq 0, v \in \mathbb{Z}\}$ introduced above, acting on the set $T = \{(t, 2^\delta) \mid t \in \mathbb{Z}[\frac{1}{2}], \delta \in \mathbb{Z}\}$ of *dyadic time-spans*, which defines a functor $S \in \mathbf{Sets}^M$. For a given $k \in \mathbb{Z}[\frac{1}{2}]$, the set $T_k = \{(t, 2^\delta) \mid t \in \mathbb{Z}[\frac{1}{2}], t \geq k, \delta \in \mathbb{Z}\}$ equipped with the same action of M is a subobject A of S. Let $\mathbb{Z}[\frac{1}{2}]_{\geq 0}$ be the set $\{p \in \mathbb{Z}[\frac{1}{2}] \mid p \geq 0\}$. The reader can verify that the subobject classifier of $\mathbf{Sets}^M$ is the union of $\mathbb{Z}[\frac{1}{2}]_{\geq 0}$ and a singleton $\{x\}$, equipped with the following action of the generators of M

$$(1,1) \cdot p = \begin{cases} p - 1 & \text{if } p \geq 1 \\ 0 & \text{otherwise} \end{cases} \tag{4}$$

and

$$(0, 1/2) \cdot p = 2p \tag{5}$$

for all $p \in \mathbb{Z}[\frac{1}{2}]_{\geq 0}$, and where the singleton $\{x\}$ is a fixed point by the action of M. The characteristic map χ_A then sends any element $(t, 2^\delta) \in T$ to $\dfrac{k - t}{2^\delta}$ if $k \geq t$, or 0 otherwise. In other words, the characteristic map measures the time period $t - k$ from a time-span $(t, 2^\delta)$ in units of 2^δ. Consider for example the PK-net (R, S, F, ϕ) defined in Example 3, and the subobject A defined from the set $T_{9/2}$. The morphism of PK-nets (id_M, χ_A) sends the PK-net (R, S, F, ϕ) to the new PK-net (R, Ω, F, ϕ'), where the natural transformation $\phi' = \chi_A \circ \phi$ sends $R(X)$ to $\{7/2\}$ and $R(Y)$ to $\{3\}$.

5   Conclusions and Perspectives

We have presented a generalized framework of Klumpenhouwer Networks based on category theory. In order to show the richness of this new framework we have chosen some pedagogical examples by focusing on the concept of set-valued PK-nets of constant form R and the category $\mathbf{PKN_R}$ they form. This construction stresses the categorical description of musical objects based on the synergy between algebra and graph-theory, as it is the case for Klumpenhouwer Networks and other constructions within transformational theory. The category $\mathbf{PKN_R}$ is the category of objects under R of the category $Diag(\mathbf{Sets})$ of diagrams in Sets which we denote by $\mathbf{PKN}$; the morphisms of $\mathbf{PKN}$ are all the PK-nets (where the diagram Δ and/or the functor R may vary). Our current research addresses the question of the musical relevance of $\mathbf{PKN}$, of $Span(\mathbf{PKN})$, and of different constructions based on $\mathbf{PKN}$, such as the construction of PK-nets of higher order (see Remark 1), or the characterization and musical applications of a notion of PK-homographies generalizing the problem of isographies.

We are also planning to better study some computational aspects underlying PK-nets, once they are integrated into some programming languages for computer-aided music theory and analysis, such the MathTools environment in OpenMusic [18]. This will probably enable one to better understand the cognitive and perceptual relevance of transformational theory and contribute to the programmatic research area of a categorical approach to creativity [19].

References

1. Lewin, D.: Transformational techniques in atonal and other music theories. Perspect. New Music **21**(1–2), 312–371 (1982)
2. Lewin, D.: Generalized Music Intervals and Transformations. Yale University Press, New Haven (1987)
3. Mazzola, G.: Gruppen und Kategorien in der Musik: Entwurf einer mathematischen Musiktheorie. Heldermann, Lemgo (1985)

4. Mazzola, G.: Geometrie der Töne. Birkhäuser, Basel (1990)
5. Mazzola, G.: The Topos of Music: Geometric Logic of Concepts, Theory, and Performance. Birkhäuser, Basel (2002)
6. Fiore, T.M., Noll, T.: Commuting groups and the topos of triads. In: Agon, C., Andreatta, M., Assayag, G., Amiot, E., Bresson, J., Mandereau, J. (eds.) MCM 2011. LNCS, vol. 6726, pp. 69–83. Springer, Heidelberg (2011)
7. Popoff, A.: Towards A Categorical Approach of Transformational Music Theory. Submitted
8. Kolman, O.: Transfer principles for generalized interval systems. Perspect. New Music **42**(1), 150–189 (2004)
9. Fiore, T.M., Noll, T., Satyendra, R.: Morphisms of generalized interval systems and PR-groups. J. Math. Music **7**(1), 3–27 (2013)
10. Nolan, C.: Thoughts on Klumpenhouwer networks and mathematical models: the synergy of sets and graphs. Music Theory Online **13**(3), 1–6 (2007)
11. Lewin, D.: Klumpenhouwer networks and some isographies that involve them. Music Theory Spectr. **12**(1), 83–120 (1990)
12. Klumpenhouwer, H.: A generalized model of voice-leading for atonal music. Ph.D. Dissertation, Harvard University (1991)
13. Klumpenhouwer, H.: The inner and outer automorphisms of pitch-class inversion and transposition. Intégral **12**, 25–52 (1998)
14. Mazzola, G., Andreatta, M.: From a categorical point of view: K-nets as limit denotators. Perspect. New Music **44**(2), 88–113 (2006)
15. Ehresmann, C.: Gattungen von lokalen Strukturen. Jahresber. Dtsch. Math. Ver. **60**, 49–77 (1957)
16. Vuza, D.: Some mathematical aspects of David Lewin's book generalized musical intervals and transformations. Perspect. New Music **26**(1), 258–287 (1988)
17. Kan, D.M.: Adjoint functors. Trans. Am. Math. Soc. **87**, 294–329 (1958)
18. Agon, C., Assayag, G., Bresson, J.: The OM Composer's Book. Collection "Musique/Sciences". IRCAM-Delatour France, Sampzon (2006)
19. Andreatta, M., Ehresmann, A., Guitart, R., Mazzola, G.: Towards a categorical theory of creativity for music, discourse, and cognition. In: Yust, J., Wild, J., Burgoyne, J.A. (eds.) MCM 2013. LNCS, vol. 7937, pp. 19–37. Springer, Heidelberg (2013)

The Spinnen-Tonnetz: New Musical Dimensions in the 2D Network for Tonal Music Analysis

Using Polarization and Tonal Regions in a Dynamic Environment

Gilles Baroin[1]($^{(\boxtimes)}$) and Hugues Seress[2]

[1] Labo MAIAA, Ecole Nationale de l'Aviation Civile,
Université de Toulouse, Toulouse, France
Gilles@Baroin.Org
[2] IReMus, Centre d'Études Supérieures de Musique de Poitiers,
University of Paris Sorbonne, Paris, France
http://www.seress-hugues.hu

Abstract. The Polarized Tonnetz developed by Hugues Seress [4] is an innovative two-dimensional representation system for visualizing statically and dynamically some triadic or post-triadic organization, harmonic and tonal paths, as well as tonal regions and musical relations. As for the traditional Tonnetz, the Polarized one relates to transformational design and parsimonious voice leading. Its originality resides however in the introduction of three differentiated criteria concerning the transformation (chromatic distance, modal orientation, polarity) and a fourth parameter: the upward or downward direction of the voice leading. We then enhanced the Polarized Tonnetz with the *Planet-4D* colorization and animation system developed by Gilles Baroin [1], and finally add a new dynamic graphical layer that illustrates the tonal regions. This paper describes the construction and features of the Spinnen-Tonnetz as well as some musical analysis performed. All videos concerning the construction of the Spinnen-Tonnetz and the musical sample studied are available online at www.MatheMusic.net.

Keywords: Tonnetz · Neo-Riemannian · Planet-4D · Transformation · Polarized Tonnetz · Animation · CGI · Mathemusical · Analysis · Tonal regions

1 The Polarized Tonnetz

1.1 Transformation Type

For each of the three neo-Riemannian elementary transformations: P, L, R [2], we define two binary relations. The first one connects all major triads to related minor triads, the second one proceeds the other way round. In parsimonious voice leading, two voices remain while the third voice moves upwards or downwards from either, one semitone or one whole tone. In order to perceive that

© Springer International Publishing Switzerland 2015
T. Collins et al. (Eds.): MCM 2015, LNAI 9110, pp. 315–320, 2015.
DOI: 10.1007/978-3-319-20603-5_32

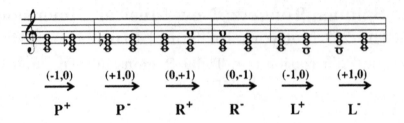

Fig. 1. Couple associated to a transformation.

the movement of the third voice occurs in opposite distances and directions, we define and represent, the couple (x,y), with x: the number of semitones and y: the number of whole tones computed. The Fig. 1 below shows the six different relations and their associated couple [2].

1.2   Modal Orientation

Each of the three transformations can be uniquely identified by the distance in semi-tones between the fundamentals ($P = 0$, $R = 3$, $L = 4$). This model is lacking the direction of the relation: the modalities of the original and the transformed chord. We name this parameter distinctive modal orientation and arbitrarily define the relation from major to minor as "original", and the one from minor to major "retrograde". We indicated an original relation with "+" and a reversed one with "-" (i.e. P^+ P^-).

1.3   Polarity

We define polarity as the direction of the relation alongside the circle of fifths, by considering both elements (argument and image) as triads. We call it subdominant (upwards) or dominant (downwards) and materialize it as a color (see Fig. 2) in our Spinnen-Tonnetz. Progressions that go down this circle, are called dominant: they correspond to a dominant progression also called fifths downward patterns. Subdominant progressions proceed the other way; neutral movements are observed when original and transformed triads have the same key signature or differ by six sharps or flats. The polarity factor introduces a diatonic consideration in a system that was originally designed in a chromatic environment. This approach is essential to understand the relationships between the different structure levels and identify traditional tonal operations.

1.4   Conclusion Concerning the Polarized Tonnetz

With the polarized Tonnetz, each transformation has now three dimensions: Intervallic distance, modal orientation and direction within the circle of fifths. By combining these elementary transformations, it is now possible to generate all the 46 possible progressions [5]. Each of them being characterized by a distance in semi-tones, a modal orientation and a polarity corresponding to the number

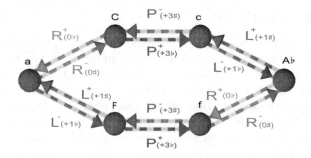

Fig. 2. The six transformations and their colors in the Polarized Tonnetz.

of steps on the circle of fifth (-6 to $+6$) that we can interpret as the differences in sharps between the two key signatures.

In Fig. 2 above, the arrow names and directions match the transformation definition. The color value has been arbitrarily chosen as purple and orange to symbolize opposite directions in the circle of fifths, the intensity indicates the amount of changes. Therefore, the polarized Tonnetz is no longer only a formal description of relationship between objects, but an attempt to capture the dynamic progressions ruling the triadic chromatic organization within the chromatic system. Because it takes into account the direction in which the transformations operate, the model enables a hierarchical structure.

2 The Spinnen-Tonnetz

2.1 Planet4D Symbols

On the *Planet Graph*, defined as the Cartesian product of the cycle graphs $C_3 \square C_4$, each of the 12 pitch classes corresponds to a vertex with a coloration defined by a complex number $(a+ib)$ with $a \in \{0,1,2\}, b \in \{0,1,2,3\}$. We translate the complex coloration into a two-parameters (color and shape) symbol and obtain a unique set of 12 symbols [1].

2.2 Triads Representation

As in the Schönbergian approach [3], we use the dual space of the Tonnetz and display the triads. Since we work with major and minor triads weve arbitrarily chosen darker colours for minor chords. We end up with four forms, three colours and two brightness levels that combine into 24 unique symbols.

2.3 Harmonic Path

Since we use the Polarized Tonnetze path, the coloured path remains visible during the movie. This path can be used in a static environment for musical analysis.

2.4   The Spider in the Web

In tonal music we do have some modulations. But, we may consider these modulations as tonal regions of the general tonality. In the same way, each triad may be part of a bigger scheme that we call the tonal region, a kind of macrostructural tonal hierarchy. In order to symbolize the tonal region we control a spider web that is centred on the tonal region, the moving sphere, symbolising the current triad, seems to be naturally attracted to the centre of the web as a spider to the centre of its web. By moving the Spiders web, we are able to display dynamically modulations and changes for tonal regions.

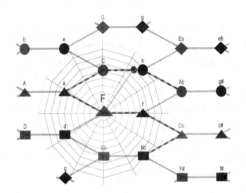

Fig. 3. The Spinnen-Tonnetz.

In the example chosen in Fig. 3 above, the tonal region is F major. It is materialized by the spider web. The spider, acting for the triad is currently moving from C major to c minor: the P transformation is included in the higher level (tonal region) of F major. The striped line reveals the followed path, its color and the nature of the diverse transformations.

3   Musical Applications

3.1   Presentation

Weve analysed 16 different musical samples and created associated video clips. The video samples are ordered in a way to demonstrate that tonal and harmonic paths that operate at different scales (piece of music or complete movement, thematic material exposition or chorus, development, introduction, transition, part of sentence or period,...), do not necessarily follow the same logic. Recursion levels in tonal structure have been long discussed in the theory of tonality. From the nineteenth century, confusion between harmonic progression and keys comes firstly from the belief in the possibility of a direct transfer from one to the other, on the other hand an imprecise definition of each level. We distinguish two hierarchical levels: the first being more stable, the second being less stable either because it is integrated into a stable category or because it is in transition between two stable categories. The main level is represented by a spider, whose web extends over an adjacent area in terms of neo-Riemannian parsimony.

3.2 About the Neo-Riemannian Approach

We consider two possible disadvantages to the Neo-Riemannian approaches. First: transformational approach often involves the mono-tonality (that is to say atonality); identification of the tonality is not a priority for transformational theories. In a sense, they allow greater permeability, but support fuzzy definitions. Second: the description of successive events often does not exceed the scope of the microstructure, that is because the underlying criteria are not directly transferable to higher levels. Reversing the path is often not possible.

3.3 Distances, Intervals and Directions

Finally, the existence of an intervallic distance is partly related to the management and modal orientation processing. These polarized and oriented distances assume discursive and formal functions and, as such, occupy a place in the structure. These transformation chains are associated with a voice leading that coincides or contradicts the polarity (circle of fifths). For minor third relationship (including augmented fourth) and mode change(s): the voice leading is in the same direction as the polarity. For the minor third, perfect fifth, major and minor seconds relations, the voice leading is done in reverse polarity. For M3, (including $P5^+$) and mode changes, the voice leading is well balanced and partially carried out in the same direction as the polarity.

4 Musical Observations

4.1 Modal Orientation

It is noted that the paths of the earlier music samples follow a direction of major triads whereas the more recent ones are most often linked with major to minor triads. In this case, it is more often the Relative (R) relationship which has more occurrences than the Leadingtone exchange (L). However even when they begin with a major key, this often leads to another major key, by compiling two transformations, either R^+P^- or P^+R^-, either L^+P^- or P^+L^-. The tonal structure of romantic musical pieces represents variants of a generic tonal path beginning with L^- before moving on to the substitutions of subdominant function. These pieces develop several types of multidirectional and transitional paths grounded in thirds and fifths (the upper minor third in the first movement of Schuberts symphony, the upper minor third for the thematic material exposition in Smetanas string quartet). Regarding recent pop music, harmonic patterns using thirds which are not inserted in a fifth relation constitute the exact transposition of classical musics tonal structure: since most of them mainly reach a minor key rather than a major key.

4.2 Polarity

If modal orientation seems to constitute a relevant criterion within third patterns, polarity thus seems less of a determining factor. This polarity depends more on the functional meaning of the harmonic structure and is linked to aesthetics, at least for art music: concretely the R⁻ progression is more often used in the early styles (18th century) as a substitution of the dominant key. This proportion seems to be progressively reversed in the 19th century (L⁻ as a substitution of the subdominant key). During the 18th century, the tonalities moved clockwise (subdominant or bottomup polarity), whereas in the 19th century, the structure goes clockwise (dominant or top-down polarity) before returning to the tonic. Both strategies seem to coexist in pop music. The very rare occurrence of the L transformation (L⁺ L⁻) seems to indicate that, whenever the piece starts from a major key, the third progression reaching a minor key is considered as a progression of a dominant polarity.

4.3 Intervallic Distance

Intervallic distance is not really relevant in third progressions: paths tend to move towards major keys by compiling two transformations, or by completing the loop. Also, we note that minor third progressions are used as frequently as major third progressions. But, the modal orientation from major key to another major key is predominant, as well as the mostly negative polarity. Whatever the level of the structure (tonality or chord), these progressions seem to be considered as substitutions of top-down fifth progressions, similar to the relation between a dominant and its own tonic.

5 Conclusion

The observation of the different criteria of relations or chains enables us to understand their hierarchy and the interaction processes between these criteria. We can add the following suggestions: even if the microstructural level is different from the macrostructural one, the comparison of the different scales of the tonal structure lead us to define generic paths.

References

1. Baroin, G.: The planet-4D model: an original hypersymmetric music space based on graph theory. In: Agon, C., Andreatta, M., Assayag, G., Amiot, E., Bresson, J., Mandereau, J. (eds.) MCM 2011. LNCS, vol. 6726, pp. 326–329. Springer, Heidelberg (2011)
2. Cohn, R.: Neo-riemannian operations, parsimonious trichords, and their "tonnetz" representations. J. Music Theor. **41**, 1–66 (1997)
3. Schönberg, A.: Structural Functions of Harmony. Faber and Faber, London (1954)
4. Seress, H.: La musique folklorique pour piano (1907–1920) de Béla Bartók: emprunt symbolique, matriau combinatoire (Doctoral dissertation, Paris 4) (2012)
5. Seress, H.: Trois exemples datonalité chez Beethoven, Bartk et Faur: une rflexion no-riemannienne sur la perte du sentiment tonal, Analyse Music Analysis Today, 315–332, Paris: Delatour (2014)

Deep Learning

Probabilistic Segmentation of Musical Sequences Using Restricted Boltzmann Machines

Stefan Lattner[1]([⊠]), Maarten Grachten[1], Kat Agres[2],
and Carlos Eduardo Cancino Chacón[1]

[1] Austrian Research Institute for Artificial Intelligence, Vienna, Austria
stefan.lattner@ofai.at
[2] Queen Mary, University of London, London, UK

Abstract. A salient characteristic of human perception of music is that musical events are perceived as being grouped temporally into structural units such as phrases or motifs. Segmentation of musical sequences into structural units is a topic of ongoing research, both in cognitive psychology and music information retrieval. Computational models of music segmentation are typically based either on explicit knowledge of music theory or human perception, or on statistical and information-theoretic properties of musical data. The former, rule-based approach has been found to better account for (human annotated) segment boundaries in music than probabilistic approaches [14], although the statistical model proposed in [14] performs almost as well as state-of-the-art rule-based approaches. In this paper, we propose a new probabilistic segmentation method, based on Restricted Boltzmann Machines (RBM). By sampling, we determine a probability distribution over a subset of visible units in the model, conditioned on a configuration of the remaining visible units. We apply this approach to an n-gram representation of melodies, where the RBM generates the conditional probability of a note given its $n - 1$ predecessors. We use this quantity in combination with a threshold to determine the location of segment boundaries. A comparative evaluation shows that this model slightly improves segmentation performance over the model proposed in [14], and as such is closer to the state-of-the-art rule-based models.

1 Introduction

Across perceptual domains, grouping and segmentation mechanisms are crucial for our disambiguation and interpretation of the world. Both top-down, schematic processing mechanisms and bottom-up, grouping mechanisms contribute to our ability to break the world down into meaningful, coherent "chunks". Indeed, a salient characteristic of human perception of music is that musical sequences are not experienced as an indiscriminate stream of events, but rather as a sequence of temporally contiguous musical groups or segments. Elements within a group are perceived to have a coherence that leads to the perception of these events as a structural unit (e.g., a musical phrase or motif).

© Springer International Publishing Switzerland 2015
T. Collins et al. (Eds.): MCM 2015, LNAI 9110, pp. 323–334, 2015.
DOI: 10.1007/978-3-319-20603-5_33

The origin and nature of this sense of musical coherence, or lack thereof, which gives rise to musical grouping and segmentation has been a topic of ongoing research. A prominent approach from music theory and cognitive psychology has been to apply perceptual grouping mechanisms, such as those suggested by Gestalt psychology, to music perception. *Gestalt principles*, such as the laws of proximity, similarity, and closure, were first discussed in visual perception [20], and have been successfully applied to auditory scene analysis [2] and inspired theories of music perception [10–12]. Narmours Implication-Realization theory [12], for example, uses measures of pitch proximity and closure that offer insight into how listeners perceive the boundaries between musical phrases. This type of theory-driven approach has given rise to various rule-based computational models of segmentation. This class of models relies upon the specification of one or more principles according to which musical sequences are grouped.

A second class of computational methods is based on statistical and information theoretic properties of musical data. Recent research in this area has used the statistical structure of sequential tonal and temporal information to compute measures of information (such as Information Content), which serve as a proxy for expectedness (see for example, [1]). Measures of expectation may then be used to calculate segmentation boundaries. For example, a highly expected musical event followed by an unexpected event is often indicative of a perceptual boundary.

A comparison of rule-based and probabilistic approaches [14] suggests the most effective segmentation methods are generally theory-based approaches. The statistical model proposed in [14] (IDyOM) is capable of much better segmentations than simpler statistical models based on digram transition probabilities and point-wise mutual information [3], but still falls slightly short of state-of-the-art rule-based models. Even if rule-based models currently outperform statistical models, there is a motivation to further pursue statistical models of melodic segmentation.

It is plausible that the rules put forth in music-theoretic and perception-based models have been induced by regularities in musical and auditory stimuli. Models that learn directly from the statistics of such stimuli are conceptually simpler than models that describe the perceptual mechanisms of human beings that have internalized the regularities of those stimuli.

In this paper, we introduce a new probabilistic segmentation method, based on a class of stochastic neural networks known as Restricted Boltzmann Machines (RBMs). We present a Monte-Carlo method to determine a probability distribution over a subset of visible units in the model, conditioned on a configuration of the remaining visible units. Processing melodies as n-grams, the RBM generates the conditional probability of a note given its n-1 predecessors. This quantity, in combination with a threshold, determines the location of segment boundaries. In Sect. 2 we give a brief overview of both rule-based and statistical models for melodic segmentation, where we restrict ourselves to an overview of the models with which we compare our approach: those evaluated in [14]. Then, we will argue that our model (explained in Sect. 3) has advantages over statistical models based on n-gram counting. In addition to this qualitative comparison of our method to other approaches (Sect. 3), we reproduce a quantitative evaluation experiment by Pearce et al. [14] (Sect. 4). The results, as reported and discussed in Sect. 5, show that our model slightly improves segmentation performance over IDyOM, and as such is closer to the state-of-the-art rule-based models. Finally, we present conclusions and directions for future work in Sect. 6.

2   Related Work

2.1   Rule Based Segmentation

One of the first models of melodic segmentation based on Gestalt rules was proposed by Tenney and Polansky in [17]. This theory quantifies rules of local detail to predict grouping judgements. However, this theory does not account for vague or ambiguous grouping judgements, and the selection of their numerical weights is rather arbitrary [10,17]. One of the most popular music theoretic approaches is Lerdahl and Jackendoff's Generative Theory of Tonal Music (GTTM) [10]. This theory pursues the formal description of musical intuitions of experienced listeners through a combination of cognitive principles and generative linguistic theory. In GTTM, the hierarchical segmentation of a musical piece into motifs, phrases and sections is represented through a *grouping structure*. This structure is expressed through consecutively numbered *grouping preference rules* (GPRs), which model possible structural descriptions that correspond to experienced listeners' hearing of a particular piece [10]. According to GTTM, two types of evidence are involved in the determination of the grouping structure. The first kind of evidence to perceive a phrase boundary between two melodic events is *local detail*, i.e. relative temporal proximity like slurs and rests (GPR 2a), inter-onset-interval (IOI) (GPR 2b) and change in register (GPR 3a), dynamics (GPR 3b), articulation (GPR 3c) or duration (GPR 3d).

The organization of *larger-level* grouping involves intensification of the effects picked out by GPRs 2 and 3 on a larger temporal scale (GPR 4), symmetry (GPR 5) and parallelism (GPR 6). While Lerdahl and Jackendoff's work did not attempt to quantify these rules, a computational model for identification of segment boundaries that numerically quantifies the GPRs 2a, 2b, 3a and 3d was proposed by Frankland and Cohen [5]. This model encodes melodic profiles using absolute duration of the notes, and MIDI note numbers for representing absolute pitch.

A related model to the quantification of the GPRs was proposed by Cambouropoulos [4]. The Local Boundary Detection Model (LBDM) consists of a *change* rule and a *proximity* rule, operated over melodic profiles that encode pitch, IOI and rests. On the one hand, the change rule identifies the strength of a segment boundary in relation to the degree of change between consecutive intervals (similar to GPR 3). On the other hand, the proximity rule considers the size of the intervals involved (as in GPR 2). The total boundary strength is then computed as a weighted sum of the boundaries for pitch, IOI and rests, where the weights were empirically selected.

Temperley [16] introduced a similar method, called Grouper, that partitions a melody (represented by onset time, off time, chromatic pitch and a level in a metrical hierarchy) into non-overlapping groups. Grouper uses three *phase structure preference rules* (PSPR) to asses the existence of segment boundaries. PSPR 1 locates boundaries at large IOIs and large offset-to-onset intervals (OOIs), and is similar to GPR 2, while PSPR 3 is a rule for metrical parallelism, analogous to GPR 6. PSPR 2 relates to the length of the phrase, and was empirically determined by Temperley using the Essen Folk Song Collection (EFSC), and therefore, may not be a general rule [13].

2.2   Statistical and Information Theoretic Segmentation

In [13], Pearce, Müllensiefen and Wiggins applied two information theoretic approaches, originally designed by Brent [3] for word identification in unsegmented speech, to construct boundary strength profiles (BSPs) for melodic events. This method relies on the assumption that segmentation boundaries are located in places where certain information theoretic measures have a higher numerical value than in the immediately neighbouring locations. The first approach constructs BSPs using transition probability (TP), the conditional probability of an element of a sequence given the preceding element, while the second method relies on pointwise mutual information (PMI), that measures to which degree the occurrence of an event reduces the model's uncertainty about the co-occurrence of another event, to produce such BSPs.

Inspired by developments in musicology, computational linguistics and machine learning, Pearce, Müllensiefen and Wiggins offered the IDyOM model. IDyOM is an unsupervised, multi-layer, variable-order Markov model that computes the conditional probability and Information Content (IC) of a musical event, given the prior context. An overview of IDyOM can be found in [14].

3   Method

The primary assumption underlying statistical models of melodic segmentation is that the perception of segment boundaries is induced by the statistical properties of the data. RBMs (Sect. 3.2) can be trained effectively as a generative probabilistic model of data (Sect. 3.5), and are therefore a good basis for defining a segmentation method. However, in contrast to sequential models such as recurrent neural networks, RBMs are models of static data, and do not model temporal dependencies. A common way to deal with this is to feed the model sub-sequences of consecutive events (n-grams) as if they were static entities, without explicitly encoding time. This n-gram approach allows the model to capture regularities among events that take place within an n-gram. With some simplification we can state that these regularities take the form of a joint probability distribution over all events in an n-gram. With Monte-Carlo methods, we can use this joint distribution to approximate the conditional probability of some of these events, given others. This procedure is explained in Sects. 3.3 and 3.4.

3.1 Relation to Other Statistical Models

Although our RBM-based method works with n-gram representations just as the statistical methods discussed in Sect. 2.2, the approaches are fundamentally different. Models such as IDyOM, TP and PMI are based on n-gram counting, and as such has to deal with the trade-off between longer n-grams and sparsity of data that is inevitable when working with longer sub-sequences. In IDyOM, this problem is countered with "back-off" a heuristic to dynamically decrease or increase the n-gram size as the sparsity of the data allows. In contrast, an RBM does not assign probabilities to n-grams based directly on their frequency counts. The non-linear connections between visible units (via a layer of hidden units) allow a much smoother probability distribution, that can also assign non-zero probability to n-grams that were never presented as training data. As a result, it is possible to work with a fixed, relatively large n-gram size, without the need to reduce the size in order to counter data sparsity.

Every computational model requires a set of basic features that describe musical events. In IDyOM, these basic features are treated as statistically independent, and dependencies between features are modelled explicitly by defining combined viewpoints as cross-products of subsets of features. An advantage of the RBM model is that dependencies between features are modelled as an integral part of learning, without the need to specify subsets of features explicitly.

Finally, the statistical methods discussed in Sect. 2 are fundamentally n-gram based, and it is not obvious how these methods can be adapted to work with polyphonic music rather than monophonic melodies. Although the RBM model presented here uses an n-gram representation, it is straight-forward to adopt the same segmentation approach using a different representation of musical events, such as the note-centred representation proposed in [7]. This would make the RBM suitable for segmenting polyphonic music.

3.2 Restricted Boltzmann Machines

An RBM is a stochastic Neural Network with two layers, a visible layer with units $\mathbf{v} \in \{0,1\}^r$ and a hidden layer with units $\mathbf{h} \in \{0,1\}^q$ [9]. The units of both layers are fully interconnected with weights $\mathbf{W} \in \mathbb{R}^{r \times q}$, while there are no connections between the units within a layer.

In a trained RBM, the marginal probability distribution of a visible configuration $\mathbf{v}$ is given by the equation

$$p(\mathbf{v}) = \frac{1}{Z} \sum_{\mathbf{h}} e^{-E(\mathbf{v}, \mathbf{h})}, \tag{1}$$

where $E(\mathbf{v}, \mathbf{h})$ is an energy function. The computation of this probability distribution is usually intractable, because it requires summing over all possible joint configurations of $\mathbf{v}$ and $\mathbf{h}$ as

$$Z = \sum_{\mathbf{v}, \mathbf{h}} e^{-E(\mathbf{v}, \mathbf{h})}. \tag{2}$$

3.3   Approximation of the Probability of v

Another way to compute the probability of a visible unit configuration $\mathbf{v}$ is to approximate it through Monte Carlo techniques. To that end, for N randomly initialized *fantasy particles*[1] $\mathbf{Q}$, we execute Gibbs sampling until thermal equilibrium. In the visible *activation vector* $\mathbf{q}_i$ of a fantasy particle i, element q_{ij} specifies the probability that visible unit j is on. Since all visible units are independent given $\mathbf{h}$, the probability of $\mathbf{v}$ based on one fantasy particle's visible activation is computed as:

$$p(\mathbf{v}|\mathbf{q}_i) = \prod_j p(v_j|q_{ij}). \tag{3}$$

As we are using binary units, such an estimate can be calculated by using a binomial distribution with one trial per unit. We average the results over N fantasy particles, leading to an increasingly close approximation of the true probability of $\mathbf{v}$ as N increases:

$$p(\mathbf{v}|\mathbf{Q}) = \frac{1}{N}\sum_i^N \prod_j \binom{1}{v_j} q_{ij}^{v_j}(1 - q_{ij})^{1-v_j}. \tag{4}$$

3.4   Posterior Probabilities of Visible Units

When the visible layer consists of many units, N will need to be very large to obtain good probability estimates with the method described above. However, for conditioning a (relatively small) subset of visible units $\mathbf{v}_y \subset \mathbf{v}$ on the remaining visible units $\mathbf{v}_x = \mathbf{v} \setminus \mathbf{v}_y$, the above method is very useful. This can be done by Gibbs sampling after randomly initializing the units $\mathbf{v}_y$ while clamping all other units $\mathbf{v}_x$ according to their initial state in $\mathbf{v}$. In (4), all $\mathbf{v}_x$ contribute a probability of 1, which results in the conditional probability of $\mathbf{v}_y$ given $\mathbf{v}_x$.

We use this approach to condition the units belonging to the last time step of an n-gram on the units belonging to preceding time steps. For the experiments reported in this paper, we found that it is sufficient to use 150 fantasy particles and for each to perform 150 Gibbs sampling steps.

3.5   Training

We train a single RBM using *persistent contrastive divergence* (PCD) [18] with *fast weights* [19], a variation of the standard *contrastive divergence* (CD) algorithm [8]. PCD is more suitable for sampling than CD, because it results in a better approximation of the likelihood gradient.

Based on properties of neural coding, sparsity and selectivity can be used as constraints for the optimization of the training algorithm [6]. Sparsity encourages competition between hidden units, and selectivity prevents over-dominance by any individual unit. A parameter μ specifies the desired degree of sparsity and selectivity, whereas another parameter ϕ determines how strongly the sparsity/selectivity constraints are enforced.

---

[1] See [18].

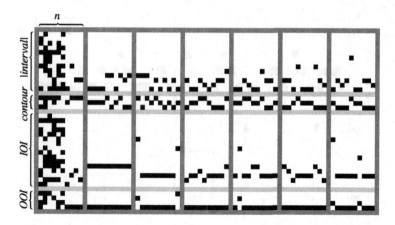

Fig. 1. Seven examples of n-gram training instances (n=10) used as input to the RBM. Within each instance (delimited by a dark gray border), each of the 10 columns represents a note. Each column consists of four *one-hot* encoded viewpoints: |*interval*|, *contour*, *IOI* and *OOI* (indicated by the braces on the left). The viewpoints are separated by horizontal light gray lines for clarity. The first instance shows an example of noise padding (in the first six columns) to indicate the beginning of a melody.

3.6   Data Representation

From the monophonic melodies, we construct a set of n-grams by using a sliding window of size n and a step size of 1. For each note in the n-gram, four basic features are computed: (1) absolute values of the pitch interval between the note and its predecessor (in semitones); (2) the contour (up, down, or equal); (3) inter-onset-interval (IOI); and (4) onset-to-offset-interval (OOI). The IOI and OOI values are quantized into semiquaver and quaver, respectively. Each of these four features is represented as a binary vector and its respective value for any note is encoded in a one-hot representation. The first n-1 n-grams in a melody are noise-padded to account for the first n-1 prefixes of the melody. Some examples of binary representations of n-grams are given in (Fig. 1).

3.7   Information Content

After training the model as described in Sect. 3.5, we estimate the probability of the last note conditioned on its preceding notes for each n-gram as introduced in Sect. 3.4. From the probabilities $p(e_t \mid e_{t-n+1}^{t-1})$ computed thus, we calculate the IC as:

$$h(e_t \mid e_{t-n+1}^{t-1}) = log_2 \frac{1}{p(e_t \mid e_{t-n+1}^{t-1})}, \tag{5}$$

where e_t is a note event at time step t, and e_k^l is a note sequence from position k to l of a melody. IC is a measure of the unexpectedness of an event given its context. According to a hypothesis of [14], segmentation in auditory perception is determined by perceptual expectations for auditory events. In this sense, the

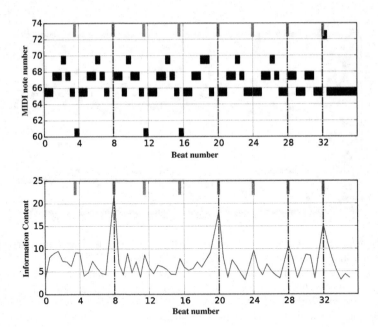

Fig. 2. A BSP calculated from 11-grams. The upper figure shows the notes of 9 measures (36 beats) of a German folk song. The lower figure shows a BSP (i.e. IC) used for segmentation. The correct segmentation (ground truth) is depicted as vertical grey bars at the top of the figures, segment boundaries found by our model are shown as dashed vertical lines. Note that the BSP has particularly high peaks at rests and at high intervals. However, the segment boundary found at beat 28 does not have any of those cues and was still correctly classified.

IC relates directly to this perceived boundary strength, thus we call the IC over a note sequence *boundary strength profile* (Fig. 2).

3.8   Peak Picking

Based on the BSP described in the previous section, we need to find a concrete binary segmentation vector. For that, we use the peak picking method described in [14]. This method finds all peaks in the profile and keeps those which are k times the standard deviation greater than the mean boundary strength, linearly weighted from the beginning of the melody to the preceding value:

$$S_n > k \sqrt{\frac{\sum_{i=1}^{n-1} (w_i S_i - \bar{S}_{w,1...n-1})^2}{\sum_1^{n-1} w_i}} + \frac{\sum_{i=1}^{n-1} w_i S_i}{\sum_1^{n-1} w_i}, \qquad (6)$$

where S_m is the m-th value of the BSP, and w_i are the weights which emphasize recent values over those of the beginning of the song (triangular window), and k has to be found empirically.

4   Experiment

4.1   Training Data

In this work, we use the EFSC [15]. This database is a widely used corpus in MIR for experiments on symbolic music. This collection consists of more than 6000 transcriptions of folk songs primarily from Germany and other European regions. The EFSC collection is commonly used for testing computational models of music segmentation, due to the fact that it is annotated with phrase markers.

In accordance with [14], we used the *Erk* subset of the EFSC, which consists of 1705 German folk melodies with a total of 78, 995 note events. Phrase boundary annotations are marked at about 12 % of the note events.

4.2   Procedure

The model is trained and tested on the data described in Sect. 4.1, with n-gram lengths varying between 1 and 11. For each n-gram length, we perform 5-fold cross-validation and average the results over all folds. Similar to the approach in [14], after computing the BSPs, we evaluate different k from the set $\{0.70, 0.75, 0.80, 0.85, 0.90, 0.95, 1.00\}$ and choose the value that maximizes F1 for the respective n-gram length. To make results comparable to those reported in [14], the output of the model is appended with an implicit (and correct) phrase boundary at the end of each melody.

Since the hyper-parameters of the model are inter-dependent, it is infeasible to exhaustively search for the optimal parameter setting. For the current experiment, we have manually chosen a set of hyper-parameters that give reasonable results for the different models tested: 200 hidden units, a batch size of 100, a momentum of 0.6, and a learning rate of 0.007 which we linearly decrease to zero during training. The fast weights used in the training algorithm (see Sect. 3.5) help the fantasy particles mix well, even with small learning rates. The learning rate of the fast weights is increased from 0.002 to 0.007 during training. The training is continued until convergence of the parameters (typically between 100 and 300 epochs). The sparsity parameters (see Sect. 3.5) are set to $\mu = 0.04$, and $\phi = 0.65$, respectively. In addition, we use a value of 0.0035 for $L2$ weight regularization, which penalizes large weight coefficients.

5   Results and Discussion

We tested three different representations for pitch, yielding the following F1 scores for 10-grams: *absolute pitch* (0.582), *interval* (0.600), and the absolute value of interval (i.e. |*interval*|) plus *contour* (0.602). The latter representation was chosen for our experiments, as it showed the best performance. Not surprisingly, relative pitch representations lead to better results, as they reduce the number of combination possibilities in the input. Event though the difference in F1 score between *interval* and |*interval*| plus *contour* representation is not significant, it still shows that it is valid to decompose viewpoints into their elementary

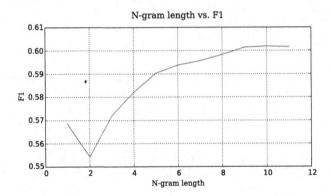

Fig. 3. Maximal F1 scores for different n-gram lengths.

Table 1. Results of the model comparison, ordered by F1 score. Table adapted from [14], with permission.

Model	Precision	Recall	F1
Grouper	0.71	0.62	0.66
LBDM	0.70	0.60	0.63
RBM (10-gram)	0.83	0.50	0.60
IDyOM	0.76	0.50	0.58
GPR 2a	0.99	0.45	0.58
GPR 2b	0.47	0.42	0.39
GPR 3a	0.29	0.46	0.35
GPR 3d	0.66	0.22	0.31
PMI	0.16	0.32	0.21
TP	0.17	0.19	0.17
Always	0.13	1.00	0.22
Never	0.00	0.00	0.00

informative parts. Such an approach, next to reducing the input dimensionality, may also support the generalization ability of a model (e.g. |interval| representation in music may help to understand the concept of inversion).

Figure 3 shows the F1 score obtained by models of different n-gram sizes. The fact that boundary detection is reasonably good even for 1-grams is likely due to the fact that the 1-gram includes the OOI, which is mostly zero, except for the relatively rare occurrence of a rest between notes. Because of this, the probability values assigned to OOI values by a trained RBM behave like an inverted rest indicator: high for $OOI = 0$, and low for $OOI > 0$. This makes the behaviour of the 1-gram RBM much like that of the GPR 2a rule (Sect. 2.1). That GPR 2a performs slightly better (see Table 1) can be explained by the fact that the RBM also detects segment boundaries at large (and unlikely) pitch intervals, which are not always correct.

Another remarkable result is that 1-grams perform better than 2-grams (see Fig. 3). Although we have no definite explanation for this yet, the difference may be related to the fact that in the 1-gram model, the probability is estimated by sampling without clamping any units. In contrast, for 2-grams half the units get clamped during sampling. Prior tests with our method for computing the conditional probability (Sect. 3.4) have revealed (unsurprisingly) that the quality of the approximation decreases with the ratio of unknown units over given (clamped) units. This phenomenon may also partly account for the steady increase of F1 scores for increasing n-grams sizes larger than one. Nevertheless, the increasing performance with increasing n-gram size demonstrates that the RBM based segmentation method is less susceptible to problems of data sparseness encountered in n-gram counting approaches.

6   Conclusion

In this paper, an RBM-based unsupervised probabilistic method for segmentation of melodic sequences was presented. In contrast to other statistical methods, our method does not rely on frequency counting, and thereby circumvents problems related to data sparsity. The method performs slightly better than IDyOM, a sophisticated frequency counting model.

The segment boundary detection capabilities of our model are still slightly lower than state-of-the-art rule based methods that rely on gestalt principles formulated for musical stimuli. This result underlines the remaining challenge to find segmentation models that correspond to human perception, based only on musical stimuli in combination with universal learning principles.

An important aspect of human perception that is missing in our current method is equivalent of short-term memory, to bias long-term expectations based on the stimuli in the direct past (see [14]). Furthermore, we wish to investigate the effect of different architectural factors on the segmentation behaviour of the model, like an increased number of hidden layers, or an increased number of hidden units per hidden layer. Lastly, the formation of boundary strength profiles may be improved by involving other information theoretic quantities, such as the entropy of conditional probability distributions.

Acknowledgements. The project Lrn2Cre8 acknowledges the financial support of the Future and Emerging Technologies (FET) programme within the Seventh Framework Programme for Research of the European Commission, under FET grant number 610859. We thank Marcus Pearce for sharing the Essen data used in [14].

References

1. Agres, K., Abdallah, S., Pearce, M.: An information-theoretic account of musical expectation and memory
2. Bregman, A.S.: Auditory Scene Analysis. MIT Press, Cambridge (1990)

3. Brent, M.R.: An efficient, probabilistically sound algorithm for segmentation and word discovery. Mach. Learn. **34**(1–3), 71–105 (1999)
4. Cambouropoulos, E.: The local boundary detection model (LBDM) and its application in the study of expressive timing. In: Proceedings of the International Computer Music Conference, San Francisco, pp. 17–22 (2001)
5. Frankland, B.W., Cohen, A.J.: Parsing of melody: quantification and testing of the local grouping rules of Lerdahl and Jackendoff's A Generative Theory of Tonal Music. Music Percept. **21**(4), 499–543 (2004)
6. Goh, H., Thome, N., Cord, M.: Biasing restricted Boltzmann machines to manipulate latent selectivity and sparsity. In: NIPS Workshop on Deep Learning and and Unsupervised Feature Learning (2010)
7. Grachten, M., Krebs, F.: An assessment of learned score features for modeling expressive dynamics in music. IEEE Trans. Multimedia **16**(5), 1211–1218 (2014). http://dx.doi.org/10.1109/TMM.2014.2311013
8. Hinton, G.E., Osindero, S., Teh, Y.: A fast learning algorithm for deep belief nets. Neural Comput. **18**, 1527–1554 (2006)
9. Hinton, G.E.: Training products of experts by minimizing contrastive divergence. Neural Comput. **14**(8), 1771–1800 (2002)
10. Lerdahl, F., Jackendoff, R.: A Generative Theory of Tonal Music. MIT Press, Cambridge (1983)
11. Meyer, L.: Emotion and Meaning in Music. University of Chicago Press, Chicago (1956)
12. Narmour, E.: The Analysis and Cognition of Basic Melodic Structures: The Implication-Realization Model. University of Chicago Press, Chicago (1990)
13. Pearce, M.T., Müllensiefen, D., Wiggins, G.: The role of expectation and probabilistic learning in auditory boundary perception: a model comparison. Perception **39**(10), 1365–1391 (2010)
14. Pearce, M.T., Müllensiefen, D., Wiggins, G.A.: Melodic grouping in music information retrieval: new methods and applications. In: Raś, Z.W., Wieczorkowska, A.A. (eds.) Adv. in Music Inform. Retrieval. SCI, vol. 274, pp. 364–388. Springer, Heidelberg (2010)
15. Schaffrath, H.: The Essen folksong collection in Kern format. In: Huron, D. (ed.) Database Containing, Folksong Transcriptions in the Kern Format and A -page Research Guide Computer Database. Menlo Park, CA (1995)
16. Temperley, D.: The Cognition of Basic Musical Structure. MIT Press, Cambridge (2001)
17. Tenney, J., Polansky, L.: Temporal gestalt perception in music. J. Music Theor. **24**(2), 205–241 (1980)
18. Tieleman, T.: Training restricted Boltzmann machines using approximations to the likelihood gradient. In: Proceedings of the 25th International Conference on Machine Learning, pp. 1064–1071. ACM, New York (2008)
19. Tieleman, T., Hinton, G.: Using fast weights to improve persistent contrastive divergence. In: Proceedings of the 26th International Conference on Machine Learning, pp. 1033–1040. ACM, New York (2009)
20. Wertheimer, M.: Laws of organization in perceptual forms. In: Ellis, W. (ed.) A Source Book of Gestalt Psychology, pp. 71–88. Harcourt, New York (1938)

¿El Caballo Viejo? Latin Genre Recognition with Deep Learning and Spectral Periodicity

Bob L. Sturm[1]([✉]), Corey Kereliuk[2], and Jan Larsen[2]

[1] School of Electronic Engineering and Computer Science,
Queen Mary University of London, Mile End Road, London E1 4NS, UK
b.sturm@qmul.ac.uk
[2] DTU Compute, Technical University of Denmark,
Richard Petersens Plads, B324, 2800 Kgs. Lyngby, Denmark
{cmke,janla}@dtu.dk

Abstract. The "winning" system in the 2013 MIREX Latin Genre Classification Task was a deep neural network trained with simple features. An explanation for its winning performance has yet to be found. In previous work, we built similar systems using the *BALLROOM* music dataset, and found their performances to be greatly affected by slightly changing the tempo of the music of a test recording. In the MIREX task, however, systems are trained and tested using the *Latin Music Dataset (LMD)*, which is 4.5 times larger than *BALLROOM*, and which does not seem to show as strong a relationship between tempo and label as *BALLROOM*. In this paper, we reproduce the "winning" deep learning system using *LMD*, and measure the effects of time dilation on its performance. We find that tempo changes of at most ±6 % greatly diminish and improve its performance. Interpreted with the low-level nature of the input features, this supports the conclusion that the system is exploiting some low-level absolute time characteristics to reproduce ground truth in *LMD*.

Keywords: Machine music listening · Genre · Deep learning · Evaluation

1 Introduction

Consider the machine music listening system that "won" the Audio Latin Genre Classification task at MIREX 2013.[1] Among the ten classes in the cleanly labeled Latin Music Database (*LMD*) [13], three systems based on deep learning of spectral periodicity features (DeSPerF) reproduced an average of 77 % of the ground truth of each class – more than any of the other systems submitted. Figure 1(a) overall figures of merit (FoM) of these three systems. These FoM, being significantly better than from just guessing, leads one to believe that these systems have successfully learned to identify a good set of musical characteristics associated with each class in *LMD* that are *general* (common to a class) and *discriminative* (distinguishing one class from another). At the heart of this claim,

---

[1] http://www.music-ir.org/nema_out/mirex2013/results/act/latin_report/summary.html.

© Springer International Publishing Switzerland 2015
T. Collins et al. (Eds.): MCM 2015, LNAI 9110, pp. 335–346, 2015.
DOI: 10.1007/978-3-319-20603-5_34

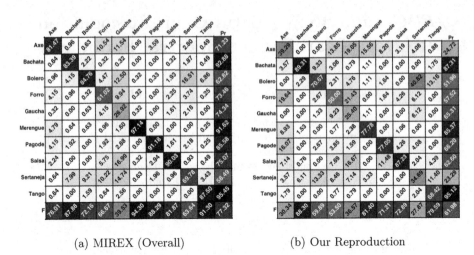

(a) MIREX (Overall)   (b) Our Reproduction

Fig. 1. Figures of merit (FoM, ×100) for the (a) three DeSPerF-based systems in MIREX 2013, and (b) our reproduced DeSPerF-LMD system. Column is "true" class, and row is selected class. Off diagonals are confusions. Precision is the right-most column, F-score is the bottom row, recall is the diagonal, and normalised accuracy (mean recall) is at bottom-right corner.

however, sits a false assumption, not to mention an unjustified confidence in the validity of this evaluation.

Despite being studied for more than 15 years, music genre recognition (MGR) still lacks an explicit, specific and reasonable definition [1,18]. The definition most commonly used is that given implicitly by, or by proxy of, some labeled dataset. Critically, the conclusions drawn about systems trained to reproduce labels in a dataset often belie the artificial and unreasonable assumptions made in creating that dataset, not to mention its specious relationship to genre in the real world [6]. Most of these conclusions also implicitly assume that there are only two possible ways to reproduce the ground truth: by chance or with music intelligence [16,18]. When a system reproduces an amount of ground truth much more than that expected from chance, success is declared, and the line of inquiry stops short of proving the outcome to be a result of *music learning*.

In earlier work [19], we sought to explain the winning performance of DeSPerF-based systems in MIREX 2013. Since we did not have access to *LMD* at that time, we used the *BALLROOM* music dataset [4]: a dataset consisting of short music audio excerpts labeled in seven classes. With a 70/30 train and test set partition of *BALLROOM*, we found that the DeSPerF-based system (DeSPerF-BALLROOM) reproduced an average of 88.8 % of the ground truth in each class of the test set. We then showed how DeSPerF-BALLROOM can perform perfectly, or no better than random, by time-stretching the test dataset recordings by at most ±6 % –effectively changing music tempo without affecting pitch. Furthermore, we showed how minor tempo changes make DeSPerF-BALLROOM label *the same music* in several different ways. For

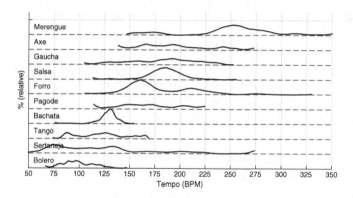

Fig. 2. Tempo distributions in *LMD* (smoothed with 3rd-order moving average).

instance, a waltz with a tempo of 87 BPM became a jive at 86 BPM, a rumba at 90 BPM, a samba at 72 BPM, and a cha cha cha at 99 BPM. The explanation for these observations comes from the fact that the labels in *BALLROOM* are highly correlated with the tempo of the excerpts – a characteristic of *BALLROOM* that has been noticed before [4, 7]. Nearest neighbour classification using *only* annotated tempo produces accuracies from 78 % [19] to 82.3 % [7]. Hence, no matter the meter of the music in the recording, no matter its rhythm, or whether a clave is involved or a bandoneon, a system that can accurately estimate tempo will *appear* quite capable, from being trained and tested in *BALLROOM*, to recognise rhythm, meter, style, instrumentation, and so on.

These results, however, are limited to DeSPerF-BALLROOM and do not explain the "winning" performance of the DeSPerF-based systems in MIREX 2013 (DeSPerF-LMD). Just what has DeSPerF-LMD learned such that it appears able to recognise Merengue with its "crisp, zippy beat, hissed and scratched out on a metal grater quira in jaunty 2/4 time" [13]; Pagode with its "[unpretentious lyrics], focusing on situations from [Brazilian] daily life" [13]; Salsa with its "essential" clave [13]; or Bachata with its standard instrumentation of guitar, maracas and bongos [13]? What has it not learned since it does not appear to recognise Gaucha with its lyrics about "respect for the women, the love for the countryside, the culture and the animals" [13]? A brief look at the characteristics of the labels in *LMD* show that some are musicological and performative, but many are topical, cultural, and geographical, which are of course outside the purview of any artificial algorithm focused exclusively on recorded musical audio. Since DeSPerF-based systems have by design features containing no information about instrumentation or lyrics [11], might DeSPerF-LMD be exploiting a strong correlation between tempo and label in *LMD*?

Recent work [5] suggests that there does not exist a very strong correlation between tempo and label in *LMD*. Figure 2 (created using data collected by Esparza et al. [5]) shows large overlaps in tempo distributions between classes. Table 1 summarises the results from 3-nearest neighbour classification with only estimated tempo, using 10-fold cross validation in *LMD*. Furthermore, the size of *LMD* is more than 4.5 times that of *BALLROOM*, and has

Table 1. FoM of 3-NN classification in LMD by tempo (10-fold CV).

Class	Recall	Precision	F-score
Axe	0.1629	0.2615	0.2008
Bachata	0.6154	0.5680	0.5908
Bolero	0.4268	0.5076	0.4637
Forro	0.1538	0.2712	0.1963
Gaucha	0.3204	0.1704	0.2225
Merengue	0.7229	0.5791	0.6431
Pagode	0.1480	0.1744	0.1601
Salsa	0.2706	0.3727	0.3136
Sertaneja	0.3156	0.4208	0.3607
Tango	0.5025	0.3764	0.4304
Mean	0.3639	0.3702	0.3582

complete songs instead of 30 s excerpts. Hence, one important question to answer is whether DeSPerF-LMD is as sensitive to "irrelevant" tempo changes as DeSPerF-BALLROOM. If it has a high sensitivity, then DeSPerF-LMD might have learned to exploit absolute temporal characteristics in LMD that are not visible from Fig. 2. If, on the other hand, DeSPerF-LMD is not as sensitive as DeSPerF-BALLROOM, then perhaps DeSPerF-LMD has learned to exploit relative temporal characteristics correlated with the labels of LMD, e.g., the rhythmic characteristics to which Pikrakis [11] alludes. Indeed, given the recordings in each of the classes of LMD have a far wider variation in tempo than BALLROOM, we expect DeSPerF-LMD should be robust to minor changes in tempo as long as the training dataset similarly displays the same variation.

In this work, we seek to answer these questions, which ultimately carry implications for the applications, limitations and improvement of DeSPerF for machine music listening. We begin by reviewing these systems, and then discuss how we create DeSPerF-LMD. We then perform two experiments to determine how time dilation affects the performance of DeSPerF-LMD. We discuss the implications of our results, and propose several avenues for future work.

2   DeSPerF-Based Systems

2.1   The Extraction of SPerF

DeSPerF-based systems combine hand-engineered features – spectral periodicity features (SPerF) – and deep neural networks (DNNs) [11]. A SPerF is generated from an audio extract of 10 s. This is broken into 100 ms frames skipped by 5 ms. The first 13 MFCCs [14] are computed for each frame, which produce a *modulation sonogram* $\mathcal{M} = (\mathbf{m}_t : 0 \leq t \leq 10)$, where $\mathbf{m}_t \in \mathbb{R}^{13}$ is a vector of the MFCCs extracted from the frame over time $[t, t+0.1]$. For *offset* $l \in [1, 4/0.005]$ define the two sequences, $\mathcal{M}_{\text{beg},l} = (\mathbf{m}_t \in \mathcal{M} : 0 \leq t \leq 10 - 0.005l)$ and

$\mathcal{M}_{\text{end},l} = (\mathbf{m}_t \in \mathcal{M} : 0.005l \leq t \leq 10)$. $\mathcal{M}_{\text{beg},l}$ are the features starting from the beginning of extract; $\mathcal{M}_{\text{end},l}$ are the features up to its end. The time overlap between features in these two sequences will always be larger than 2 s.

Now, define the *distance* between the sequences for an offset l as

$$d[l] = \frac{\|\mathbf{M}_{\text{beg},l} - \mathbf{M}_{\text{end},l}\|_F}{|\mathcal{M}_{\text{beg},l}|} \tag{1}$$

where the columns of $\mathbf{M}_{\text{beg},l}$ and $\mathbf{M}_{\text{end},l}$ are the sequences $\mathcal{M}_{\text{beg},l}$ and $\mathcal{M}_{\text{end},l}$, and $\|\cdot\|_F$ is the Frobenius norm. The denominator is the number of columns in both matrices. The sequence $d[l]$ is then filtered $y[l] = ((d * h) * h)[l]$, where

$$h[n] = \begin{cases} \frac{1}{n}, & -0.1/0.005 \leq n \leq 0.1/0.005, n \neq 0 \\ 0, & \text{otherwise} \end{cases} \tag{2}$$

and adapting $h[n]$ around the end points of $d[l]$ (shortening its support to a minimum of two). The sequence $y[l]$ then is an approximation of the second derivate of $d[l]$. Finally, a SPerF is created by a non-linear transformation:

$$x[l] = [1 + \exp\left(-(y[l] - \hat{\mu})/\hat{\sigma}\right)]^{-1} \tag{3}$$

where $\hat{\mu}$ is the mean of $y[l]$ and $\hat{\sigma}$ is its standard deviation. The function of $\hat{\sigma}$ is to remove the influence of energy in the modulation sonograms computed from many audio extracts, thereby making them comparable.

From this derivation, we clearly see that SPerF describe temporal periodicities of modulation sonograms. If the sequence $\mathcal{M}$ is periodic with period T seconds, then the sequence $d[l]$ should be small, and $y[l]$ should be large positive, for all $l \approx kT/0.005$ with k a positive integer. At some of these offsets $x[l]$ will be close to 1. The hope is that $x[l]$ provides insight into musical characteristics such as tempo, meter and rhythm, when they exist over durations of at most 10 s. An estimate of a multiple of the tempo can come from a spectral analysis of $x[l]$, i.e., the amplitudes and frequencies of its harmonics. Predicting meter requires approaches that are more heuristic, e.g., deciding on the beat level and then grouping peaks of $x[l]$. Describing rhythm from $x[l]$ should involve even more heuristics, not to mention information SPerF does not contain, e.g., instrumentation. By using SPerF as input to deep learning systems, one hopes that it automatically develops such heuristics meaningful for music listening.

2.2 The Construction and Operation of DeSPerF-Based Systems

The deep neural network (DNN) used in DeSPerF-based systems specifically use feedforward architectures, whereby the input data is propagated through one or more hidden layers. This forward propagation is achieved via a series of cascaded operations consisting of a matrix multiplication followed by a non-linear function (e.g., a logistic sigmoid or hyperbolic tangent). Since each DNN layer computes a feature representation of the data in the previous layer, the hidden layers are said to compute "features-of-features." The hierarchical nature of DNNs is a commonly cited motivation for choosing to work with them [3,9]. For instance, it might be argued that music rhythm perception is hierarchical in nature, e.g., beat-level, measure-level, and so on, which motivates the application of DNNs to recognising rhythmic qualities in recorded music.

Several efficient DNN training techniques have been developed [2,8,15]. A DeSPerF-based system employs a DNN trained using a common two-phase process: unsupervised pre-training followed by supervised fine-tuning. Pre-training initializes the network with a 'good' set of weights, which can be critical for achieving learning times that are independent of depth [12]. The pre-training phase is accomplished by greedily training a stack of restricted Boltzmann machines using 1-step contrastive divergence [8]. In the subsequent fine-tuning step, back-propagation is used to adjust the network parameters in order to minimize the expected misclassification error on the labeled training data. The DeSPerF-based systems in MIREX 2013 have five hidden layers with 400 units each.

The final layer of a DeSPerF-based system involves a softmax unit, the output of which can be interpreted as the posterior probability of the classes for an input SPerF. The class of the largest posterior is thus applied to the unlabelled observation. In *LMD*, however, observations are longer than 10 s, and so a single music recording can produce many SPerF. Since the classification problem implicit in *LMD* is to classify whole recordings and not 10 s excerpts, the DeSPerF-based systems in MIREX employs majority vote. In other words, for each SPerF extracted from the same music recording, a vote is recorded for the class of the maximum posterior probability. Once all SPerF have been processed, the class with the most votes is selected.

2.3   Evaluating DeSPerF-LMD

Though we have access to *LMD* we do not have access to the specific folds used in this MIREX task. We thus reverse engineer the folds given the results of MIREX 2013,[2] and using the claim that the task employs artist filtering. Table 2.3 shows the number of tracks of each class appearing in each fold. We create an approximately 70/30 train/test partition by combining the numbers in the coloured cells. Our copy of LMD contains 3229 excerpts (1 extra each in Pagode and Sertaneja). We compose the train and test folds using the blocks of artists in each class. Since more than 81 % (334/408) of the tracks in LMD Tango are by one artist (Carlos Gardel), we have to violate artist filtering by including 41 excerpts of his in the test set. We use his first 41 excerpts listed by filename.

Figure 1(b) shows the FoM of our DeSPerF-LMD system. Comparison with Fig. 1(a) shows some possible discrepancies. First, the normalised accuracies differ by more than 15 points; however, the FoM in Fig. 1(a) is the overall FoM for three systems tested on the three folds summarised by Table 2. In fact, the three normalised accuracies measured in the MIREX 2013 folds for each DeSPerF-LMD system are reported 73.34, 65.81 and 51.75.[3] Hence, our observation of 61.98 is not alarming. With the exception of Bachata, the FoM of our system

---

[2] http://www.music-ir.org/nema_out/mirex2013/results/act/latin_report/files.html.

[3] The fold composition in the MIREX task is problematic. Table 2 shows folds 1 and 2 are missing examples of 2 classes, and fold 1 has only one example in another.

Table 2. An overview of the composition of the three folds used in the 2013 MIREX Audio Latin Genre Classification Train-test Task. We construct an approximately 70/30 split in each class by combining the shaded numbers of tracks to the test partition.

Fold Class	1	2	3	Total	Proportion in our test
Axe	257	14	42	313	17.9 %
Bachata	1	131	181	313	41.9 %
Bolero	68	172	75	315	23.8 %
Forro	183	0	130	313	41.5 %
Gaucha	0	126	186	312	40.4 %
Merengue	224	80	11	315	28.9 %
Pagode	60	246	0	306	19.6 %
Salsa	75	217	19	311	30.2 %
Sertaneja	0	272	49	321	15.3 %
Tango	114	0	294	408	27.9 %
Totals	982	1258	987	3227	28.7 %

is worse than those seen in MIREX 2013. We see large differences in the recall and precision for Axe, Merengue, Sertaneja, Tango and Bolero. Again, looking over the FoM for the individual systems in MIREX 2013, these are not alarming. Of all DeSPerF-LMD systems tested in MIREX 2013, the one built using folds 1 and 2 performed the worst in these classes. For Axe, its recall and precision was 0.43 and 0.23, respectively. For Merengue, these were 0.72 and 0.35; for Sertaneja: 0.43 and 0.15; for Bolero: 0.71 and 0.37; and for Tango: 0.82 and 0.95. Hence, we conclude that our DeSPerF-LMD system is working comparably to those built in MIREX 2013 with respect to their FoM.

3   Measuring the Sensitivity to Tempo of DeSPerF-LMD

Given the above results, we now attempt to inflate and deflate its FoM by the method of irrelevant transformation [17] through pitch-preserving time stretching using the RubberBand library.[4] We then attempt to make DeSPerF-LMD apply different labels to the same music by the same transformation.

3.1   Inflation and Deflation of FoM

By the same deflation and inflation procedures we applied in [19], we find that DeSPerF-LMD obtains the FoM shown in Fig. 3 with changes of at most ±6 % (i.e., a dilation factor 0.94 or 1.06). Comparison with Fig. 1(b) shows severe harm or great benefit to the FoM of DeSPerF-LMD. If we change the tempo by at most ±10 %, we find the normalised classification accuracy reaches 0.11 with deflation, and 0.94 with inflation. Figure 4 shows how even for small tempo changes the F-scores for all classes are dramatically affected.

3.2   Picking Any Class in LMD

We now randomly select one excerpt of each label from the test set, and attempt to make DeSPerF-LMD classify them in every way. Table 3 shows the new

---

[4] http://breakfastquay.com/rubberband/.

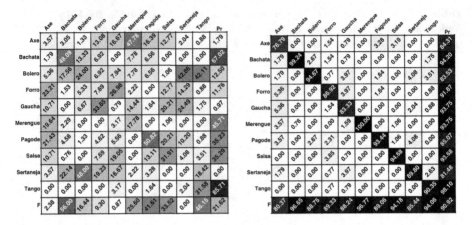

Fig. 3. FoM (×100) resulting from deflation (left) and inflation (right) for DeSPerF-LMD. The maximum change in tempo here is ±6 %. Interpretation as in Fig. 1.

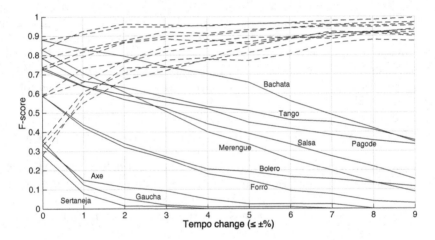

Fig. 4. Inflation (dashed) and deflation (solid) of F-score of DeSPerF-LMD in each label of *LMD* as a function of the maximum change in tempo. We label only the deflation, but inflation for each class begins from the same F-score.

tempo of each track (using data from [5]), and the resulting classifications.[5] From Fig. 1(b) we see that Bachata receives the highest recall and F-score, and second highest precision. DeSPerF-LMD classifies the Bachata tracks in 6 categories, with the largest change of 0.79. It classifies as Bachata six not-Bachata tracks, with the largest change of 1.23. In Merengue, DeSPerF-LMD has the second highest F-score, and third highest precision. It labels the Merengue excerpt eight different ways. The hardest classification to force was Tango, where only one not-Tango track was classified Tango with a change 1.25.

---

[5] Audition this table at http://www.eecs.qmul.ac.uk/~sturm/research/DeSPerFtable2/exp.html.

Table 3. *LMD* tracks (left column) are classified in a number of different ways by time-stretching. Resulting estimated tempi are shown.

Label *Title* (tempo)	*Class*	*Axe*	*Bachata*	*Bolero*	*Forro*	*Gaucha*	*Merengue*	*Pagode*	*Salsa*	*Sertaneja*	*Tango*
Axe *Batom Na Cueca, Arrasta* (168.36)		168.36	141.48	134.69	160.34	177.22	155.89	181.03	154.46	157.34	
Bachata *Aventura, Ensename A Olvidar* (124.71)			125.97	118.77	145.01	113.37			164.09	124.71	
Bolero *Emilio Santiago - Dilema* (101.92)			118.52	101.92	80.89			141.56	113.25	102.95	
Forro *Trio Forrozao - Ze do Rock* (163.85)		176.18			163.85	180.05			195.06	142.48	
Gaucha *Alma Serrana, O Garrafao* (178.82)		178.82	155.50		173.61	177.05		158.25	184.35		
Merengue *Ronny Moreno - Matame* (279.49)		288.14	268.74		321.26	234.87	279.49	297.33	221.82	220.07	
Pagode *Grupo Sensacao, Sorriso de marfin* (142.11)		175.45	129.19	122.51				142.11	194.67	124.66	
Salsa *Eddie Santiago, Hagamoslo* (168.67)		168.67	165.36		170.37	191.67		167.00	172.11	137.13	
Sertaneja *Leandro & Leonardo, Eu Juro* (87.04)				87.04		106.15	70.77			83.70	69.64
Tango *A Passion For Tango, Milonga de Mis Amores* (112.29)		155.96		113.43	111.18	129.07		142.14		112.29	118.20

3.3 Pick Any Class Outside of LMD

We now attempt to make DeSPerF-LMD classify in every way time-stretched versions of the ten music recording excerpts used in [17]. Table 4 shows that we were able to do this for most labels and with minor time-stretching factors.

4 Discussion

The results of our experiments show the performance of DeSPerF-LMD to be strongly dependent upon some characteristic related to *absolute time*. Figure 3 shows the normalised accuracy of DeSPerF-LMD drops 40 points or increases 30 with tempo changes of at most ±6 %. Figure 4 shows that small tempo changes greatly impact the reproduction of ground truth of all labels. Table 3 shows DeSPerF-LMD can be made to classify several *LMD* excerpts in most ways it has learned; and Table 4 shows the same result for music excerpts that are not a part of *LMD*. Though Fig. 1 is evidence that DeSPerF-LMD has certainly learned something about *LMD*, the results of our experiments show that what it has learned may not be of much use when it comes to identifying or discriminating Latin music genre or style. An impressive precision in Bachata inspires hope that DeSPerF-LMD has automatically learned why something does or does not "sound like" Bachata. By the results in Table 3, DeSPerF-LMD says slightly speeding up the Bolero excerpt makes it sound more like Bachata than Bolero; and slightly slowing down the Bachata excerpt make it sound more like Bolero

Table 4. Not-*LMD* tracks (left column) are classified in a number of different ways by time-stretching. Resulting tempi (found manually) are shown.

"Original" (tempo)	Axe	Bachata	Bolero	Forro	Gaucha	Merengue	Pagode	Salsa	Sertaneja	Tango
Little Richard *Last Year's Race Horse* (82.00)	96.47	110.81	82.00	79.61	78.10		128.12	113.89	81.19	70.09
Rossini *William Tell Overture* (164.00)	165.66	146.43	164.00	157.69	160.78	133.33	298.18	182.22	150.46	140.17
Willie Nelson *A Horse Called Music* (63.00)	70.79		68.48	66.32	56.25		75.00	92.65	70.00	63.00
Simian Mobile Disco *10000 Horses Can't Be Wrong* (130.00)	128.71			106.56	113.04	130.00	149.43	111.11	114.04	
Rubber Bandits *Horse Outside* (114.00)	110.68	121.28	109.62	142.50	112.87	114.00		193.22	106.54	
Leonard Gaskin *Riders in the Sky* (95.00)	84.07	120.25	95.00	82.61	66.43	68.84	148.44	102.15	95.96	74.22
Jethro Tull *Heavy Horses* (113.00)	97.41	124.18	114.14		113.00	221.57	137.80	166.18	108.65	125.56
Echo and The Bunnymen *Bring on the Dancing Horses* (120.00)	118.81	127.66	104.35	146.34	114.29	120.00		110.09	115.38	
Count Prince Miller *Mule Train* (91.00)	95.79	121.33	91.00	86.67	105.81	88.35		110.98	94.79	
Rolling Stones *Wild Horses* (70.00)	51.09		71.43	54.26	75.27				70.00	68.63

then Bachata. Table 4 shows how for DeSPerF-LMD the "original" excerpt of the "William Tell Overture" sounds most like Bolero, but slowing it down 11 % makes it sound more like Bachata, slowing it down by 15 % makes it become Tango, and slowing it down 19 % creates Merengue. This is not good behaviour.

In their brief musicological descriptions of the music labels in *LMD*, Silla et al. [13] allude to tempo only twice: Merengue has a "zippy" beat, and Axe is "energetic." Supported by Fig. 2, minor changes in tempo should be insignificant to *LMD*. For the mode of the narrowest distribution (Bachata, 130 BPM), a tempo change of $\pm 6\%$ is a difference of about 8 BPM. For the mode of the Merengue tempo distribution (255 BPM), such a change is a difference of about 15 BPM. Since these intervals are well within the spreads of each distribution, one hopes DeSPerF-LMD would not be so sensitive to these changes. While the input SPerF are by construction intimately connected to absolute time characteristics (Sect. 2.1), the results of our experiments suggest that the deeper features produced by deep learning are sensitive to changes of a characteristic that has minor importance for the designation of a recording of music as any *LMD* label (Tables 3 and 4).

From the size of *LMD*, the distribution of tempi of its excerpts, and the fact that the FoM in Fig. 1 are produced using artist filtering, it is hard to believe there to be a specific absolute time characteristic confounded with the labels. In our previous experiments [19], we found the mechanism introducing such a confound into the *BALLROOM* dataset. So, we must discover the cue used by DeSPerF-LMD to produce an illusion of music understanding. An opportunity for this is given in Figs. 3(b) and 4. Analysing the SPerF extracted from the set of time-stretched test signals inflating these FoM might reveal the cues learned by DeSPerF-LMD. While these are negative results, they are also opportunities to improve assumptions and models, as well as machine music listening systems and approaches to their evaluation. Our work motivates in particular the transformation of SPerF (3) to be time-relative rather than time-absolute, and then to measure the impact of this change by performing the experiments above with the new system. Our work also suggests new ways to evaluate systems submitted to the MIREX LMD task, and in fact any of its train-test tasks.

5   Conclusion

DeSPerF-LMD appears to be quite adept at a complex human feat, in spite of the fact that it does not have access to many of the most significant characteristics identifying and distinguishing the labels in *LMD* (e.g., topical, geographical, instrumental). When one claims the only two explanations for such an outcome are either by chance or by music learning, it is easy to see why one would accept the conclusion that the system has learned something general and useful about music. Along with our results in [19], there is however little to support the claim that DeSPerF-based systems trained in *BALLROOM* and in *LMD* have learned anything general about music, meter or rhythm. The story of Clever Hans [10,17] provides a third and much more reasonable explanation: the old horse (*el caballo viejo*) has learned to exploit cues hidden by the lack of control over the independent variables of the evaluation. Once these cues are removed, e.g., giving Clever Hans a private office in which to solve the firm's accounting, or slightly adjusting the tempo of a music recording, the horse reveals its shenanigans.

Speaking more broadly, the 2013 MIREX victory of DeSPerF-LMD, and indeed any victory in the current MIREX train-test tasks, is hollow. When an experimental design lacks an accounting for all independent variables, then one cannot conclude a system has learned to solve some problem implicitly defined by a labeled dataset *no matter how good is its FoM* [16,18]. A machine music listening system can appear to be solving a complex listening task merely by exploiting irrelevant but confounded factors [17,19]. "Solutions" will freely masquerade as advancements until evaluation methods are required to possess the relevance and validity to make the desired conclusions. The development of these valid methods is impossible as long as the problem remains undefined; but we have shown in this paper that it is possible to test claims such as: "System X is performing significantly better than random because it has learned something general about music." It just requires thinking outside the stable.

Acknowledgments. We greatly appreciate Aggelos Pikrakis for making his code available for analysis and testing. CK and JL were supported in part by the Danish Council for Strategic Research of the Danish Agency for Science Technology and Innovation under the CoSound project, case number 11-115328. This publication only reflects the authors' views.

References

1. Aucouturier, J.J., Pachet, F.: Representing music genre: A state of the art. J. New Music Res. **32**(1), 83–93 (2003)
2. Bengio, Y., Lamblin, P., Popovici, D., Larochelle, H.: Greedy layer-wise training of deep networks. Adv. Neural Inf. Process. Syst. **19**, 153 (2007)
3. Deng, L., Yu, D.: Deep Learning: Methods and Applications. Now Publishers, Hanover (2014)
4. Dixon, S., Gouyon, F., Widmer, G.: Towards characterisation of music via rhythmic patterns. In: Proceedings of the ISMIR, pp. 509–517 (2004)
5. Esparza, T., Bello, J., Humphrey, E.: From genre classification to rhythm similarity: Computational and musicological insights. J. New Music Res. **44**, 39–57 (2014)
6. Frow, J.: Genre. Routledge, New York (2005)
7. Gouyon, F., Dixon, S., Pampalk, E., Widmer, G.: Evaluating rhythmic descriptors for musical genre classification. In: Proceedings of the Audio Engineering Society Conference, pp. 196–204 (2004)
8. Hinton, G., Osindero, S., Teh, Y.W.: A fast learning algorithm for deep belief nets. Neural Comput. **18**(7), 1527–1554 (2006)
9. Humphrey, E., Bello, J., LeCun, Y.: Feature learning and deep architectures: New directions for music informatics. J. Intell. Info. Syst. **41**(3), 461–481 (2013)
10. Pfungst, O.: Clever Hans (The horse of Mr. Von Osten): A Contribution to Experimental Animal and Human Psychology. Henry Holt, New York (1911)
11. Pikrakis, A.: A deep learning approach to rhythm modeling with applications. In: Proceedings of International Workshop Machine Learning and Music (2013)
12. Saxe, A.M., McClelland, J.L., Ganguli, S.: Exact solutions to the nonlinear dynamics of learning in deep linear neural networks. CoRR abs/1312.6120 (2013)
13. Silla, C.N., Koerich, A.L., Kaestner, C.A.A.: The Latin music database. In: Proceedings of ISMIR (2008)
14. Slaney, M.: Auditory toolbox. Technical report, Interval Research Corporation (1998)
15. Srivastava, N., Hinton, G., Krizhevsky, A., Sutskever, I., Salakhutdinov, R.: Dropout: A simple way to prevent neural networks from overfitting. J. Mach. Learn. Res. **15**(1), 1929–1958 (2014)
16. Sturm, B.L.: Classification accuracy is not enough: On the evaluation of music genre recognition systems. J. Intell. Info. Syst. **41**(3), 371–406 (2013)
17. Sturm, B.L.: A simple method to determine if a music information retrieval system is a "horse". IEEE Trans. Multimedia **16**(6), 1636–1644 (2014)
18. Sturm, B.L.: The state of the art ten years after a state of the art: Future research in music information retrieval. J. New Music Res. **43**(2), 147–172 (2014)
19. Sturm, B.L., Kereliuk, C., Pikrakis, A.: A closer look at deep learning neural networks with low-level spectral periodicity features. In: Proceedings of the International Workshop on Cognitive Information Processing (2014)

Scales

Can a Musical Scale Have 14 Generators?

Emmanuel Amiot$^{(\boxtimes)}$

Classes Préparatoire aux Grandes Ecoles,
Perpignan, France
manu.amiot@free.fr

Abstract. A finite arithmetic sequence of real numbers has exactly two generators: the sets $\{a, a + f, a + 2f, \ldots a + (n - 1)f = b\}$ and $\{b, b - f, b - 2f, \ldots, b - (n - 1)f = a\}$ are identical. A different situation exists when dealing with arithmetic sequences modulo some integer c. The question arises in music theory, where a substantial part of scale theory is devoted to generated scales, i.e. arithmetic sequences modulo the octave. It is easy to construct scales with an arbitrary large number of generators. We prove in this paper that this number must be a totient number, and a complete classification is given. In other words, starting from musical scale theory, we answer the mathematical question of how many different arithmetic sequences in a cyclic group share the same support set. Extensions and generalizations to arithmetic sequences of real numbers modulo 1, with rational or irrational generators and infinite sequences (like Pythagorean scales), are also provided.

Keywords: Musical scale · Generated scale · Generator · Interval · Interval vector · DFT · Discrete fourier transform · Arithmetic sequence · Music theory · Modular arithmetic · Cyclic groups · Irrational

Foreword

In the cyclic group $\mathbb{Z}_c$, arithmetic sequences are not ordered as they are in $\mathbb{Z}$. For instance, the sequence (0 7 14 21 28 35 42) reduces modulo 12 to (0 7 2 9 4 11 6), whose support is (rearranged) $\{0, 2, 4, 6, 7, 9, 11\} \subset \mathbb{Z}_{12}$. In music theory, such arithmetic sequences in a cyclic group are the basis of studies on Maximally Even Sets, Well-Formed Scales [5] and other major topics of scale theory[1], but also of many rhythms, e.g. the Tresillo 0 3 6 mod 8 which is so prominent in Latin America's music.

In the example above, the G major scale is generated by $f = 7$, of backwards by $-f = 5$ using the convention $C = 0$. All twelve diatonic scales can be generated thusly. It seems natural to infer that, as in $\mathbb{Z}$, such arithmetic sequences have exactly two possible (and opposite) generators. But such is not the case.

---

[1] I am indebted to a reviewer for reminding me of Mazzola's 'circle chords' which provide 'a generative fundament for basic chords in harmony', cf. [12] pp. 514 for a reference in English.

© Springer International Publishing Switzerland 2015
T. Collins et al. (Eds.): MCM 2015, LNAI 9110, pp. 349–360, 2015.
DOI: 10.1007/978-3-319-20603-5_35

1. The 'almost full' scale, with $c-1$ tones, is generated by any interval f coprime with c: such a scale is $f, 2f, \ldots (d-1)f \bmod c^2$. For $c = 12$ for instance we have $\Phi(12) = 4$ different generators, Φ being Euler's totient function. *Idem* for the full aggregate, i.e. $\mathbb{Z}_c$ itself.

2. When f generates a scale which is a subgroup/coset of $\mathbb{Z}_c$, then kf generates the same scale as f, for any k coprime[3] with c, that is to say the generators are all those elements of the group $(\mathbb{Z}_c, +)$ whose order is equal to some particular divisor of c. Consider for instance a 'whole-tone scale' in a 14-tone chromatic universe, i.e. a 7-element sequence generated by 2 modulo 14. It exhibits 6 generators, which are the elements of $\mathbb{Z}_{14}$ with order 7, namely the even numbers: the set $\{0, 2, 4, 6, 8, 10, 12\}$ is produced by either of the 6 following sequences

$$(024681012)(048122610)(061241028)(082104126)(010621284)(012108642)$$

Mathematically speaking, the subgroup of $\mathbb{Z}_c$ with d elements where d divides c, is generated by precisely $\Phi(d)$ intervals. Conversely, the subgroup of $\mathbb{Z}_c$ generated by f is the (one and only) cyclic subgroup with $c/\gcd(c, f)$ elements.

3. One last example: the 'incomplete whole-tone scale' in 10-tone chromatic universe $\{1, 3, 5, 7\}$ has 4 different generators, namely 2, 4, 6, 8.

The present paper studies all possible arithmetic sequences (a.k.a. generated scales) $\{a, a + f, a + 2f, \ldots a + (d-1)f\}$ in $\mathbb{Z}_c$. We will prove that the above examples cover all possible cases, hence the number of generators is always a 'totient number' $\Phi(n)$.[4]

– When a generator f is coprime with c, there are two generators only, except in the cases of the full and 'almost full scale' (when $d = c$ or $d = c - 1$) which admit $\Phi(c)$ generators.

– When some generator of the scale is not coprime with c, it will be seen that the number of generators can be arbitrarily large. At this point the whole classification shall be obtained.

– Then we will endeavour to bring the question into a broader focus, considering partial periodicity and its relationship to Discrete Fourier Transform.

– Lastly, these results will be generalized for scales with non-integer generators.

Several subcases and developments had to be removed because of the size limits.

Notations and Conventions

– Unless otherwise mentioned, computations take place in $\mathbb{Z}_c$, the cyclic group with c elements whose elements are 'tones'.

---

[2] This was mentioned to Norman Carey by Mark Wooldridge, see [4], Chap. 3.

[3] This case was suggested by David Clampitt in a private communication; it also appears in [13].

[4] Sloane's integer sequence A000010.

- $\mathbb{Z}$ (respectively $\mathbb{N} = \mathbb{Z}^+$) stands for the integers (respectively the non-negative integers).
- $a \mid b$ means that a is a divisor of b in the ring of integers.
- The symbol $\lfloor t \rfloor$ denotes the floor function, i.e. the greatest integer lower than, or equal to t.
- The word 'scale' is used, incorrectly but according to custom, for 'pc-set', i.e. an unordered subset of $\mathbb{Z}_c$. Some authors have provided more correct definitions but I shall use this one since the main point of this paper is the possibility of different sequencings.
- A 'generated scale' is a subset of $\mathbb{Z}_c$ that can be built from the values of some finite arithmetic sequence (mod c), e.g. $A = \{a, a + f, a + 2f \ldots\} \subset \mathbb{Z}_c$.
- 'ME set' stands for Maximally Even Set',[5] 'WF' means 'Well Formed', 'DFT' is 'Discrete Fourier Transform'.
- Φ is Euler's totient function, i.e. $\Phi(n)$ is the number of integers smaller than n and coprime with n.
- We will say that $A \subset \mathbb{Z}_c$ is generated by f if A can be written as $A = \{a, a + f, a + 2f \ldots\}$ for some suitable starting point $a \in A$. Notice that sets are denoted using curly braces, while sequences are given between parentheses (somewhat salvaging the sloppiness of the definition of a 'scale').

1   Results

The different cases discussed have no one-to-one relationship with the different cases in the conclusion.

1.1   The Simpler Cases

For any generated scale *where the generator f is coprime with c*, there are only two generators f and $c - f$, except for the extreme cases mentioned in the foreword:

Proposition 1. *Let $1 < d < c - 1$; the scales $A = \{0, a, 2a, \ldots (d - 1)a\}$ and $B = \{0, b, 2b, \ldots (d - 1)b\}$ with d tones, generated in $\mathbb{Z}_c$ by intervals a, b, where one at least of a, b is coprime with c, cannot coincide up to translation, unless $a = b$ or $a + b = c$ (i.e. $b = -a$ mod c). In other words, A admits only the two generators a and $-a$. But when $d = c - 1$ or $d = c$, then there are $\Phi(c)$ generators. (All non-trivial proofs at the end of the paper.)*

Another simple case allows multiple generators:

Proposition 2. *A regular polygon in $\mathbb{Z}_c$, i.e. a translate of some subgroup $f\mathbb{Z}_c$ with d elements, has exactly $\Phi(d)$ generators (Fig. 1).*

Clearly in that case f will not be coprime with c (except in the extreme case of $d = c$ when the polygon is the whole aggregate $\mathbb{Z}_c$).

---

[5] These sets were introduced by Clough and Myerson [7], they can be seen as scales where the elements are as evenly spaced as possible on a number of given sites and comprise the diatonic, whole-tone, pentatonic and octatonic scales among others. See also [1,6,8].

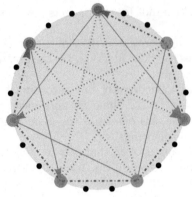

Generators of a whole polygon

Fig. 1. Many generators for a regular polygon

1.2   The Remaining Cases

Theorem 1. *A scale generated by f **not** coprime with c, with a cardinality $1 < d < c$, has*

1. *one generator when the scale is (a translate of) $\{0, c/2\}$ (a 'tritone');*
2. *two generators (not coprime with c) when d is strictly between 1 and $c' - 1 = c/m - 1$ where $m = \gcd(c, f)$;*
3. *$\Phi(d)$ generators when $d = c' = c/m$, i.e. when A is a regular polygon;*
4. *$\Phi(d + 1)$ generators when $d = c' - 1$, and A is a regular polygon minus one point; all generators share the same order in the group $(\mathbb{Z}_c, +)$;*

The third case is actually that of Proposition 1. The last, new case, features *incomplete regular polygons*, i.e. regular polygons with one point removed. For any divisor c' of c, let $f = c/c'$ and $d = c' - 1$:

$$\{f, 2f, \ldots df \pmod{c}\} \qquad \text{where } df = c - f = -f$$

is the simplest representation of such a scale.[6] Consideration of the 'gap' in such scales yields an amusing fact, whose proof will be left to the reader:

Proposition 3. *For each different generator of an 'incomplete polygon', there is a different starting point of the arithmetic sequence. Hence...*

Theorem 2. *There are three cases of generated scales: regular polygons, regular polygons minus one tone, and 'diatonic-like' scales, i.e. affine images of a chromatic segment, as summarized in Fig. 2. The number of generators of such a scale can be any totient number.*

---

[6] A reviewer sums up nicely the two 'plethoric' cases by identifying them with multiple orbits of affine endomorphisms, i.e. different affine maps generating the same orbit-sets. See also [2] about orbits of affine maps modulo n.

Conversely, *nontotient numbers*, that is to say numbers that are not a $\Phi(n)$, can never be the number of generators of a scale. All odd numbers (apart from 1) are nontotient, the sequence of even nontotients begins with 14, 26, 34, 38, 50, 62, 68, 74, 76, 86, 90, 94, 98...[7]

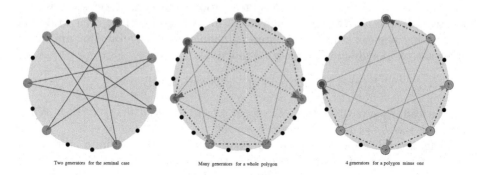

Two generators for the seminal case   Many generators for a whole polygon   4 generators for a polygon minus one

Fig. 2. Three cases

1.3 Partial Periodicities and Fourier Transform

In the seminal case, the values of the two generators can be conveniently recognized as the indexes of the maximum Fourier coefficients of the scale. Informally, a generated scale A is at least partially periodic, and this is apparent on the Fourier transform $\mathcal{F}_A : t \mapsto \sum_{k \in A} e^{-2i\pi\, k\, t/c}$.

More precisely, in the online supplementary of [1] it is established that

Theorem 3. *For c, d coprime, a scale with d tones is generated by an interval f, coprime with c, if and only if the semi-norm*

$$\|\mathcal{F}_A\|^* = \max_{t\ coprime\ with\ c} |\mathcal{F}_A(t)| = \max_{t\ coprime\ with\ c} \left|\sum_{k \in A} e^{-2i\pi\, k\, t/c}\right|$$

is maximum among all d-element scales, i.e. for any scale B with d elements, $\|\mathcal{F}_A\|^* \geq \|\mathcal{F}_B\|^*$. Moreover, if $\|\mathcal{F}_A\|^* = |\mathcal{F}_A(t_0)|$, then $t_0^{-1} \in \mathbb{Z}_c$ is one generator of scale A, the only other being $-t_0^{-1}$.

This extends the discovery made by Ian Quinn [13]:

Quinn's Fourier Characterization 1. *A is Maximally Even with d elements if, and only if, $|\mathcal{F}_A(d)|$ has maximum value among d-subsets.*

The connexion being that the scales involved in Theorem 3 are affinely equivalent to some ME sets, which are themselves affine images of chromatic clusters (see also proof of Proposition 1). Informally it vindicates Quinn's idea of *saliency* or chord quality: a pc-set with a large magnitude of its kth coefficient is c/k-ish (or k^{-1}-ish in $\mathbb{Z}_c$ as the case may be), e.g. (for $c = 12$) a large third coefficient means *major-thirdish*, a large fifth coefficient means *fifthish*.

---

[7] Sloane's sequence A005277 in his online encyclopedia of integer sequences [15]. For the whole sequence including odd numbers, see A007617.

1.4 Non-integer Generation

The present paper covered exhaustively the case of scales of the form

$$\{a + k\,f \pmod{c}, k = 0 \ldots d - 1\}, a, c, f \in \mathbb{Z}$$

but in other music-theoretic models, generation from non-integer steps is also common. There are two important cases: for a non-integer or even irrational generator f one can still retrieve integers modulo some c by rounding up, which gives rise to Clough and Douthett's J-functions:

$$J_\alpha(k) = \lfloor k\alpha \rfloor \pmod{c}, \quad \text{wherein usually } \alpha = \frac{c}{d}$$

Let us call J_α-sets the sets of the form $\{J_\alpha(0); \ldots J_\alpha(d-1)\} \subset \mathbb{Z}_c.$[8] When $\alpha = c/d$, the J_α-set is a Maximally Even set.

The second important case adresses finite or even infinite arithmetic sequences in the continuous circle, with the seminal notion of $\mathcal{P}_x$-sets, made up with consecutive values of the maps $k \mapsto \mathcal{P}_x(k) = k\,x \pmod{1}$, that express pythagorean-style scales (e.g. $x = \log_2(3/2)$). Some of them are the Well-Formed Scales [4,5]. The question of different generators can be formulated thus: *If two sets of values of J_α, J_β (resp. $\mathcal{P}_x, \mathcal{P}_y$) are transpositionally equivalent, do we have necessarily $\alpha = \pm\beta$ (resp. $x = \pm y$)? In other words, do such scales have exactly 2 generators and no more?*

We have already seen counter-examples, e.g. in Proposition 2 when $\alpha \in \mathbb{N}, d\,\alpha = c$ and the J_α-set is a regular polygon (or similarly $d\,x \in \mathbb{N}$ for the $\mathcal{P}_x$ case). Other cases are worth investigating. Let us first consider the values of a J function with a random multiplier, e.g. $J_\alpha(k) = \lfloor k\,\alpha \rfloor \pmod{c}$ with some $\alpha \in \mathbb{R}$. These values have been mostly scrutinized when $\alpha = c/d$. Since the floor function is locally right-constant, $J_\alpha(k)$ does not change for a small increase of α and hence

Proposition 4. *A J_α-set (up to translation) does not characterize the pair $\pm\alpha$: there are infinitely many real α's that generate the same sequence (the set of those α's has strictly positive measure).*

Secondly, we state a result when the generator x, in a finite sequence of values of $\mathcal{P}_x$, is irrational:

Theorem 4. *If the sets $\mathcal{P}_x^d = \{0, x, 2x, \ldots (d-1)x\}$ and $\mathcal{P}_y^d = \{0, y, 2y, \ldots (d-1)y\}$ (both mod 1) are transpositionally equivalent, with x irrational and $d > 0$, then $x = \pm y \pmod{1}$.*

This has been implicitly known in the case of Well-Formed Scales in non-tempered universes [4], but this theorem is more general. Lastly, we characterize the generators of the 'infinite scales'[9] $\mathcal{P}_x^\infty = \{nx \pmod{1}, n \in \mathbb{Z}\}$:

---

[8] Some degree of generalization is possible, see [6] for instance, but results merely in translations of the set.

[9] I have to stress the musical interest of such bizarre objects, actively researched both in the domain of word/scale theory [4,5,9] and aperiodic rhythms [3] and providing compositional material.

Theorem 5. *Two infinite generated scales are equal up to translation, that is $\exists \tau, \mathcal{P}_x^\infty = \tau + \mathcal{P}_y^\infty$, if and only if they have the same generator up to a sign, i.e. $x = \pm y \pmod 1$, when x is irrational.*

In the case where x is rational and $x = a/b, \gcd(a, b) = 1$, there are $\Phi(b)$ different possible generators (just as in the finite case, see Theorem 2).

In other words, an infinite Pythagorean scale has 2 or $\Phi(b)$ generators, according to whether the number of actually different tones is infinite or finite. The special (tritone) case of one generator already mentioned in Proposition 1 also occurs for $x = 1/2$.

2 Proofs

2.1 Proof of Proposition 1

The following proof relies on the one crucial concept of (oriented) interval vector,[10] that is to say the multiplicities of all intervals inside a given scale. This is best seen by transforming A, B into segments of the chromatic scale, by way of affine transformations.

Proof. All computations are to be understood modulo c. The extreme cases $d = c - 1$ and $d = c$ were discussed before: any f coprime with c generates the whole $\mathbb{Z}_c$, hence $\{f, 2f, \ldots (d-1)f\}$ is always equal to $\mathbb{Z}_c$ deprived of 0. Up to a change of starting point, we can generate with such an f any subset with cardinality $d - 1$.

Furthermore, a generator f **not** coprime with c would only generate (starting with 0, without loss of generality) a part of the strict subgroup $f\mathbb{Z}_c \subset \mathbb{Z}_c$, hence a subset with stricly less than $c - 1$ elements. So we are left with the general case, $1 < d < c - 1$. Without loss of generality we take b invertible modulo c (a and b are interchangeable, and we assumed that one of them is invertible modulo c, i.e. coprime with c).

Let $D = \{0, 1, 2, \ldots d - 1\}$; assume $A = B + \tau$, as $A = aD$ and $B = bD$, then D must be its own image ($\varphi(D) = D$) under the following affine map:

$$\varphi : x \mapsto b^{-1}(a\,x - \tau) = b^{-1}ax - b^{-1}\tau = \lambda x + \mu.$$

We now elucidate the different possible multiplicities of intervals between two elements of D.

Lemma 1. *Let $1 < d < c - 1$, we define the oriented interval vector $\boldsymbol{IV}_D$ by $\boldsymbol{IV}_D(k) = \#\{(x, y) \in D^2, y - x = k\}$. Then $\boldsymbol{IV}_D(k) < d - 1 \; \forall k = 2 \ldots c - 2$; to be more precise,*

$$\boldsymbol{IV}_D = [_D(0), \boldsymbol{IV}_D(1), \ldots, \boldsymbol{IV}_D(c-1)] = [d, d-1, d-2, \ldots, d-2, d-1]$$

---

[10] A famous concept in music theory, see for instance [14]. I had initially found an alternative proof based on majorizations of the Fourier Transform, omitted here in favour of a shorter one.

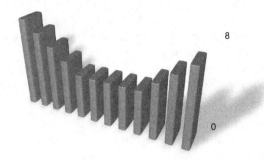

Fig. 3. Interval vector of (0 1 2 3 4 5 6 7) mod 12

i.e. interval 1 and its opposite $c-1$ are the only ones with multiplicity $d-1$.

For instance, one computes $\boldsymbol{IV}_D = [8,7,6,5,4,4,4,4,4,5,6,7]$ with $c = 12$, $d = 8$.[11] See a picture of this interval vector on Fig. 3.

Proof. This can proved by induction, or by direct enumeration (left as an exercise). Figure 4 may help to distinguish the two different ways of getting interval k, either between consecutive elements (dashed) or across the gap (dotted).

As we will see independently below, if one generator is invertible, then so are all others.[12] Hence $\lambda = b^{-1}a$ is invertible in $\mathbb{Z}_c$ and the map $\varphi : x \mapsto \lambda x + \mu$ above is one to one; it multiplies all intervals by λ mod c:

$$\varphi(j) - \varphi(i) = (\lambda j + \mu) - (\lambda i + \mu) = \lambda.(j-i),$$

which turns the interval vector $\boldsymbol{IV}_D$ into $\boldsymbol{IV}_{\varphi(D)}$ wherein $\boldsymbol{IV}_D(\lambda i) = \boldsymbol{IV}_{\varphi(D)}(i)$: the same multiplicities occur, but for different intervals.[13]

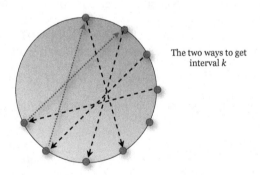

The two ways to get
interval k

Fig. 4. Double origin of one interval

---

[11] The minimum value of $\boldsymbol{IV}$ and the number of its repeated occurrences could be computed – it is 0 for $d < c/2$ – but are irrelevant to the discussion.

[12] This could also be proved directly from $\varphi(D) = D$.

[13] It is well known that affine transformations permute interval vectors.

Most notably, the *only*[14] two intervals with multiplicity $d-1$ in $\textbf{IV}_{\varphi(D)}$ are λ and $-\lambda$. Hence, if $\textbf{IV}_{\varphi(D)} = \textbf{IV}_D$, the maximal multiplicity $d-1$ must appear in positions 1 and $c-1$, which compels λ to be equal to ± 1. Finally, as $\lambda = a\,b^{-1}$ mod c, we have indeed proved that $a = \pm b$, qed.

2.2   Proof of Proposition 2

A regular polygon is a translate of a subgroup with d elements, and such a subgroup has $\Phi(d)$ generators.

2.3   Proof of Theorem 1

The first case is obvious, when c is even the only generator of $\{0, f = c/2\}$ is $c/2$. Case 3 was studied in Proposition 2. There remains the case of a scale generated by some f **not** coprime with c, when that scale is not a regular polygon. For the end of this discussion, let $\gcd(f,c) = m > 1$, and assume $d > 1$ and (up to translation) $0 \in A$.

Lemma 2. *If f, g are two generators of a same scale A, then $m = \gcd(c, f) = \gcd(c, g)$.*

Proof. Consider the group D_A of differences of elements of A, generated by iteration (here a finite number thereof) of the operator $\Delta : A \mapsto A - A$. This is a subgroup of $\mathbb{Z}_c$, hence a cyclic group, i.e. some $m\mathbb{Z}_c$.[15] Now for any generator f of A, it is obvious that D_A is also generated by f (any difference of elements of A being a multiple of f). The order of f is the cardinality c/m of D_A, hence the Lemma. (NB: this lemma was initially proved using DFT.)

As a special case, this proves what we had advanced during the proof of Proposition 2, namely that if one of the generators of a scale is invertible modulo c, then so are all others. From there, one can divide A by m and assume without loss of generality that $f' = f/m$ and $c' = c/m$ coprime. We are dealing now with a scale $A' = A/m$ in $\mathbb{Z}_{c'}$, generated by f' and $g' = g/m$, both coprime with c': then Proposition 1 provides two cases, either $\#A' = d < c' - 1$ or not. In the latter case, we have $\Phi(c')$ generators for an almost full, or full, regular polygon; in the former, only two, like for the generic 'diatonic-like' scale.

As for instance, $f' = \pm g' \pmod{c}' \iff f = \pm g \pmod{c}$, we have exhausted all possible cases when a generator is not coprime with c, and proved Proposition 1, hence Theorem 2.

2.4   Proof of Theorem 4

Proof. It is the same the idea as the proof of Proposition 1, using the interval vector; but since the universe is so different (infinite and continuous) I feel the need for an independent proof.

---

[14] Because φ is one to one.

[15] This is the group generated by $A - A$, in all generality, cf. [12], 7.26.

This works in both cases because the affine map $\mathcal{P}_x$ is one to one again: since x is irrational, we have the immediate.

Lemma 3. $\forall a, b \in \mathbb{Z}$, $ax \equiv bx \pmod 1$ $\iff$ $a = b$. (Else x equals some $\frac{k}{b-a}$.)

Consider now all possible intervals in $\mathcal{P}_x^d$, i.e. the $(i-j)x \pmod 1$ with $0 \le i, j < d$. By our hypothesis, these intervals occur with the same multiplicity in $\mathcal{P}_x^d$ and $\mathcal{P}_y^d$. Let us have a closer look at these intervals (computed modulo 1), noticing first that

- There are d different intervals from 0 to kx, with k running from 0 to $d-1$. They are distinct because x is irrational, see Lemma 3. Their set is $\mathcal{I}_0 = (0, x, 2x \ldots (d-1)x)$.
- From x to $x, 2x, 3x, \ldots (d-1)x$ and 0, there are $d-1$ intervals common with $\mathcal{I}_0$, and a new one, $0 - x = -x$. It is new because x is still irrational. For the record, their set is $\mathcal{I}_x = (0, x, 2x \ldots (d-2)x, -\boldsymbol{x})$.
- From $2x$ to the others, $d-1$ intervals are common with $\mathcal{I}_x$ and only $d-2$ are common with the $\mathcal{I}_0$.
- Similarly for $3x, 4x \ldots$ until
- Finally, we compute the intervals from $(d-1)x$ to $0, x, \ldots, (d-2)x, (d-1)x$. One gets $\mathcal{I}_{(d-1)x} = (0, -x, -2x \cdots - (d-1)x)$.

The table Fig. 5, not unrelated to Fig. 3, will make clear the values and coincidences of the different possible intervals. So only two intervals (barring 0) occur $d-1$ times in $\mathcal{P}_x^d$ (resp. $\mathcal{P}_y^d$), namely x and $-x$. Hence $x = \pm y$, qed.

Intervals starting in									
0					0	x	2x	...	(d-2)x	(d-1)x
x				-x	0	x	2x	...	(d-2)x
2x			-2x	-x	0	x	2x	...
...									
(d-2)x		-(d-2)x	...	-x	0	x			
(d-1)x	-(d-1)x	...	...	-x	0				

Fig. 5. The different intervals from each starting point

2.5 Proof of Theorem 5

Notice that $\mathcal{P}_x^\infty = \{kx \pmod 1, k \in \mathbb{Z}\}$ is a subgroup of the circle (or one-dimensional torus) $\mathbb{R}/\mathbb{Z}$, quotient group of the subgroup of $\mathbb{R}$ generated by 1 and x. All computations are to be understood modulo 1. If we consider a translated version $\mathcal{S} = a + \mathcal{P}_x^\infty$, then the group can be retrieved by a simple difference:

$$\mathcal{P}_x^\infty = \mathcal{S} - \mathcal{S} = \{s - s', (s, s') \in \mathcal{S}^2\}$$

So the statement of the theorem can be simplified, without loss of generality, as "if $\mathcal{P}_x^\infty = \mathcal{P}_y^\infty$ then $x = \pm y$ (mod 1)" (and similarly in the finite case).

Proof. We must distinguish the two cases, whether x is rational or not.

- The case x rational is characterized by the finitude of the scale. Namely, when $x = a/b$ with a, b coprime integers (we will assume $b > 0$), then $\mathcal{P}_x^\infty$ is the group generated by $1/b$: one inclusion is clear, the other one stems from Bezout relation: there exists some combination $au + bv = 1$ with u, v integers, and hence

$$1/b = u\,a/b + v = u\,a/b \pmod 1 = \underbrace{a/b + \ldots a/b}_{u \text{ times}}$$

  is an element of $\mathcal{P}_x^\infty$. As $1/b \in \mathcal{P}_x^\infty$, it contains the subgroup generated by $1/b$, and finally these two subgroups are equal. The subgroup $< 1/b >$ (mod 1) is cyclic with b elements, hence it has $\Phi(b)$ generators, which concludes this case of the theorem.[16]
- Now assume x irrational and $\mathcal{P}_x^\infty = \mathcal{P}_y^\infty$. An element of $\mathcal{P}_x^\infty$ can be written as ax (mod 1), with $a \in \mathbb{Z}$. Since $y \in \mathcal{P}_x^\infty$ then $y = ax$ (mod 1) for some a. Similarly, $x = by$ for some b. Hence

$$x = abx \pmod 1 \quad \text{that is to say in } \mathbb{Z}, \quad x = abx + c$$

  where a, b, c are integers. This is where we use the irrationality of x: $(1 - ab)x = c$ implies that $ab = 1$ and $c = 0$. Hence $a = \pm 1$, i.e. $x = \pm y$ (mod 1), which proves the last case of the theorem.

Remark 1. *The argument about retrieving the group from the (possibly translated) scale applies also if the scale is just a semi-group, e.g. $\mathcal{P}_x^{+\infty} = \{kx$ (mod 1)$, k \in \mathbb{Z}, k \geq 0\}$ (or even to many subsets of $\mathcal{P}_x^\infty$), which are perhaps a less wild generalization of the usual Pythagorean scale. So the theorem still holds for the half-infinite scales.*

Conclusion

Apart from the seminal case of diatonic-like generated scales, it appears that many scales can be generated in more than two ways. This is also true for more complicated modes of 'generation'. Other simple sequences appear to share many different generation modes. It is true for instance of geometric sequences, like the *powers* of 3, 11, 19 or 27 modulo 32 which generate the same 8-note scale in $\mathbb{Z}_{32}$, namely $\{1, 3, 9, 11, 17, 19, 25, 27\}$ – geometric progressions being interestingly dissimilar in that respect from arithmetic progressions.[17] I hope the above discussion will shed some light on the mechanics of scale construction.

---

[16] The different generators are the k/b where $0 < k < b$ is coprime with b.
[17] Such geometric sequences occur in Auto-Similar Melodies [2], like the famous initial motive in Beethoven's Fifth Symphony, autosimilar under ratio 3.

Acknowledgments. I thank David Clampitt for fruitful discussions on the subject, and Ian Quinn whose ground-breaking work edged me on to explore the subject in depth, and my astute and very helpful anonymous reviewers.

References

1. Amiot, E.: David lewin and maximally even sets. J. Math. Music Taylor Fr. **1**(3), 152–172 (2007)
2. Amiot, E.: Self similar melodies. J. Math. Music Taylor Fr. **2**(3), 157–180 (2008)
3. Callender, C.: Sturmian canons. In: Yust, J., Wild, J., Burgoyne, J.A. (eds.) MCM 2013. LNCS, vol. 7937, pp. 64–75. Springer, Heidelberg (2013)
4. Carey, N.: Distribution Modulo 1 and musical scales. Ph.D. thesis, University of Rochester (1998). Available online
5. Carey, N., Clampitt, D.: Aspects of well formed scales. Music Theory Spectr. **11**(2), 187–206 (1989)
6. Clough, J., Douthett, J.: Maximally even sets. J. Music Theory **35**, 93–173 (1991)
7. Clough, J., Myerson, G.: Variety and multiplicity in diatonic systems. J. Music Theory **29**, 249–270 (1985)
8. Douthett, J., Krantz, R.: Maximally even sets and configurations: common threads in mathematics, physics, and music. J. Comb. Optim. (Springer) **14**(4), 385–410 (2007). http://www.springerlink.com/content/g1228n7t44570442
9. Carey, N.: Lambda words: a class of rich words defined over an infinite alphabet. J. Integer Seq. **16**(3), 13.3.4 (2013)
10. Lewin, D.: Re: intervalic relations between two collections of notes. J. Music Theory **3**, 298–301 (1959)
11. Lewin, D.: Generalized Musical Intervals and Transformations. Yale University Press, New Haven (1987)
12. Mazzola, G., et al.: Topos of Music. Birkhäuser, Basel (2002)
13. Quinn, I.: A unified theory of chord quality in chromatic universes. Ph.D. Dissertation, Eastman School of Music (2004)
14. Rahn, D.: Basic Atonal Theory. Longman, New York (1980)
15. Sloane, N.J.A., Online encyclopedia of integer sequences. https://oeis.org/

On the Step-Patterns of Generated Scales that are Not Well-Formed

Marco Castrillón López[1] and Manuel Domínguez Romero[2](✉)

[1] ICMAT (CSIC-UAM-UC3M-UCM)
Geometria y Topologia, Universidad Complutense de Madrid,
28040 Madrid, Spain
mcastri@mat.ucm.es
[2] Departamento de Matemática Aplicada
E.T.S. de Arquitectura, Universidad Politécnica de Madrid,
28040 Madrid, Spain
mdrmanuel@gmail.com

Abstract. It is well-known that generated scales (with irrational generator) may have two or three different steps. It is also known that the scale has exactly two steps precisely if the number of notes coincides with the denominator of a (semi-)convergent of the generator. Moreover, the step-pattern is a Christoffel word: a mechanical word with rational slope. In this article we investigate the *bad* case: generated scales with three different steps. We will see that their step-patterns share some properties with the Christoffel case: they are Lyndon words and their right Lyndon factorization is determined by the generator. Some conjectures on their left Lyndon factorization are also given.

Overview

In the present work we connect notions from three different fields: the theory of numbers (convergents of a real number and Farey numbers), the theory of scales (generated and well-formed scales) and the combinatorics on words (Christoffel and Lyndon words). We define below the main concepts that will be needed from these three domains. In Sect. 1 the notion of non well-formed word (NWF word) is defined as the word that encode a generated scale that is not well-formed. These words will center our attention in the whole paper. In Sect. 2 it is proven that every NWF word is a Lyndon word. Finally, in Sect. 3 we describe the right Lyndon factorization of a NWF word, which has a very nice musical interpretation: as in the case of well-formed scales, it coincides with the factorization determined by the generator of the scale. We end up by presenting a conjecture that describes the scales for which right and left Lyndon factorizations coincide.

Let $\theta = [0; a_1, a_2, \ldots]$ be the continued fraction expansion of a real number in the interval $I = [0, 1)$. Every rational number $\frac{M}{N} = [0; a_1, a_2, \ldots, a_{k-1}, b+1]$ with $0 < b \leq a_k$ is called *semi-convergent* of θ.

Let us denote by $\mathfrak{F}$ the set of all fractions $\frac{M}{N}$ such that $0 \leq M < N$ and $gcd(M, N) = 1$. $\mathfrak{F}$ is called the set (or tree) of *Farey numbers*. It can be displayed

© Springer International Publishing Switzerland 2015
T. Collins et al. (Eds.): MCM 2015, LNAI 9110, pp. 361–372, 2015.
DOI: 10.1007/978-3-319-20603-5_36

as the left half of the binary *Stern-Brocot tree* whose nodes comprise all positive rational numbers $r \in \mathbb{Q}^+$. Every Farey number $\frac{M}{N}$ is determined by a path down the tree starting with a move to the left, i.e. by a word of the semi-ideal $L \cdot \{L, R\}^*$ (set of words in two letters, L and R, starting with letter L). The set of N-*Farey numbers* is the set $\mathfrak{F}_N = \{\frac{p}{q} \in \mathfrak{F} \text{ such that } q \leq N\}$.

The monoid $SL_2(\mathbb{N})$ of 2×2-matrices $\left(\begin{smallmatrix} a & b \\ c & d \end{smallmatrix}\right)$ with natural number entries $a, b, c, d \in \mathbb{N}$ and determinant $ad - bc = 1$ is freely generated by the matrices $L = \left(\begin{smallmatrix} 1 & 0 \\ 1 & 1 \end{smallmatrix}\right)$ and $R = \left(\begin{smallmatrix} 1 & 1 \\ 0 & 1 \end{smallmatrix}\right)$. Notice that we use the same notation as for the branches in the Stern-Brocot tree. This is due to the following number-theoretic facts about the conversion between nodes of the Stern-Brocot tree and paths from the root leading to these nodes are well-known [5]:

Lemma 1. *The following three facts are equivalent:*

1. $\frac{M}{N} = [0; a_1, \ldots, a_k + 1]$.
2. $\frac{M}{N}$ *is the node associated with the path* $L^{a_1} R^{a_2} \cdots A^{a_k}$ *in the Stern-Brocot tree, where* $A = L$ *if* k *is odd and* $A = R$ *otherwise.*
3. *If* $L^{a_1} R^{a_2} \cdots A^{a_k} = \left(\begin{smallmatrix} M_2 & M_1 \\ N_2 & N_1 \end{smallmatrix}\right)$, *then* $\frac{M_1}{N_1} < \frac{M}{N} < \frac{M_2}{N_2}$ *are three N-Farey numbers, where* $\frac{M}{N} = \frac{M_1 + M_2}{N_1 + N_2}$.

If θ is an irrational number in the interval $(0, 1)$, the set

$$\Gamma(\theta, N) = \{\{k\theta\} \text{ with } k = 0, 1, \ldots, N - 1\}$$

is called generated scale of N notes. It can be represented geometrically as a *regular polygonal chain* inscribed in the circle (see Fig. 1). For example, if we take as generator the pure fifth $\theta = \log_2\left(\frac{3}{2}\right)$ then we are building scales in Phytagorean tuning.

The three gap theorem (see, for example, [9]) asserts that any generated scale

$$\Gamma(\theta, N) = \{0 = x_0 < x_1 < \cdots < x_{N-1}\}$$

has at most three different step with lengths $x_1 = \alpha$, $1 - x_{N-1} = \beta$ and $\alpha + \beta$. The third step may not appear if the scale has just two steps. As one generates

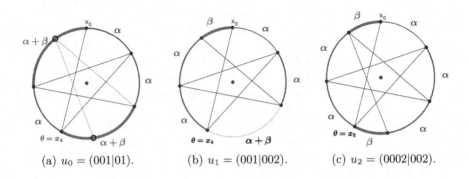

(a) $u_0 = (001|01)$.   (b) $u_1 = (001|002)$.   (c) $u_2 = (0002|002)$.

Fig. 1. Three scales generated by the pure fifth $\log_2\left(\frac{3}{2}\right)$ and their step patterns.

each note, one step of length $\alpha + \beta$ splits in two of lengths α and β. The scale has just one step if and only if the generator is a rational number $\frac{M}{N}$ and we generate at least N notes. To avoid this *degenerated* case we will suppose from now on that the generator is an irrational number.

A finite word w of length N over an alphabet $\mathcal{A}$ is an application

$$w : \{0, 1, \dots, N-1\} \to \mathcal{A}.$$

The set of finite words over the alphabet $\mathcal{A}$ is denoted by $\mathcal{A}^*$. Thus, the step-pattern of Γ may be encoded with the word $w_\Gamma \in \{0, 1, 2\}^*$ as follows:

$$w_\Gamma = w[0]w[1]\dots w[N-1] \in \{0, 1, 2\}^* \text{ with } w[k] = \begin{cases} 0 & \text{if } x_{k+1} - x_k = \alpha \\ 2 & \text{if } x_{k+1} - x_k = \beta \\ 1 & \text{if } x_{k+1} - x_k = \alpha + \beta \end{cases}$$

Following [2], the step-pattern w_Γ of a scale $\Gamma(\theta, N)$ is in $\{0, 2\}^*$ if and only if N is the denominator of a (semi-) convergent $\frac{M}{N}$ of θ. In this case it holds that the order in which the notes of the scale are generated is a multiplicative permutation of the natural order. More precisely, it holds that

$$x_k = \{(M^* \cdot k) \bmod N \cdot \theta\} \quad \text{for } k = 0, 1, \dots, N-1$$

where $M^* = M^{-1} \bmod N$. These scales are called well-formed scales (WF scales) with multiplier M and were introduced by N. Carey and D. Clampit in [3].

We introduce now some notions of Combinatorics on words. Recall that a Christoffel word of slope $\frac{p}{q}$ is the binary word (in $\{0, 1\}^*$) given by:

$$w[k] = \left\lceil \frac{p(k+1)}{q} \right\rceil - \left\lceil \frac{pk}{q} \right\rceil \quad \text{with } k = 0, 1, \dots, N-1.$$

A Christoffel word can be written in any other ordered binary alphabet $\{x < y\}$ via the transformation $(0, 1) \to (x, y)$. One of the first links between Combinatorics on Words and the Theory of Scales was introduced in [4], where it was proven that the step-pattern, w_Γ, of a WF scale of N notes and generator θ is a Christoffel word of slope $\frac{M^{-1} \bmod N}{N}$ whenever $\frac{M}{N}$ is a convergent or semi-convergent of the generator θ. The aim of the present work is to describe some combinatorial properties of the generated scales *that do not verify well-fomedness*. Lyndon words, that are defined just below, will be crucial for this purpose.

If the alphabet $\mathcal{A}$ is ordered, one can define a complete ordering, called *lexicographic ordering*, over the set of finite words $\mathcal{A}^*$: given two words $u, v \in \mathcal{A}^*$, we say that $u < v$ if and only if either $u \in v\mathcal{A}^*$ or

$$u = ras \quad v = rbs \quad \text{with } a < b, \quad a, b \in \mathcal{A} \quad r, s \in \mathcal{A}^*.$$

A word is called primitive if it cannot be written as a power of a word of smaller length. For every $u, v \in \mathcal{A}^*$, the words uv and vu are said conjugated. Conjugation is an equivalence relation: two words are conjugated if they are equivalent modulo rotations. A word $u \in \mathcal{A}^*$ is called a Lyndon word if it is primitive and minimal in its conjugation class. Equivalently, a word w is Lyndon if and only if it holds that:

$$\text{for all } u, v \in \mathcal{A}^* \text{ such that } w = uv \Rightarrow w < vu.$$

1   NWF Words: Definition and Construction

In this section we describe the words codifying generated scales that are not well-formed.

Definition 1. *A word $w \in \{0,1,2\}^*$ is NWF if it is the step-pattern of a generated scale that is not well-formed.*

We recall now some facts on Christoffel words. A word $c \in \{0,2\}^*$ is called central if $0c2$ is a Christoffel word. A positive integer p is a period of w if $w[i] = w[i+p]$ for all $1 \leq i \leq |w| - p$. The following classical two characterizations of central words (see [7, Corollary 2.2.9] and [1, Theorem 4.6]) will be useful in the study of NWF words:

Proposition 1. *The set of central words coincides with*

$$0^* \cup 2^* \cup (PAL \cap PAL20PAL)$$

where PAL is the set of palindrome words. The decomposition of a central word c as $c = p20q$ with p, q palindrome words is unique.

Every (non-trivial) Christoffel word w has a unique factorization, called *standard factorization*, as $w = (w'|w'')$, where w' and w'' are Christoffel words. If $w = 0c2$ with $c \in PAL$, then $w' = 0p2$ and $w'' = 0q2$, where $c = p20q$ is the unique decomposition of the central word c, as in the Proposition above.

Proposition 2. *If $w = (w'|w'')$ is the standard factorization of a Christoffel word and $w = 0c2$, the central word c has the periods $|w'|$ and $|w''|$.*

Lemma 2. *If $w \in \{0,2\}^*$ is a Christoffel word of slope $\frac{M^*}{N}$, then it holds:*

1. $w[k] = 0 \Leftrightarrow k = s \cdot M \mod N$ for a certain $s \in \{0, 1, \ldots, |w|_0 - 1\}$.
2. $w[k] = 2 \Leftrightarrow k = N - 1 + s \cdot M \mod N$ for a certain $s \in \{0, 1, \ldots, |w|_2 - 1\}$.

Proof. Notice first that $w = (w'|w'')$ with $|w'| = M$ and $|w''| = N - M$ is the standard factorization of w. It follows that

$$w[0] = 0 = w[M] \quad \text{and} \quad w[N-1] = 2 = w[N - M - 1].$$

But $w[i] = w[i+M]$ for $i = 1, \ldots, N - M - 1$ and $w[i] = w[i + N - M]$ for $i = 1, \ldots, M-1$, since M and $N - M$ are periods of the central word c associated with w (that is $w = 0c2$). Then $w[i] = [i + M \mod N]$ for $i = 1, \ldots, N - 1$, which concludes the proof.

Let $\frac{M}{N} \in \mathfrak{F}$ and $\frac{M_1}{N_1} < \frac{M}{N} < \frac{M_2}{N_2}$ three consecutive $N-$ Farey numbers in $\mathfrak{F}_N$. As we previously saw, given $\theta \in \left(\frac{M_1}{N_1}, \frac{M_2}{N_2}\right)$, the scale $\Gamma(\theta, N) = \{x_i\}$ is WF and its step-pattern is the Christoffel word w_Γ of slope $\frac{M^*}{N}$. Let us denote by ψ and ξ the morphisms of words in $\{0, 1, 2\}^*$ given by $\psi(0, 1, 2) = (0, 1, 1)$ and $\xi(0, 1, 2) = (1, 1, 2)$. Furthermore, let $u_0 = \psi(w_\Gamma)$ and $v_0 = \xi(w_\Gamma)$. Following Proposition describes the NWF words corresponding with the step-patterns of the scales $\Gamma(\theta, N + k)$, being $\Gamma(\theta, N)$ a WF scale:

Proposition 3. *With the previous notation, let us consider the sequences $\{l_k\}$ and $\{r_k\}$ given by:*

$$l_k = (N - 1 + k \cdot M) \bmod N \quad with \quad k = 0, 1, \ldots N_1 - 1,$$

$$r_k = k \cdot M \bmod N \quad with \quad k = 0, 1, \ldots N_2 - 1.$$

If $\theta \in \left(\frac{M_1}{N_1}, \frac{M}{N}\right)$, the step-pattern of the scale $\Gamma(\theta, N + k)$ is given by

$$u_k = u_{k,0} \cdot u_{k,1} \cdots u_{k,N-1} \quad where \quad u_{k,i} = \begin{cases} 02 & if\ i \in \{l_0, l_1, \ldots, l_{k-1}\} \\ u_0[i] & if\ i \notin \{l_0, l_1, \ldots, l_{k-1}\} \end{cases}$$

for every $k = 1 \ldots, N_1$. In a parallel way, if $\theta \in \left(\frac{M}{N}, \frac{M_2}{N_2}\right)$ and $k = 1, \ldots, N_2$, the step-pattern of the scale $\Gamma(\theta, N + k)$ is given by

$$v_k = v_{k,0} \cdot v_{k,1} \cdots v_{k,N-1} \quad where \quad v_{k,i} = \begin{cases} 02 & if\ i \in \{r_0, r_1, \ldots, r_{k-1}\} \\ v_0[i] & if\ i \notin \{r_0, r_1, \ldots, r_{k-1}\} \end{cases}$$

Proof. If $\theta \in \left(\frac{M_1}{N_1}, \frac{M}{N}\right)$, then the first step $\alpha = x_1$ is smaller than the last one $\beta = 1 - x_{N-1}$. When we generate the $(N + 1)$-the note $\{N\theta\}$, it will lay between x_{N-1} and x_0 and the step of length β will have split into two steps: α and $\beta - \alpha$. Following Lemma 2, the order in which the steps of length β split in two of lengths α and $\beta - \alpha$ is given by the sequence $\{l_s\}$. It follows that the step-pattern of $\Gamma(\theta, N + k)$ will be built from u_0 exchanging $u_0[l_i] = 1$ with 02 for all $i = 0, 1, \ldots, k - 1$. If $\theta \in \left(\frac{M}{N}, \frac{M_2}{N_2}\right)$, $\{N\theta\}$ will lay between x_0 and x_1, the step of length α will have split into two steps: β and $\alpha - \beta$. The step-pattern v_k of $\Gamma(\theta, N + k)$ will be built from v_0 exchanging $v_0[r_i] = 1$ with 02 for all $i = 0, 1, \ldots, k - 1$.

Notice that u_{N_1} coincides with the Christoffel word $G(w_\Gamma)$, where G stands for the Christoffel morphism $G(0, 2) = (0, 02)$. The words $\{u_1, \ldots, u_{N_1-1}\}$ given in previous Proposition will be called *left*-NWF words. In a parallel way, $v_{N_2} = \tilde{D}(w_\Gamma)$, where $\tilde{D}$ stands for the *right* Christoffel morphism $\tilde{D}(0, 2) = (02, 2)$ and the words $\{v_1, \ldots, v_{N_2}\}$ will be called *right*-NWF words.

Example 1. We illustrate previous Proposition with the example of Fig. 1. The step pattern of the pentatonic scale $\Gamma(\log_2(\frac{3}{2}), 5)$ is $w = (002|02)$. Since the last

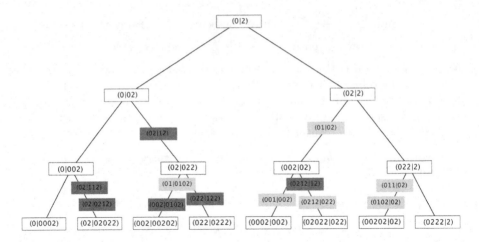

Fig. 2. The Christoffel tree (white nodes) completed with the NWF words (grey nodes) and their standard factorization.

step is greater than the first one (Fig. 1(a)) the next WF scale will be the diatonic $\Gamma(\log_2(\frac{3}{2}), 7)$ with step pattern $G(w) = (0002|002)$ (each long step will split into two that are labeled with 02 (Fig. 1(c)) and following the ordering given by the sequence $\{r_k\} = \{4 + 3 \cdot k \bmod 5\} = \{4, 2\}$). By Proposition 3, the sequence of left-NWF words is $u_0 = \psi(w)$, $u_1 = (001|002)$ (which corresponds with the step pattern of the Guidonian Hexachord) and at last $u_2 = G(w) = (0002|002)$.

The set of standard factorizations of all Christoffel words is usually displayed in an infinite, complete binary tree whose root is $(0|1)$ and the nodes of which are labeled with Christoffel words $(w'|w'')$ subjected to the rules G and $\tilde{D}$. This tree is called Christoffel tree. If $\frac{M_1}{N_1} < \frac{M}{N} < \frac{M_2}{N_2}$ are consecutive fractions in $\mathfrak{F}_N$ and $M^* = M^{-1} \bmod N$, between the Christoffel word w of slope $\frac{M^*}{N}$ and its left son $G(w)$ (resp. its right son $\tilde{D}(w)$) the $N_1 - 1$ NWF words $\{u_k\}_{k=1,\ldots,N_1-1}$ (resp. the $N_2 - 1$ NWF words $\{v_k\}_{k=1,\ldots,N_2-1}$) given in Proposition 3 can be inserted. The tree obtained in that way will display the set of step-patterns of every generated scale (see Fig. 2) and it will be called *extended Christoffel tree*.

The standard factorization of a Christoffel word $w = (w'|w'')$ determines a standard factorization of its NWF sons:

The standard factorization $w = w' \cdot w''$ determines a standard factorization of its NWF sons:

Definition 2. *If w is a Christoffel word of slope $\frac{M^*}{N}$ and z_k is an associated left or right NWF word, we call standard factorization of z_k to the decomposition given by:*

$$z_k = (z_k'|z_k'') = (z_{k,0} \cdot z_{k,1} \cdots z_{k,M-1}|z_{k,M} \cdots z_{k,N-1})$$

where $z_{k,j}$ are the factors of z_k given by Proposition 3.

The NWF words can be computed recursively, as the following proposition asserts:

Proposition 4. *Let u_{i-1} and u_i two consecutive NWF words with $1 < i \leq N_r$ ($r = 1$ for left-NWF words and $r = 2$ for right-NWF words). If the standard factorization of u_i is*

$$u_i = (u_i'|u_i'') = (0x|y2) \quad \text{with } x, y \in \{0, 1, 2\}^*,$$

then the word u_{i-1} can be written as $u_{i-1} = y1x$.

Proof. Let us suppose first that u_{i-1} and u_i are left-NWF words, sons of a Christoffel word w of slope $\frac{M^*}{N}$. If $w = 0c2$ is the corresponding Christoffel word with $(0p2|0q2)$ the standard factorization of w, with c, p, q palindrome words in $\{0, 2\}^*$, it follows that $0q$ is a prefix of w of length $N - M$. LEt $u_0 = \psi(w)$, as in Proposition 3. Every time that $u_0[k] = 1$ is replaced by 02 in a NWF word u_s with $k \geq M$, previously the letter $u_0[k - M] = 1$ had been replaced by 02 in the immediately preceding NWF word u_{s-1}. It follows that y must be a prefix of u_{i-1}. A similar argument shows that x is a suffix of u_{i-1}

To show that $y1$ is a prefix of u_{i-1} as well, notice that the last 1 of u_0 to be replaced by 02 takes place at the position:

$$M \cdot (N_2 - 1) \bmod N = M \cdot (N - M^* - 1) \bmod N = N - M - 1 = |0q|.$$

Since $u_{i-1} \neq u_{N_2}$, $y1$ must be a prefix of u_{i-1}.

A similar argumentation works if u_{i-1} and u_i are right-NWF words. y ad x are respectivelly a prefix and a suffix of u_{i-1} and the last 1 to be replaced by 02 in $v_0 = \xi(w)$ takes place at the position:

$$N - 1 + M \cdot (N_1 - 1) \bmod N = N - 1 + M \cdot (M^* - 1) \bmod N = N - M$$

2   NWF Words are Lyndon Words

If one take a closer look to the extended Christoffel tree of Fig. 2, one can see that the factors u' and u'' of any NWF word $u = (u'|u'')$ are themselves nodes of the tree (in case they are not a single letter or a Christoffel word). To prove this fact we need the following technical lemma:

Lemma 3. *Let $\frac{M}{N}$ be an irreducible fraction. Let moreover $\frac{M_1}{N_1} < \frac{M}{N} < \frac{M_2}{N_2}$ be three consecutive N–Farey fractions and $M^* = M^{-1} \pmod{N}$. The following equalities hold:*

1. $M^* - N \bmod M^* = M_1^{-1} \bmod N_1$.
2. $M^* \bmod (N - M^*) = M_2^{-1} \bmod N_2$.

Proof. We present just a sketch of the proof. Recall that if $M^{-1} \bmod N = M^*$, then the associated matrices of $\frac{M^*}{N}$ and $\frac{M}{N}$ as in Lemma 3 can be written as

$$\begin{pmatrix} M_2 & N_2 \\ M_1 & N_1 \end{pmatrix} = \begin{pmatrix} 1 & 0 \\ 1 & 1 \end{pmatrix} \cdot \begin{pmatrix} a & b \\ c & d \end{pmatrix} \quad \begin{pmatrix} M_2^* & N_2^* \\ M_1^* & N_1^* \end{pmatrix} = \begin{pmatrix} 1 & 0 \\ 1 & 1 \end{pmatrix} \cdot \begin{pmatrix} d & b \\ c & a \end{pmatrix}$$

where $\frac{M^*}{N} = \frac{M_2^* + M_1^*}{N_2^* + N_1^*}$. We have to compute three consecutive N_1-Farey fractions $\frac{A}{B} < \frac{M_1}{N_1} < \frac{C}{D}$ (and the same for $\frac{M_2}{N_2}$ and $\frac{M}{N}$) and then *flip* the main diagonals of the corresponding matrices. For that purpose, we need to distinguish two cases, namely $M_2 > M_1$, $N_2 > N_1$ or $M_2 < M_1$, $N_2 < N_1$. If $M_2 > M_1$ and $N_2 > N_1$ hold one can show that

$$\frac{A}{B} < \frac{M_1}{N_1} < \frac{C}{D} \qquad \frac{M_1}{N_1} < \frac{C + \lambda M_1}{D + \lambda N_1} = \frac{M_2}{N_2} < \frac{C + (\lambda - 1)M_1}{D + (\lambda - 1)N_1} \quad \text{and}$$

$$\frac{M_1}{N_1} < \frac{C + (\lambda + 1)M_1}{D + (\lambda + 1)N_1} = \frac{M}{N} < \frac{M_2}{N_2}$$

are consecutive (semi-convergents). If we take

$$\begin{pmatrix} C & A \\ D & B \end{pmatrix} = \begin{pmatrix} 1 & 0 \\ 1 & 1 \end{pmatrix} \cdot \begin{pmatrix} c & a \\ d & b \end{pmatrix} \implies \frac{M_1}{N_1} = \frac{a + c}{a + b + c + d},$$

and $\frac{M_2}{N_2}$ and $\frac{M}{N}$ are also written depending on a, b, c, d, one gets after an easy computation the modular equalities:

$$M^* - N \bmod M^* = a + b = M_1^{-1} \bmod N_1$$

$$M^* \bmod (N - M^*) = a + b + c + d = M_2^{-1} \bmod N_2,$$

which complete the proof. The case $M_2 < M_1$, $N_2 < N_1$ is symmetrical.

Proposition 5. *If $u = (u'|u'')$ is a word in the extended Christoffel tree, each of the words u' and u'' verify one of the following three conditions:*

1. *It is a single letter.*
2. *It is a Christoffel word within $\{0, 1\}^*$, $\{0, 2\}^*$ or $\{1, 2\}^*$.*
3. *It is a NWF word.*

Proof. Notice first that every Christoffel word verifies the statement. Let $w = (w'|w'')$ be a Christoffel word of slope $\frac{M}{N}$ and $\frac{M_1}{N_1} < \frac{M}{N} < \frac{M_2}{N_2}$ are three consecutive N-Farey numbers. We need to prove that if $u = (u'|u'')$ is a NWF word laying between w and $\tilde{D}(w)$ –resp. between w and $G(w)$– both u' and u'' meet the conditions of the Proposition. We will prove that u' is under the word w' and u'' is under the word w'' in the extended Christoffel tree.

If u is a NWF word that lays between w and $\tilde{D}(w)$, it is obtained from $v_0 = \xi(w)$ with the substitution $1 \rightarrow 02$ of some 1's following the ordering $i \cdot M^* \bmod N$, where $M^* = M^{-1} \bmod N$ (see Proposition 3). For $i = 1$ we obtain M^*, the place were w is divided in its standard factorization $(w'|w'')$. Notice that

$$|w| = N = \left[\frac{N}{M^*}\right] \cdot M^* + N \bmod M^*.$$

It follows that the smaller i satisfying $i \cdot M^* \bmod M^* < M^*$, verifies the condition $i \cdot M^* \bmod M^* = M^* - N \bmod M^*$ (see Fig. 3). The word u' will be a node of the extended Christoffel tree if the Christoffel word w' has a standard factorization that splits as $((w')'|(w')'')$ with $|(w')'| = M^* - N \bmod M^*$. But we know that $|(w')'| = M_1^{-1} \bmod N_1$. We conclude due to Lemma 3.

Something similar (rather symmetrical) can be said if u is a NWF word that lays between w and $G(w)$.

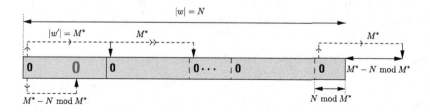

Fig. 3. Graphical motivation for the modular equalities of Lemma 3.

Proposition 6. *If $u = (u'|u'')$ is a word in the extended Christoffel tree, then it holds that $u' < u''$.*

Proof. Notice first that the result holds for every Christoffel word. We proceed by induction on the depth of the Christoffel tree. Let $w = (w'|w'')$ be a Christoffel word and let $\{u_s\}_{s=1,2,\ldots,N_1-1}$ and $\{v_t\}_{t=1,2,\ldots,N_2-1}$ the NWF words that lay between w and, respectively, $G(w)$ and $\tilde{D}(w)$.

We will prove that the result holds first for every NWF word u_i, with standard factorization $(u_i'|u_i'')$, between w and $G(w)$ by induction over i. Let

$$\psi_{21}(w'|w'') = 0p1 \cdot 0q1 = 0x1, \quad \text{with } p, q \text{ and } x \text{ palindrome words in } \{0,1\}^*.$$

By construction, $u_1 = 0x02 = 0p1 \cdot 0q02 = u_1' \cdot u_1''$. As $x \in PAL(\{0,1\}^*)$, the factor $q0$ must be a prefix of x. If $|0q0| < |u_1'|$, then $u_1' \cdot u_1'' = 0q0y \cdot 0q02$, with $y \in \{0,1\}^*$. If $|0q0| > |u_1'|$, then u_1' is a prefix of u_1''. In both cases $u_1' < u_1''$. Thus, the statement holds for $i = 1$.

In each step of the induction $u_{i-1}' \cdot u_{i-1}'' \to u_i' \cdot u_i''$ a 1 is replaced by a 02. If the replaced 1 is in u_{i-1}', there is nothing to prove since, in that case, by induction, $u_i' < u_{i-1}' < u_{i-1}'' = u_i''$.

We can suppose that we are in the following situation: $u_{i-1}' = u_i'$ and $u_{i-1}'' = x1y2 \to x02y2 = u_i''$ with $i < N_1$. Following Proposition 4, $x02y1$ is a prefix of u_{i-1}.

Now, if $|x02y1| > |u_{i-1}'|$, then $u_{i-1}' = u_i'$ is a prefix of u_i'' and it follows that $u_i' < u_i''$. If $|x02y1| < |u_{i-1}'| = |u_i'|$, then $u_i' = x02y1z$ and $u_i'' = x02y2$ for some $z \in \{0,1,2\}^*$ and $u_i' < u_i''$ holds as well.

Finally, let v_i be a NWF word with standard factorization $v_i' \cdot v_i''$, that lays between w and $\tilde{D}(w)$. Again, we proceed by induction over i. The case $i = 1$ is trivial now, since $v_1' = 0x$ and $v_1'' = y$ with $x, y \in \{1,2\}^*$ implies that $v_1' < v_1''$

holds. We use a similar argument as above. In each step $v'_{i-1} \cdot v''_{i-1} \to v'_i \cdot v''_i$ of the induction a 1 is replaced by a 02. If the replaced 1 is in v'_{i-1}, there is nothing to prove since, by induction one has that $v'_i < v'_{i-1} < v''_{i-1} = v''_i$. In the other case, namely $v'_{i-1} = v'_i$ and $v''_{i-1} = x1y2 \to x02y2 = u''_i$, then, following Proposition 4, the word $x02y1$ is a prefix of $v'_{i-1} \cdot v''_{i-1}$. It holds then necessarily that $v'_i < v''_i$.

Following [6, Proposition 5.1.3], a word w is Lyndon if and only if it is a single letter or it can be written as $w = w' \cdot w''$ with $w' < w''$ Lyndon words. This characterization will be used in the following

Theorem 1. *If u is a word of the extended Christoffel tree, it is a Lyndon word.*

Proof. It is enough to show the assertion for the NWF words since Christoffel words are Lyndon words. If $u = u' \cdot u''$ is a NWF word, following Proposition 5, u' is a single letter, a Christoffel word (in both cases, a Lyndon word) or it is a NWF word that appears upper in the extended Christoffel tree. The same holds for u''. By induction, u' and u'' are Lyndon words. Since $u' < u''$ (Proposition 6) u must be a Lyndon word.

3 Lyndon Factorizations of NWF Words

Given w a Lyndon word that is not a letter, if u is its longest proper prefix that is Lyndon, the factorization $w = u \cdot v$ is called left Lyndon factorization. It holds that v is a Lyndon word as well and $u < v$. In a parallel way, one can define the right Lyndon factorization of w as $w = x \cdot y$ where y is the longest suffix of w that is Lyndon. As in the left case, x is Lyndon and it holds $x < y$ (see [6, Sect. 5.1]).

There is a tight connection between Christoffel words and Lyndon factorization: the left and right Lyndon factorization of a Christoffel word w coincide with the standard factorization $w = w_1 \cdot w_2$ and the same happens if we factorize recursively w_1 and w_2. This fact is used in [8] to propose a generalization of Christoffel words for non binary alphabets defining the so-called Lyndon-Christoffel words.

In this section we study the right and left Lyndon factorization of NWF words. As we will prove, the right Lyndon factorization coincides with the standard factorization, as in the Christoffel case, but the left Lyndon factorization does not. We describe the nodes of the extended tree for which left and right Lyndon factorization coincide. These nodes are painted in green in Fig. 2.

We begin by studying the right Lyndon factorization of NWF words. The following result [6, Proposition 5.1.4] states a characterization of this factorization:

Proposition 7. *Let w be a Lyndon word different from a single letter and let $w = u' \cdot u''$ its right Lyndon factorization. For any Lyndon word v with $w < v$ the right Lyndon factorization of wv is $w \cdot v$ if and only if $u'' \geq v$.*

Theorem 2. *If u is a word of the extended Christoffel tree, its standard factorization $u' \cdot u''$ coincides with its right Lyndon factorization. That is, u'' is the longest proper suffix of u that is Lyndon.*

Proof. The assertion is already known for Christoffel words. Let us suppose, thus, that u_i is a NWF word, and let $u_i' \cdot u_i''$ be its standard factorization. By induction and following Proposition 7, if $u_i' = (u_i')' \cdot (u_i')''$ is the standard factorization of u_i', it is enough to show that $(u_i')'' > u_i''$. We will proceed by induction over i. If $i = 0$ the statement is clear, since u_0 is a Christoffel word.

Let us suppose that u_{i-1} is a NWF word, with $i \geq 1$ verifying the Theorem. Now we can distinguish two cases:

1. If the slope verifies $\frac{M^*}{N} < \frac{1}{2}$, the standard factorization of w is $w = w_1 \cdot (w_1 \cdot w_2)$. Let $w_1 = x_1 \cdot x_2$ be the standard factorization of w_1. Following [6, Propositon 5.1.3], $x_1 < x_1 x_2 < x_2$. If $u_{i-1} = (y_1 \cdot y_2 \,|\, y_3 \cdot y_4 \cdot z)$ with $y_2 \geq y_3 \cdot y_4 \cdot z$, where the factors y_1, y_2, y_3, y_4 and z lay respectively below the factors of the decomposition of $w = (x_1 \cdot x_2 \,|\, x_1 \cdot x_2 \cdot w_2)$, the 1 that is replaced with 02 in the trasformation $u_{i-1} \to u_1$ can be in y_1, y_2, y_3, y_4 or z. If the replaced 1 comes from y_3, y_4 or z, there is nothing to prove, since

$$(u_i')'' = (u_{i-1}')'' = y_2 \geq y_3 \cdot y_4 \cdot z > u_i''.$$

If the replaced 1 comes from y_1 then

$$(u_i')'' = (u_{i-1}')'' = y_2 \geq y_3 \cdot y_4 \cdot z = u_i''.$$

It remains to check only the case when the 1 that is replaced comes from y_2. Let $u_{i-1}' = y_1 \cdot y_2 = 0a \cdot b1c2$, $u_i' = 0a \cdot b02c2$ with $y_1 = 0a$ and $y_2 = b1c2$. Since u_i' is a NWF word, following Proposition 4, one has that $u_{i-1}' = b02c1a$. Notice now that $u_i'' = u_{i-1}''$ and also that u_{i-1}' is a prefix of u_{i-1}''. One may conclude, since

$$(u_i')'' = b02c2 > b02c1a \geq u_{i-1}'' = u_i''.$$

2. If the slope $\frac{M^*}{N}$ of w is greater than $\frac{1}{2}$, then w is on the right side of the Christoffel tree and $w = w_1 w_2 \cdot w_2$ and the proof is similar to the previous case.

As of yet, the left Lyndon factorization of a NWF words is still a conjecture:

Conjecture 1. *Let $u_{i-1} = u_{i-1}' \cdot u_{i-1}''$ and $u_i = u_i' \cdot u_i''$ be the standard factorization of two consecutive NWF words u_{i-1} and u_i with $i \geq 1$. The left and right Lyndon factorizations of u_{i-1} coincide if and only if $u_{i-1}'' = u_i''$.*

We will say that a NWF word verifies LR property if its left and right Lyndon factorization coincide. In Fig. 2, NWF words verifying LR property are displayed with lighter grey. Last conjecture can be translated into the terminology of generated scales as follows:

Conjecture 2. *The step-pattern of a generated scale $\Gamma(\theta, N)$ verifies the LR property if and only if it holds that $0 < \{N\theta\} < \theta$.*

As a consequence, the following propositions are easy to prove:

Corollary 1. *If* $\{u_i\}_{i=0,\ldots,N_1-1}$ *and* $\{v_j\}_{j=0,\ldots,N_2-1}$ *are the left and right NWF words associated with a Christoffel word* w*, the following list of facts hold:*

1. *The first NWF word on the left side,* u_1*, verifies LR Property and the first one on the rights,* v_1*, does not.*
2. *If* w *is on the left side of the Christoffel tree (if its slope is smaller than* $\frac{1}{2}$*), then the last NWF words* u_{N_1-1} *and* v_{N_2-1} *do not verify LR property.*
3. *On the contrary, if* w *is on the right side of the Christoffel tree,* u_{N_1-1} *and* v_{N_2-1} *do verify LR property.*
4. *If two NWF words* x *and* y *are symmetric in the extended Christoffel tree, then* x *verifies LR property if and only if* y *does not. LR property is thus verified by half of the NWF words.*

Remark 1. In [8], the class of Christoffel-Lyndon words were introduced in a recursive way: a word w is Lyndon-Christoffel if it is a single letter or its left and right standard factorizations coincide, say $w = (w_1|w_2)$, and if moreover w_1 and w_2 are Christoffel-Lyndon. NWF words are not, in general, Christoffel-Lyndon. For example, the NWF word $(0212|12)$ verifies LR property, but the left standard factorization of $0212 = (021|2)$ does not coincide with its right Lyndon factorization $(02|12)$.

References

1. Berstel, J., Lauve, A., Reutenauer, C., Saliola, F.: Combinatorics on words: Christoffel words and repetition in words. CRM-AMS, Canada (2008)
2. Carey, N.: Distribution modulo 1 and musical scales. Ph.D. thesis, University of Rochester (1998)
3. Carey, N., Clampitt, D.: Aspects of well-formed scales. Music Theory Spectr. **11**(2), 187–206 (1989)
4. Domínguez, M., Clampitt, D., Noll, T.: Well-formed scales, maximally even sets and Christoffel words. In: Proceedings of the MCM2007 (2009)
5. Graham, R.L., Knuth, D.E., Patashnik, O.: Concrete Mathematics: a Foundation for Computer Science. Addison-Wesley Longman Publishing Co., Boston (1994)
6. Lothaire, M.: Combinatorics on Words. Cambridge University Press, Cambridge (1983)
7. Lothaire, M.: Algebraic Combinatorics on Words. Cambridge University Press, Cambridge (2002)
8. Melançon, G., Reutenauer, C., et al.: On a class of Lyndon words extending Christoffel words and related to a multidimensional continued fraction algorithm. J. Integer Sequences **16**(2), 3 (2013)
9. Slater, N.B.: Gaps and steps for the sequence nr mod1. Proc. Camb. Phil Soc. **63**, 1115–1123 (1967)

Triads as Modes within Scales as Modes

Thomas Noll[(✉)]

Departament de Teoria, Composició i Direcció, Escola Superior de Música de Catalunya, C. Padilla, 155 - Edifici L'Auditori, 08013 Barcelona, Spain
thomas.mamuth@gmail.com

Abstract. The paper revisits results from scale theory through the study of modes. Point of departure is the nested hierarchy of triads embedded into diatonic modes embedded into the chromatic scale. Generic diatonic triads can be described as stacks of thirds or as stacks of triple-fifths. These two possibilities lead to different generalizations and it is argued that the latter possibility is better suited for the understanding of harmonic tonality. Three types of well-formed modes are investigated: Simple Modes, Chain Modes, and Diazeuctic Modes. Chain Modes and Diazeuctic modes are the modal counterparts of the two types of reduced scales with the Partitioning Property in Clough and Myerson (1985), as well as of Agmon's (1989) types A and B of generalized diatonic systems.

Keywords: Diatonic modes · Well-formed words · Algebraic combinatorics on words

1 Harmonic Tonality and Formal Concepts of Diatonicity

Mathematical models of tone relations in harmonic tonality are being investigated in mathematical music theory as well as in music cognition. In both disciplines proposals have been made to consider hierarchically nested levels of tone systems, where the interaction of triadic, diatonic and chromatic tone relations may be studied in combination. In [16] Marek Žabka quite recently emphasizes the importance of this subject. The present paper aims at an extension and refinement of the mathematical approach by taken by John Clough and Jack Douthett in the second and third sections of their seminal paper [8] on maximal even sets. In these sections, entitled *2. Ambiguities, Tritones and Diatonic Sets* and *3. Generated Sets, Complements; Interval Circles and Second-Order ME Sets* Clough and Douthett revisit concepts from preceding papers by Clough and Myerson [7] and Eytan Agmon [1]. In order to track the logical dependencies between the involved concepts, it is useful to consult the taxonomic study by Clough, Engebretsen and Kochavi [9]. Figure 1 visualizes its results in terms of a formal concept lattice (see [13] for an introduction to Formal Concept Analysis). The eponymous concept of [8] is represented by the node labeled ME. The aforementioned Sects. 2 and 3 of [8] investigate tightened concepts, above all the node labeled DT. Theoretical interest and appreciation for this node has been articulated in several contributions. In their investigation of the partitioning property

© Springer International Publishing Switzerland 2015
T. Collins et al. (Eds.): MCM 2015, LNAI 9110, pp. 373–384, 2015.
DOI: 10.1007/978-3-319-20603-5_37

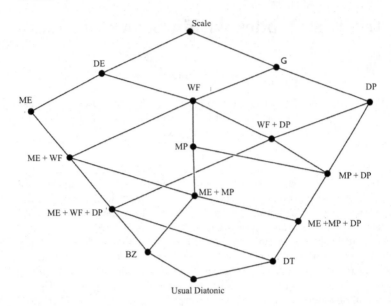

Fig. 1. Each node in this lattice stands for a scale concept, whose extension is a set of scales and whose intension is a set of attributes. The labels denote combinations of attributes, which uniquely determine the concept in question. The abbreviations stand for the following terms: DE = distributionally even, G = generated, ME = maximally even, WF = well-formed, DP = deep property, MP = Myhill property, BZ = Balzano property, DT = diatonic propery. Each descending edge — and by concatenation each descending path — represents a conceptual tightening, where extensions become smaller and intensions larger.

for reduced scales, Clough and Myerson ([7] Theorem 5 on p. 267) distinguish two main cases, namely $c = 2d - 1$, $c = 2d - 2$. The latter one corresponds to the *diatonic property* DT. Agmon [1] distinguishes the same two cases as candidates for diatonic systems (type A and type B). His closer examination of type B (corresponding to the DT-node) suggests the high relevance of this concept for the study of harmonic tonality. Clough and Douthett ([8], Theorem 2.2, p. 138 - 39) establish a set of no less than 10 equivalent characterizations for this concept.

A programmatic statement towards the end of [8] is the point of departure for the present paper. There Clough and Douthett give a conceptual characterization of the diatonic and of the inner (harmonic) levels of a nested hierarchy.

> The structure observed here is remarkable for what it shows regarding relationships between three nested *levels* of pc sets, which might be labeled chromatic, diatonic, and harmonic. The celebrated unique multiplicity property of the diatonic set with its consequences for hierarchical structuring of transpositions of that set within the 12-tone chromatic, as first observed, we believe, by Milton Babbitt [2], is present as well in triads, and seventh chords, considered as subsets within a seven-note world,

where the consequences of that property for hierarchical structuring of transpositions may reasonably be regarded as an important feature of tonal harmony.

To generalize the above, chromatic universes which support diatonic sets also support nested arrays of first-order, second-order, and high-order ME sets that are deep scales with respect to their parent collections. We suggest that the concept of diatonic system might well include, as an essential feature, the hierarchical aspects discussed here. Thus, it would encompass the diatonic system as defined by Agmon (our diatonic set), plus one or more additional levels of structure including, in the usual case, the *harmonic* level. ([8], p. 172)

In his article [12] Jack Douthett successfully accomplishes several of the theoretical challenges of this programmatic statement. In terms of dynamical voice leadings he brings the idea of an iteration of the ME-concept to fruition. The pivotal point for the innovations is Douthett's idea to turn the *mode index*, a parameter of the *J-function* (from [8], definition 1.8, p. 100) into a continuous variable. For fixed diatonic and chromatic cardinalities d and c and any given value $m \in \mathbb{R}/c\mathbb{Z}$ the *J*-function is a function in one variable k, namely $J_{c,d}^m : \mathbb{Z}_d \to \mathbb{Z}_c$ with $J_{c,d}^m(k) := \left\lfloor \dfrac{ck+m}{d} \right\rfloor$. It can reasonably be noticed that Douthett's approach already involves a plausible realization of the promise behind the title of the present paper, i.e. *Triads as Modes within Scales as Modes*. The 21 functions $J_{7,3}^m$ for $m = 0, ..., 20$ can be interpreted as inversions of generic diatonic triads on the seven diatonic scale degrees. This is in perfect analogy with the interpretation of the 84 functions $J_{12,7}^m$ for $m = 0, ..., 83$ as diatonic modes on the 12 chromatic notes:

$$
\begin{array}{l|l}
I \quad = J_{7,3}^0 = (0,2,4) & C_{Locrian} \ = J_{12,7}^0 = (0,1,3,5,6,8,10) \\
VI^6 = J_{7,3}^1 = (0,2,5) & C_{Phrygian} = J_{12,7}^1 = (0,1,3,5,7,8,10) \\
\quad\quad \ldots & \quad\quad\quad \ldots \\
III^{\overset{6}{4}} = J_{7,3}^{20} = (6,2,4) & B_{Lydian} \ \ = J_{12,7}^{83} = (11,1,3,5,6,8,10)
\end{array}
$$

For every pair of (affine) functions $(m_1 : \mathbb{R} \to \mathbb{R}/c\mathbb{Z}, m_2 : \mathbb{R} \to \mathbb{R}/d\mathbb{Z})$ one obtains a locally constant function: $t \mapsto (J_{12,7}^{m_1(t)} \circ J_{7,3}^{m_2(t)}, J_{12,7}^{m_1(t)})$. And for any discrete (arithmetic) subsequence $(....t_0, t_1, t_2, ...)$ one obtains a corresponding sequence of inverted diatonic triads, each embedded into respective diatonic modes. Besides its theoretical merits Douthett's underlying visual metaphor of rotating concentric wheels compels through its illustrative plasticity. Its descriptive range varies with the combinatorial freedom in the choice of mode functions m_1 and m_2. It seems, however, that the most convincing analytical examples belong to the area of 19th-century pan-triadic tonality. With respect to the area of common practice harmonic tonality there are several theoretical issues that find no straightforward counterpart in the mathematical theory of higher order maximally even sets and their dynamical voice leadings.

It is therefore helpful to remember that the *J*-function corresponds to a quite general concept on the concept lattice in Fig. 1, the node *ME*. In general the cardinalities d and c do not need to be coprime, while for the purpose of the present paper this additional condition is crucial. It is useful to bring also other scale attributes into play which contribute to the tightening of the node DT.

2   Chains of Thirds Versus Chains of Triple-Fifths

An important feature of the diatonic scale for the understanding of its role in harmonic tonality regards the structure of the embedding of the three major and three minor triads. The roots of the three triads of each of the two species form a chain of fifths: an embedded structural scale. The left side of Fig. 2 displays two representations of the generic triad $(0, 2, 4)$ within the usual diatonic scale. The upper heptagon represents the seven steps of the scale in the scalar order: $(0, 1, 2, 3, 4, 5, 6)$ and the lower heptagon represents them in fifth-order $(0, 4, 1, 5, 2, 6, 3)$. The segments of the inner circles in these drawings stand for the triadic intervals and measure them in steps and fifths, respectively. The interval cycle $(2, 2, 3)$ in the upper drawing characterizes the triad as a chain of two thirds (double-steps), which is closed by a single fourth. The interval cycle $(3, 3, 1)$ in the lower drawing characterizes the triad as a chain of two triple-fifths which is closed by a single fifth. Actually, both interval cycles are composed of the same three generic dyads $\{0, 2\}, \{2, 4\}, \{4, 0\}$. What differs, is their measurement.

In accordance with the Structure-Implies-Multiplicity property [7] the interval cycle $(1, 3, 3)$ also represents the multiplicities of the three triadic species. When the diminished fifth occupies the place of the fifth $\{0, 4\}$ of the triad, it becomes a diminished triad, when it occupies the place of one of the three fifths $\{4, 1\}, \{1, 5\}, \{6, 2\}$ filling the upper third $\{4, 2\}$ of the triad, it becomes a minor triad, and when it occupies the place of one of the three fifths $\{2, 6\}, \{6, 3\}, \{3, 0\}$ filling the lower third $\{0, 2\}$ of the triad, it becomes a major triad. The interval structure $(1, 3, 3)$ does not only explain the multiplicity 3 of the major triads, but also the fact that their roots form a chain of fifths; analogously for the minor triads.

In the search for extended triads in larger DT-scales, it is not possible to inherit both interval structures at the same time. Either one can opt for the chain-of-thirds structure or for the chain-of-triple-fifths structure. This conceptual bifurcation proves to be useful for a better understanding of the properties of the usual triads. In particular, the pseudo-diatonic 13-note scale, with 7 as a generator, is an interesting playground for a comparison of the two options. The right side of Fig. 2 displays the corresponding solutions. The middle column shows the circle-of steps and circle-of-fifths representations of an hexad, which is constructed as a chain of thirds $(2, 2, 2, 2, 2, 3)$. Its interval structure along the circle-of-fifths is $(3, 1, 3, 1, 4, 1)$ involves three different values: 1, 3 and 4. The right column shows the circle-of steps and circle-of-fifths representations of a pentad, which is constructed as a chain of triple-fifths $(3, 3, 3, 3, 1)$. Its interval structure along the circle-of-steps is $(3, 2, 3, 2, 3)$. It is no longer a chain, but exemplifies the step pattern of a well-formed scale. This point is interesting in

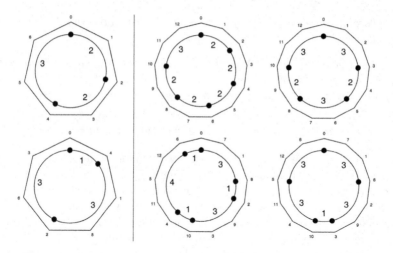

Fig. 2. The left side of the figure displays the interval cycles of the generic diatonic triad $\{0, 2, 4\}$ on the circle of seven steps (top) and the circle of seven fifths (bottom). The right side of the figure (middle and right columns) displays two alternative choices for an analogous "pseudo-triad" within the generic 13-note scale. The middle column represents a hexad forming a chain of thirds (on the circle-of-steps), the right column represents a pentad forming a chain of triple fifths (on the generalized circle-of-fifths).

connection with Clough and Douthett's argumentation in the citation of Sect. 1. They recall the importance of the unique interval multiplicity and its combinatorial consequences for the role of triads and seventh chords in tonal harmony. But unique interval multiplicity is already a consequence of generatedness alone.[1] Consequently, there is no reason to disregard pentads as candidates for the role of extended triads.

3   Refinement of Myhill's Property: Well-Formed Modes

The formal concept behind the Myhill property (labeled MP in the concept lattice of Fig. 1) entails a surprisingly rich theory about the structure of the modes associated with these scales. The extensions of tighter concepts, such as DT, are contained in the extension of MP and so this modal theory applies to these scales as well. It is then natural to ask, whether the relations between the formal concepts can be traced within the theory of the modes. To answer this question in the following section, a brief synopsis of the theory of well-formed modes is needed as a prerequisite.

The theory emerges naturally from a meanwhile classical result by Carey and Clampitt (c.f. [4,5]), which can be paraphrased as follows:

Proposition 3.1. *Consider an n-tone scale $S = \{k \cdot g \bmod 1 \mid k = 0, ..., n-1\} \subset \mathbb{R}/\mathbb{Z}$, generated by $g \in (0, 1)$. Let $q, s : \mathbb{Z}_n \to S$ denote the associated generation*

---

[1] In this regard Jacques Handschin's discussion of the unique interval multiplicities of the diatonic intervals in [14] p. 18 predates Babbitt's observation.

order and scalar order encodings of S, respectively. The following three conditions are equivalent:

1. *S is non-degenerate well-formed, i.e. $s^{-1}(q(k)) = m \cdot k \mod n$ for a suitable $m \in \mathbb{Z}_n$ and $S \neq \frac{1}{n}\mathbb{Z}/\mathbb{Z}$.*
2. *(Myhill property) Each non-zero generic interval comes in precisely two specific sizes.*
3. *The ratio $\dfrac{m}{n}$ is a semiconvergent of the generator g with $\dfrac{m}{n} \neq g$.*

To refine the map from condition 1 above, we include octaves and augmented primes into the consideration. It is useful to have abstract symbols x and y, representing the generator and its octave complement (the co-generator) as well as symbols a and b representing the two step intervals, namely the (sharpward oriented) step and the (flatward oriented) co-step, respectively. Step and co-step have both unique expressions $a = lx - iy$ and $b = -jx + ky$ with $i, j, k, l \in \mathbb{N}$ and $k, l > 0$. What distinguishes the step from the co-step are the positions of the minus-signs in these expressions. The two expressions define a linear map $\begin{pmatrix} l & -j \\ -k & i \end{pmatrix}$: $\mathbb{Z}[a, b] \to \mathbb{Z}[x, y]$ converting step/co-step coordinates into generator/co-generator coordinates. The middle arrow in the diagram below shows the inverse linear map $\begin{pmatrix} i & k \\ j & l \end{pmatrix}$: $\mathbb{Z}[x, y] \to \mathbb{Z}[a, b]$ converting generator/co-generator coordinates into step/co-step coordinates. It lifts the linear map $m : \mathbb{Z}_n \to \mathbb{Z}_n$ (from proposition 3.1, condition 1) to the free commutative group $\mathbb{Z}[x, y]$, generated by the generator x and the co-generator y. This lift is a refinement insofar as the matrix also acts on octaves $x + y$ and augmented primes $(l + j)x - (i + k)y$. These intervals are factored out in $\mathbb{Z}_n$.

$$0 \to \mathbb{Z}[x + y, (l + j)x - (i + k)y] \hookrightarrow \mathbb{Z}[x, y] \longrightarrow \mathbb{Z}_n \to 0$$

$$\downarrow \qquad\qquad\qquad \downarrow \begin{pmatrix} i & j \\ k & l \end{pmatrix} \qquad \downarrow \cdot m$$

$$0 \to \mathbb{Z}[(i + j)a + (k + l)b, a - b] \hookrightarrow \mathbb{Z}[a, b] \longrightarrow \mathbb{Z}_n \to 0$$

The next level of refinement is achieved through the transition from a commutative principle of interval concatenation (i.e. addition of group elements within $\mathbb{Z}[x, y]$ or $\mathbb{Z}[a, b]$) to a non-commutative one. Let $\{a, b\}$ denote a two-letter alphabet and let $\{a, b\}^*$ denote the free monoid of finite words over these two letters. A two-letter word, such as $abaa$ can be considered as a refinement of an ordered pair of natural numbers. In this case the number pair in question would be $(3, 1)$. In fact, the words $aaab, aaba, abaa$ and $baaa$ share the same multiplicities of the letters a and b, and thus one may consider them as refinements of that ordered number pair $(3, 1)$. The idea of a refinement can be expressed in terms of a surjective monoid homomorphism: $V : \{a, b\}^* \to \mathbb{N}^2$, with $V(w) = (|w|_a, |w|_b)$, where $|w|_a$ and $|w|_b$ denote the letter multiplicities of a and b in the word w, respectively.

In our musical interpretation the words stand for species of musical intervals, i.e. musical intervals being interpreted as chains of other musical intervals. For example, $abaa, baaa, aaab$, and $aaba$ may represent the Dorian, Phrygian, Lydian and Mixolydian species of the perfect fifth, respectively, each consisting of a specific ordering of three major steps and one minor step.

In the same way as 2×2-matrices act as linear transformations on number pairs, one may study associated refined transformations of two-letter words. The action of the monoid St_0 of the *special Sturmian morphisms* on two-letter words thereby refines the action of the monoid $SL(2, \mathbb{N})$ of special linear transformations on pairs of natural numbers.

St_0 is generated by the four transformations $G, \tilde{G}, D, \tilde{D} : \{a, b\}^* \to \{a, b\}^*$, acting on $\{a, b\}^*$ in terms of parallel rewriting rules as follows:

$$G(a) = a, G(b) = ab, D(a) = ba, D(b) = b$$
$$\tilde{G}(a) = a, \tilde{G}(b) = ba, \tilde{D}(a) = ab, \tilde{D}(b) = b$$

Also in the case of transformations the idea of a refinement can be expressed through a surjective monoid homomorphism. To that end we consider the map $M_- : St_0 \to SL(2, \mathbb{N})$, with $M_f = \begin{pmatrix} |f(a)|_a & |f(b)|_a \\ |f(a)|_b & |f(b)|_b \end{pmatrix}$, sending Sturmian morphisms to their *incidence matrices*. For the generators we obtain the triangular matrices $M_G = M_{\tilde{G}} = \begin{pmatrix} 1 & 1 \\ 0 & 1 \end{pmatrix}$ and $M_D = M_{\tilde{D}} = \begin{pmatrix} 1 & 0 \\ 1 & 1 \end{pmatrix}$.

The definition of a *Well-formed Mode* is based on a transformational interpretation of the pseudo-classical modes. All three intervals of the perfect octave $P8$, perfect fifth $P5$, and perfect fourth $P4$ are substituted by their associated species. In the case of the authentic modes the species of the octave is a concatenation of the species of the fifth with a species of the fourth. In the case of the plagal modes the order is reversed and mathematically the "non-special" exchange morphism $E : \{a, b\}^* \to \{a, b\}^*, E(a) = b, E(b) = a$ would come into play. In the present article we stick to authentic modes only. The odd one out — the Locrian mode — has an underlying species of the octave, but it is not properly divided into a species of the perfect fifth and fourth, respectively. From a transformational point of view it is amorphous. The situation is analogous for the modes of any non-degenerate well-formed scale.

Definition 3.2. *Elements of the submonoid $\langle G, D \rangle \subset St_0$ are called* special standard morphisms *and elements of the submonoid $\langle G, \tilde{D} \rangle \subset St_0$ are called* Special Christoffel Morphisms. *The images of the word ab under special standard morphisms and special Christoffel model are called* standard modes *and* Christoffel modes, *respectively.*

The monoid $\{a, b\}^*$ is a submonoid of the free group F_2 over the two generators a and b. Two words $w, w' \in \{a, b\}^*$ are said to be conjugated, if they are conjugated in F_2. The following definition generalizes the concept of a species of the octave.

Definition 3.3. *Conjugates of special standard words shall be called* well-formed modes.

The Locrian mode is included in the scope of this concept. First one finds the Ionian mode in the role of a standard mode $GGD(ab) = aabaaab$ and which is then conjugate to $baabaaa = b(aabaaab)b^{-1}$. All the species of the octave are conjugate to the Ionian mode in this sense. The concept of an *authentic mode* involves a transformational perspective on the division:

Definition 3.4. *With every special Sturmian morphism $f \in St_0$ we associate the triple $(f(a), f(b), f(ab))$, and call it the* authentically divided well-formed mode *(or briefly: the* authentic mode*) associated with f. For convenience we use the following notation with the divider-sign $f(a)|f(b)$, identifying $f(a)$ as the* divider-prefix *and $f(b)$ as the* divider-suffix *of $f(ab)$. The length $|f(a)|$ of the divider prefix is called the* winding number *of the mode.*

Conjugacy of the authentic modes extends to their divider prefixes and suffixes. Furthermore it can be expressed through *conjugacy* of their generating morphisms, as the following proposition states. The special Sturmian monoid is a submonoid of the automorphism group $Aut(F_2)$ of the free group (see [15] p. 85). Two special Sturmian morphisms are conjugate, if they are conjugate with respect to a suitable element of $Aut(F_2)$. The following proposition compiles results from [3] (Sect. 4).

Proposition 3.5. *Two authentic modes are conjugate iff their generating morphisms are conjugate. The conjugacy class of an authentic mode has $n - 1$ elements, where $n = |f(ab)|$ is the length of the underlying well-formed mode $f(ab)$. In other words, all but one conjugates of the word $f(ab)$ have authentic divisions. The exceptional one is called the* bad conjugate.

The authentic well-formed mode $a|b$ associated with the identity map, shall be called the *essential mode*. If we interpret the letters a and b in terms of the perfect fifth $P5$ and perfect fourth $P4$ respectively, the essential mode becomes the authentic division of the octave.

4   Simple Modes, Chain Modes and Diazeuctic Modes

Definition 4.1. *The four authentic modes, which are associated with the generators $G, \tilde{G}, D, \tilde{D}$ of the special Sturmian monoid, shall be called* Simple (Authentic) Modes*:*

$$G(a)|G(b) = a|ab, \tilde{G}(a)|\tilde{G}(b) = a|ba, \quad D(a)|D(b) = ba|b, \tilde{D}(a)|\tilde{D}(b) = ab|b.$$

The corresponding bad conjugates $b|aa$ (of $G(a|b)$ and $\tilde{G}(a|b)$) and $bb|a$ (of $D(a|b)$ and $\tilde{D}(a|b)$) will also be subsumed under the simple modes. The simple modes are the simplest instances of two families of modes which shall be called *Chain Modes* and *Diazeuctic Modes*.

Definition 4.2. *Consider a natural number $k \in \mathbb{N}$ and the special standard morphisms G^k, and D^k. The associated well-formed modes $G^k(a|b) = a|a^k b$ and $D^k(a|b) = b^k a|b$ shall be called* standard chain modes*. All the conjugates thereof, including the bad conjugates $b|a^{k+1}$ and $b^{k+1}|a$ are called* chain modes.

The hexad $\{0, 2, 4, 6, 8, 10\} \subset \mathbb{Z}_{13}$ (see Sect. 2) corresponds to the standard chain mode $G^4(a|b) = a|aaaab$ and its complement $\{1, 3, 5, 7, 9, 11, 0\}$ to $G^5(a|b) = a|aaaaab$. Diatonic scales (node DT in Fig. 1) are of odd cardinality $d = 2k + 1$. Clough and Douthett [8] suggest to choose third chains $\{0, 2, 4, ..., d - 3\}$ and $\{0, 2, 4, ..., d - 1\}$ from Agmon's type A in the role of generalized triads and seventh chords. Their step interval patterns $(2, 2, ..., 2, 3)$ and $(2, 2, ..., 2, 1)$ are instances of the chain modes.

Definition 4.3. *Consider a natural number $k \in \mathbb{N}$ and the special standard morphisms $G^k D$, and $D^k G$. The associated well-formed modes $G^k D(a|b) = G^k(ba|b) = a^k ba|a^k b$ and $D^k G(a|b) = D^k(a|ab) = b^k a|b^k ab$ shall be called disjunct and conjunct standard diazeuctic modes, respectively. All the conjugates thereof, including the bad conjugates $b|a^{k+1}$ and $b^{k+1}|a$ are called (disjunct or conjunct) diazeuctic modes.*

The factors $a^k b$ and $b^k a$ are interpreted as generalizations of the tetrachord, which motivates the pseudo-classical attributes *disjunct* and *conjunct*. The incidence matrices for these morphisms are $\begin{pmatrix} k+1 & k \\ 1 & 1 \end{pmatrix}$ and $\begin{pmatrix} 1 & 1 \\ k & k+1 \end{pmatrix}$, respectively and they both correspond to the formal concept DT. Thus the diazeuctic modes comprise, on the one hand, the modal refinements of the generalized diatonic scales in terms of disjunct modes. On the other hand, the conjunct modes comprise the pentads from Sect. 2 and generalizations thereof.

5   Diazeuctic Segmentations of Diazeuctic Modes

A word-theoretic variation of the concept of *second order maximally even sets* can be obtained through an iteration of well-formed modes as follows:

Definition 5.1. *Consider a well-formed mode $w \in \{a, b\}^*$ of length $n > 1$, two positive integers $k, l \in \mathbb{N}$ and a well-formed mode $v \in \{c, d\}^*$ of length $m = |v|$ with letter multiplicities $|v|_c, |v|_d > 0$, such that $k \cdot |v|_c + l \cdot |v|_d = n$. The segmentation $seg[v](w)$ of w by v divides the word w into m consecutive factors*

$$seg[v_i](w), (i = 1, ..., m) \text{ in such a way that } |seg[v_i](w)| = \begin{cases} k & \text{if } v_i = c, \\ l & \text{if } v_i = d. \end{cases}$$

Segments shall be designated by means of parentheses. For example, the segmentation of the Ionian mode $w = aabaaab$ by means of the generic triadic mode $v = ccd$ with segment lengths $c \mapsto 2, d \mapsto 3$, would be $seg[v](w) = (aa)(ba)(aab)$. If authentic divisions come into play, different divider symbols have to be used. The segmentation of the authentic Ionian mode $w = aaba|aab$ by means of the authentic generic triadic $v = c \wr cd$, would be $seg[v](w) = (aa) \wr (ba)|(aab)$.

Definition 5.2. *The segmentations of two conjugate well-formed modes $w_1, w_2 \in \{a, b\}^*$ by the same segmenting mode $v \in \{c, d\}^*$ and segment lengths are said to belong to the same species, iff the commutative images of corresponding segments coincide, i.e. iff $V(seg[v_i](w_1)) = V(seg[v_i](w_2))$, for all $i = 1, ..., m$.*

The goal is now to modally refine and generalize the situation of the pentads in Fig. 2 on the basis of these definitions. Thereby we concentrate on a situation, where the segmented modes as well as the segmenting modes are instances of diazeuctic modes. Fix $j \in \mathbb{N}$ together with $k = 3j + 2$ and consider the disjunct diazeuctic standard mode $w = G^k D(a|b) = a^k ba|a^k b$ of length $n = 2k + 3 = 2(3j + 2) + 3 = 6j + 7$ together with the conjunct diazeuctic Christoffel mode $v_1 = \tilde{D}^j G(c \wr d) = cd^j \wr cd^{j+1}$ as well as with the associated standard mode $v_2 = D^j G(c \wr d) = d^j c \wr d^j cd$ of length $m = 2j + 3$. With segment lengths $c \mapsto 2$ and $d \mapsto 3$ both modes v_1 and v_2 define segmentations of w, as they satisfy the condition $2|v|_c + 3|v|_d = 2 \cdot 2 + 3 \cdot (2j + 1) = 6j + 7 = n$. For $j = 0$ we have $seg[v_1](w) = seg[v_2](w) = (aa) \wr (ba)|(aab)$, and so it is not clear from the outset, which of the two segmentations is the canonical generalization of the triad. For $j > 0$ we obtain

$$seg[v_1](w) = (aa)(aaa)^j \wr (ba)|(aaa)^j(aab),$$
$$seg[v_2](w) = (aaa)(aaa)^{j-1}(aa) \wr (ba|a)(aaa)^{j-1}(aa)(aab).$$

It turns out that the Christoffel segmentation $seg[v_1](w)$ behaves much better than the standard segmentation $seg[v_2](w)$ with regard to the division of w. First of all it is eye-catching that the divider of w coincides with a segmentation point of the Christoffel segmentation. Even more striking is the following proposition, which states that the divider of w occupies the last generated position in the segmentation mode.

Proposition 5.3. *For $j \in \mathbb{N}$ consider the segmentation $seg[v](w)$ of the diazeuctic standard mode $w = G^{3j+2}D(a|b)$ through the diazeuctic Christoffel mode $v = \tilde{D}^j G(c \wr d)$ with the segment lengths $c \mapsto 2$ and $d \mapsto 3$. The segmentation points can be indexed by a well-formed (diazeuctic) scale $s, q : \mathbb{Z}_{2j+3} \to \mathbb{Z}_{6j+7}$, whose specific step intervals are given by the sequence $(2, 3, ..., 3, 2, 3, ..., 3, 3)$. The authentic division point of w is matched by the segmentation point $s(j + 2)$ which coincides with the last generated point $q(2j + 2)$.*

Proof. The incidence of $s(j+2)$ with the authentic divider of w follows directly from the structure of $seg[v](w)$. The well-formed (diazeuctic) scale of the segmentation points is given by $s(t) = \begin{cases} 0 & \text{for } t = 0 \\ 2 + 3(t - 1) & \text{for } t = 1, ..., j + 1, \\ 4 + 3(t - 2) & \text{for } t = j + 2, ..., 2j + 2. \end{cases}$ and
$q(t) = t \cdot (3j + 2)$ for $t = 0, ..., 2j + 2$. The last generated segmentation point is $g(2j+2) = (2j+2) \cdot (3j+2) = 6j^2 + 10j + 4 = j \cdot (6j+7) + 3j + 4 = 4 + 3j = s(j+2)$.

This finding reconciles to some degree the modal concept of the triad with the older derivation of the major and minor thirds as the result of a division of the fifth (see for example [11]). The proposition implies that the upper half of the authentic division of w (the standard species $G^{3j+2}D(b)$ of the co-generator) is a species of the closing interval $3j + 3 \in \mathbb{Z}_{6j+7}$ of the segmentation mode. For the usual case this means, that the specific perfect fourth $P4 = V(aab) = (2, 1)$ of the authentic division of the Ionian mode is a species of the generic "augmented third" 3 of the generic triad $\{0, 2, 4\} \subset \mathbb{Z}_7$.

The explicit compilation below confirms, that the same is true for the Christoffel segmentations $seg[v](\tilde{w})$ of all authentic conjugates $\tilde{w}$ of w and gives a detailed portrait about the species of these segmentations in terms of Definition 5.2.

$$
\begin{aligned}
major & \begin{cases} (aa)(aaa)...(aaa)(ab)(aaa)...(aaa)(aab) & Lydian \\ (aa)(aaa)...(aaa)(ba)(aaa)...(aaa)(aab) & Ionian \\ (aa)(aaa)...(aaa)(ba)(aaa)...(aaa)(aba) & Mixolydian \end{cases} \\
minor & \begin{cases} (aa)(aaa)...(aab)(aa)(aaa)...(aaa)(aba) & SuperIonian \\ (aa)(aaa)...(aab)(aa)(aaa)...(aaa)(baa) & SuperMixolydian \\ (aa)(aaa)...(aba)(aa)(aaa)...(aaa)(baa) & SuperSuperIonian \end{cases}
\end{aligned}
$$

$$\vdots \qquad\qquad \vdots$$

$$
\begin{aligned}
ronim & \begin{cases} (aa)(aba)...(aaa)(aa)(aab)...(aaa)(aaa) & SubSubAeolian \\ (aa)(baa)...(aaa)(aa)(aab)...(aaa)(aaa) & SubDorian \\ (aa)(baa)...(aaa)(aa)(aba)...(aaa)(aaa) & SubAeolian \end{cases} \\
rojam & \begin{cases} (ab)(aaa)...(aaa)(aa)(aba)...(aaa)(aaa) & Dorian \\ (ab)(aaa)...(aaa)(aa)(baa)...(aaa)(aaa) & Aeolian \\ (ba)(aaa)...(aaa)(aa)(baa)...(aaa)(aaa) & Phrygian \end{cases} \\
dimin & \{ \; (ba)(aaa)...(aaa)(ab)(aaa)...(aaa)(aaa) \quad\; Locrian
\end{aligned}
$$

Each conjugation along this generalized cycle-of-fifths ordering corresponds to a single switch of the form $ab \mapsto ba$. Which of the two bs undergoes the switch, depends on the number of a's between them on both sides of the cycle. These numbers must always differ by 1. Observe that each instance of a particular species of authentic modes starts with one segment of the form $(ab...)$ with the larger number of as left to it on the cycle and another segment of the form $(...ab)$ with the larger number of as right to it on the cycle. So precisely two switches may take place without changing the species. The following proposition summarizes this observation.

Proposition 5.4. *Under the assumptions of proposition 5.3 consider the segmentations $seg[v](\tilde{w})$ for all conjugates of w (including w). Every segmentation species of an authentic mode contains precisely three segmented modes. The only exception is the "diminished" segmentation of the bad conjugate of w, the generalized locrian mode, which forms a singleton species.*

In the usual case the traditional major, minor and diminished species of the generic triad correspond to three species of segmentations of the associated modes. In general the number of species coincides with the number m of notes in the generalized triad in accordance with the *cardinality equals variety*-property. For $j > 0$ the dualistic and the parsimonious derivations of the traditional minor mode bifurcate into different species. The term "minor" stands for the parsimoniously related species, while the retrogradation "rojam" of the term "major" stands for a dualistic relation. Consequently there is also a dualistic counterpart "ronim" of the parsimoniously related "minor" species.

Proposition 5.4 actually shows that three central elements of harmonic tonality interact as different instances of this concept within a hierarchy: (1) the diatonic modes, (2) the triads as their segmentations and (3) the *structural modes* parametrizing the species of the segmentations by authentic modes [10].

References

1. Agmon, E.: A mathematical model of the diatonic system. J. Music Theory **33**(1), 1–25 (1989)
2. Babbitt, M.: The stucture and function of music theory: I. Coll. Music Symp. **5**, 49–60 (1965)
3. de Berthé, V.A., Luca, C., Reutenauer, C.: On an involution of christoffel words and sturmian morphisms. Eur. J. Comb. **29**(2), 535–553 (2008)
4. Carey, N., Clampitt, D.: Aspects of well formed scales. Music Theory Spectr. **11**(2), 187–206 (1989)
5. Carey, N., Clampitt, D.: Self-similar pitch structures their duals, and rhythmic analogues. Perspect. New Music **34**(2), 62–87 (1996)
6. Clampitt, D., Noll, T.: Modes, the height-width duality, and handschin's tone character. Music Theory Online, **17/1** (2011)
7. Clough, J., Myerson, G.: Variety and multiplicity in diatonic systems. J. Music Theory **29**(2), 249–270 (1985)
8. Clough, J., Douthett, J.: Maximally even sets. J. Music Theory **35**, 93–173 (1991)
9. Clough, J., Engebretsen, N., Kochavi, J.: Scales, sets, and interval cycles: a taxonomy. Music Theory Spectr. **21**, 74–104 (1999)
10. de Jong, K., Noll, T.: Fundamental passacaglia: harmonic functions and the modes of the musical tetractys. In: Agon, C., Andreatta, M., Assayag, G., Amiot, E., Bresson, J., Mandereau, J. (eds.) MCM 2011. LNCS, vol. 6726, pp. 98–114. Springer, Heidelberg (2011)
11. Descartes, R.: 1656 = 1978. In: Descartes, R. (ed.) Musicae Compendium - Leitfaden der Musik. Wissenschaftliche Buchgesellschaft, Darmstadt (1978)
12. Douthett, J.: Filtered Point-Symmetry and Dynamical Voice-Leading. In: Douthett, J., et al. (eds.) Music Theory and Mathematics: Chords, Collections, and Transformations. University of Rochester Press, Rochester (2008)
13. Ganter, B., Wille, R.: Formal Concept Analysis Mathematical Foundations. Springer, Heidelberg (1999)
14. Handschin, J.: Der Toncharakter: Eine Einführung in die Tonpsychologie. Atlantis, Zurich (1948)
15. Lothaire, M.: Algebraic Combinatorics on Words. Cambridge University Press, Cambridge (2002)
16. Žabka, M.: Editorial. J. Math. Music **7**(2), 83–88 (2013)

Greek Ethnic Modal Names
vs. *Alia Musica*'s Nomenclature

David Clampitt[✉] and Jennifer Shafer

The Ohio State University, Columbus, OH, USA
{clampitt.4,shafer.212}@osu.edu

Abstract. The ethnic names associated with the diatonic modes (e.g., Dorian, Lydian, et al.) were assigned in one way by the ancient Greeks and in a different way in the anonymous medieval treatise *Alia musica*. Music historians usually say that this renaming was the result of a confusion, and leave it at that, but Edward Gollin showed that there was a logic here that could be captured in transformational terms. In this paper we add/uncover another layer to the palimpsest with the observation that the respective Greek and medieval nomenclatures are correlated with one of the fundamental distinctions in mathematical scale theory/word theory, the distinction between plain and twisted adjoint folding patterns of modes, as represented by conjugacy classes of Christoffel words.

Keywords: *Alia musica* · Aristoxenus · Boethius · Combinatorics on words · Modes

1 Introduction

In present-day parlance we refer to the diatonic modes, in the sense of octave species, by ancient Greek ethnic or topical names: Dorian, Phrygian, etc. For example, Phrygian refers to the octave species STTTSTT, where T and S represent Tone and Semitone step intervals, respectively (to within some reasonable tuning of these intervals). (Octave species is the only aspect of mode—a complicated topic—under consideration in this paper.) The ultimate source for these names goes back to the ancient Greek music theorists, possibly to Aristoxenus in the 4th century BCE or to the earlier Harmonicists, whom he criticized; but in any case, later Aristoxenians, such as Cleonides, of the 2nd century CE, identified these ethnic names with the species of the octave [2]. Greek music theory was transmitted to the Middle Ages primarily by Boethius, in his early 6th-century *De institutione musica* (see Bower's discussion in [3]). From Boethius, the late 9th- or early 10th-century *Alia musica* authors took the names, but reassigned them to medieval modes associated generally with completely different octave species [4]. David Cohen writes: "It is on the basis of a misreading of this complicated, utterly alien system that our Author 2 imports the octave species and the ethnic Greek names into the system of the ecclesiastical *toni* (the church modes)..." [5] (pp. 333–334). Explanations of this misreading are

© Springer International Publishing Switzerland 2015
T. Collins et al. (Eds.): MCM 2015, LNAI 9110, pp. 385–390, 2015.
DOI: 10.1007/978-3-319-20603-5_38

explored in [5] and [1], and Gollin in particular suggests historically plausible rationales for a succession of renamings and reorderings of the modes (generally in terms of octave species). The present paper demonstrates a correlation of the respective naming conventions with the plain and twisted adjoints of mathematical scale theory, discussed below. This is a correlation that is generalizable, because the respective adjoints hold for all classes of modes of non-degenerate well-formed scales (e.g., pentatonic, chromatic, as well as diatonic), as represented by conjugacy classes of Christoffel words [6]. This correlation is in the spirit of Gollin's transformational description, but distinct from his observation of diatonic inversion.

The respective Greek and medieval associations of octave species with the ethnic names are shown in Table 1. The ranges of white-key octave species are given using medieval letter notation, for uniformity. (Note that the range differs only in the case of Hypodorian, the one instance where the same name is applied.) The step-interval patterns read the scales as ascending. Many modern discussions of ancient Greek theory (following Boethius's Latin treatise) presume that scales should be conceptualized as descending, but according to Thomas Mathiesen, perhaps the foremost authority on Greek music theory, there is no evidence for this in the Greek treatises (private communication at the 2001 Mannes Institute for Advanced Studies in Music Theory). Whether we read all scales ascending or descending is purely conventional and makes no difference to the analysis here, but we do require a consistent approach.

Table 1. Greek octave species names; *Alia musica* octave species names

Name	Range	Step-interval pattern	Range	Name
Mixolydian	B-b	S-T-T-S-T-T-T	B-b	Hypophrygian
Lydian	C-c	T-T-S-T-T-T-S	C-c	Hypolydian
Phrygian	D-d	T-S-T-T-T-S-T	D-d	Dorian
Dorian	E-e	S-T-T-T-S-T-T	E-e	Phrygian
Hypolydian	F-f	T-T-T-S-T-T-S	F-f	Lydian
Hypophrygian	G-g	T-T-S-T-T-S-T	G-g	Mixolydian
Hypodorian	a-aa	T-S-T-T-S-T-T	A-a	Hypodorian

The relationship between the octave species and what the ancient Greeks referred to as *tonoi*, on the one hand, and that between the octave species and the medieval church modes, on the other, are highly complex matters, and are not the subject of this paper. The prefix "hypo-" means "under" or "below." In the case of the three Greek modes with this prefix the reference is likely to locations on a stringed instrument (lower down or shorter string lengths, thus higher in pitch). In the medieval modal system "hypo-" denotes a plagal mode, with range or *ambitus* beginning a perfect fourth below the final (*finalis*). The authentic modes have their ranges entirely above their finals, the octave divided into a perfect fifth plus a perfect fourth. There are thus eight church modes, with finals D, E, F, and G, with authentic forms given the names Dorian, Phrygian,

Lydian, and Mixolydian, respectively, and the plagal forms named Hypodorian, Hypophrygian, Hypolydian, and Hypomixolydian, respectively, in *Alia musica*. Boethius, in one of his presentations, included an eighth mode that he attributed to Ptolemy, Hypermixolydian, for the A-a octave. The coincidence that there are eight church modes and eight modes found in Boethius is one of the reasons for the medieval misreading. If we associate the eight church modes with the (seven) diatonic octave species, the octave from D to d is the one that is counted twice: once for authentic Dorian, once for plagal Hypomixolydian. The authentic modes are the principal or unmarked forms, so we choose Dorian as the mode associated with the *Alia musica* octave species TSTTTST.

2   Combinatorics on Words and Folding Patterns

The two naming conventions displayed in Table 1 above may be correlated with the folding patterns (defined below) arising from the application of combinatorics on words (word theory) to the study of the diatonic modes. We follow the exposition of this application in [6]. In the word-theoretical context that follows, the symbols T and S are replaced by a and b, respectively. In the context of folding patterns we employ the letters x and y (to be consistent with the literature).

In word theory, one considers a finite alphabet A (here, specialized to a two-letter alphabet) and the set of words over A, $A^* = \{w = w_1 \ldots w_n | w_i \in A, n \in \mathbb{N}\}$. A^* is a free monoid (semi-group with identity) where the monoid operation is concatenation of words, and one understands that the empty word ε is in A^*; for all words w in A^*, $\varepsilon w = w = w\varepsilon$. Cyclic permutations (rotations) of words are called *conjugates* and the set of all conjugates of a given word is its *conjugacy class*. See Lothaire [7] for an exposition of word theory.

Let us identify the octave species of the diatonic scale with their step-interval patterns, and thus with the seven members of the conjugacy class of $w = aaabaab$. Mathematical music theory investigates properties of the modal octave species by employing a duality, defined by a unique word over a binary alphabet that encodes the (forward or backward) *folding pattern* for that mode. The forward folding patterns are defined musically in the following way: consider the pattern, departing from F, of rising perfect fifths and falling perfect fourths such that all notes lie within the modal octave, i.e., including the modal final and the notes of the mode within the octave above it. The unique forward folding pattern for the given mode is determined by encoding the pattern of rising and falling intervals as a word over a two-letter alphabet. For example, consider the diatonic mode on C, for which the step-interval pattern is encoded by the word $aabaaab$. The forward folding pattern that has C as its lowest note and includes the notes above C and within the C-c octave, with x and y representing rising fifths and falling fourths, respectively, is: F y C x G y D x A y E x B y (F#). Note that just as in the scale C-c one includes a note not strictly in the scale, the note c an octave above the final, in order to determine the step intervals of the step-interval pattern, so one also includes in the folding a note excluded from the scale, F#, in order to determine the folding pattern. The word that encodes the forward folding for the C mode is thus $yxyxyxy$. The backward folding patterns are defined musically in an analogous way: consider the pattern, departing from B, of falling perfect fifths and rising perfect fourths such that all notes lie within the modal octave, i.e., including the modal final and the notes of the mode within the octave above it. The unique backward folding pattern for the given mode is determined by encoding the pattern of falling and rising intervals as a word over a two-letter alphabet. For example, consider the diatonic mode on D, which determines the word $abaaaba$. The backward folding pattern that has D as its lowest note and includes the notes above D and within the D-d octave, with x and y now representing falling fifths and rising fourths, respectively, is: B x E y A x D y G y C x F y (Bb). The word that encodes the backward folding for the D mode is thus $xyxyyxy$. Table 2 collects the diatonic modes and their forward and backward foldings.

Table 2. The diatonic modes and their forward and backward foldings

Forward folding	Range	Scale-step pattern	Backward folding
$yxyyxyx$	A-a	$abaabaa$	$yxyyxyx$
$yxyxyyx$	B-b	$baabaaa$	$yyxyxyx$
$yxyxyxy$	C-c	$aabaaab$	$xyxyxyy$
$xyyxyxy$	D-d	$abaaaba$	$xyxyyxy$
$xyxyyxy$	E-e	$baaabaa$	$xyyxyxy$
$xyxyxyy$	F-f	$aaabaab$	$yxyxyxy$
$yyxyxyx$	G-g	$aabaaba$	$yxyxyyx$

An important set of endomorphisms of the monoid A^* are the following, which map A^* to itself by replacing each letter of $w \in A^*$ by:

$$G(a) = a, G(b) = ab \qquad (1)$$

$$\tilde{G}(a) = a, \tilde{G}(b) = ba \qquad (2)$$

$$D(a) = ba, D(b) = b \qquad (3)$$

$$\tilde{D}(a) = ab, \tilde{D}(b) = b \qquad (4)$$

The set of all compositions of these morphisms forms the monoid St_0 of *special Sturmian morphisms*, under composition of mappings. If $F \in St_0$, F is a morphism by construction: for $u, v \in A^*, F(uv) = F(u)F(v)$, since in particular for any word $w = w_1 \ldots w_n$ for letters w_i in A, $F(w) = F(w_1)F(w_2) \ldots F(w_n)$ by definition.

Words derived by the application of members of St_0 to the root word ab (or xy) are privileged elements of A^*, called conjugates of Christoffel words [7]. As shown in [6], and as one can compute, the scale-step patterns of six of the seven octave species are formed by the application to the root word ab of the morphisms $GGD, G\tilde{G}D, \tilde{G}\tilde{G}D, GG\tilde{D}, G\tilde{G}\tilde{D}, \tilde{G}\tilde{G}\tilde{D}$. The scale-step pattern that is not the image of ab under a morphism, the B-b pattern $baabaaa$, is called the *bad conjugate*.

The conjugacy class of folding patterns is obtained by taking the reversals of the morphisms DGG, etc., applied to the root word xy. But which ones correspond to which scales depends on whether they are to represent forward foldings or backward foldings. In the case of the forward foldings, the bad conjugate is the folding for the G-g mode. In the special cases of standard words (compositions of G and D morphisms), and of Christoffel words (compositions of G and $\tilde{D}$ morphisms), the forward folding patterns are obtained by precisely reversing the morphisms: DGG for the C-c forward folding and $\tilde{D}GG$ for the F-f forward folding. Clampitt and Noll call this set of relationships the *plain adjoint* [6].

The correspondence between scale-step pattern morphisms and backward folding morphisms is much cleaner mathematically: it is an involution obtained by reversing the morphism, while replacing every D by $\tilde{D}$ and vice versa, and leaving fixed every G and $\tilde{G}$. The bad conjugate for the backward folding thus corresponds to the bad conjugate for the scale, those corresponding to the B-b octave species. Clampitt and Noll call this set of relationships the *twisted adjoint*.

For example, the morphism that produces the word $abaaaba$ corresponding to the step-interval pattern of the D-d octave species is $\tilde{G}GD$: $\tilde{G}GD(ab) = \tilde{G}G(bab) = \tilde{G}(abaab) = abaaaba$. Under the twisted adjoint the morphism for the backwards folding is thus $\tilde{D}G\tilde{G}$, and $\tilde{D}G\tilde{G}(xy) = \tilde{D}G(xyx) = \tilde{D}(xxyx) = xyxyyxy$. One may easily confirm that, departing from a note B and interpreting x as falling perfect fifth and y as rising perfect fourth, D is the lowest note reached.

As Table 3 shows, the relationship between the plain and twisted adjoints corresponds to that between the octave species that share the same Greek and medieval name. That is, reinterpreting a folding pattern from backwards to forwards or vice versa changes the corresponding octave species such that the name remains unchanged: note that, in a given row, the forward folding pattern (second column) and backward folding pattern (sixth column) are represented by the same word. A diatonic symmetry about the pitch class A obtains, as the Greek ranges are displayed descending from the a-aa octave while medieval church mode ranges are displayed ascending from the A-a octave.

This correlation is not uniquely confined to the usual diatonic scale and its modes. An analogous relationship may be contrived for any class of modes of a

Table 3. Plain adjoint (Greek name) twisted adjoint (medieval name)

Scale	Folding	Range	Name	Range	Folding	Scale
$\tilde{G}\tilde{G}\tilde{D}$	$D\tilde{G}\tilde{G}$	a-aa	Hypodorian	A-a	$D\tilde{G}\tilde{G}$	$\tilde{G}\tilde{G}\tilde{D}$
$\tilde{G}G\tilde{D}$	bad conj.	G-g	Hypophrygian	B-b	bad conj.	bad conj.
$GG\tilde{D}$	$\tilde{D}GG$	F-f	Hypolydian	C-c	$\tilde{D}GG$	GGD
$\tilde{G}\tilde{G}D$	$\tilde{D}G\tilde{G}$	E-e	Dorian	D-d	$\tilde{D}G\tilde{G}$	$\tilde{G}GD$
$G\tilde{G}D$	$\tilde{D}\tilde{G}\tilde{G}$	D-d	Phrygian	E-e	$\tilde{D}\tilde{G}\tilde{G}$	$\tilde{G}\tilde{G}D$
GGD	DGG	C-c	Lydian	F-f	DGG	$GG\tilde{D}$
bad conj.	$DG\tilde{G}$	B-b	Mixolydian	G-g	$DG\tilde{G}$	$\tilde{G}G\tilde{D}$

Table 4. Labelings of the Christoffel word *aabab*

Scale	Folding	Range	Number	Range	Folding	Scale
$\tilde{G}D$	$\tilde{D}\tilde{G}$	d-dd	1	D-d	$\tilde{D}\tilde{G}$	$\tilde{G}D$
GD	DG	C-c	2	F-f	DG	$G\tilde{D}$
bad conj.	$D\tilde{G}$	A-a	3	G-g	$D\tilde{G}$	$\tilde{G}\tilde{D}$
$\tilde{G}\tilde{D}$	bad conj.	G-g	4	A-a	bad conj.	bad conj.
$G\tilde{D}$	$\tilde{D}G$	F-f	5	C-c	$\tilde{D}G$	GD

non-degenerate well-formed scale, represented by a conjugacy class of a Christoffel word. For example, the octave species of the usual anhemitonic pentatonic scale, e.g., FGACd(f), represented by the Christoffel word *aabab*, could be labeled (numbered) according to the same principle, as shown in Table 4.

These examples prompt some concluding thoughts. The status of the plain and twisted adjoints remains a question in mathematical music theory. Unlike the decision as to whether to read scales as ascending or descending, the choice is not just a matter of convention. On the one hand, the twisted adjoint is clearly more mathematically satisfactory: it is an involution on St_0. The plain adjoint, however, affords musical insights—it would take us too far afield to specify them here, but they involve the modern major and minor modes. The twisted adjoint is more consonant with the medieval modes, privileging in particular authentic Dorian. The correspondence adduced here between the Greek and medieval nomenclatures on the basis of the adjoints is clearly ahistorical (in contrast to the work of Gollin and Cohen), but reaffirms the historical logic traced by those authors, and is potentially of interest in exploring the musical significance of the plain and twisted adjoints in the wider context of mathematical scale theory.

References

1. Gollin, E.: From tonoi to modi: a transformational perspective. Music Theor. Spectr. **26**, 119–129 (2004)
2. Mathiesen, T.: Greek music theory. In: Christensen, T. (ed.) The Cambridge History of Western Music Theory, pp. 109–135. Cambridge University Press, Cambridge (2002)
3. Bower, C.: The transmission of ancient music theory into the middle ages. In: Christensen, T. (ed.) The Cambridge History of Western Music Theory, pp. 136–167. Cambridge University Press, Cambridge (2002)
4. [Anonymi]: Alia musica, Chailley, J. (ed. and Fr. trans.). Centre de documentation universitaire, Paris (1965)
5. Cohen, D.: Notes, scales, and modes in the earlier middle ages. In: Christensen, T. (ed.) The Cambridge History of Western Music Theory, pp. 307–363. Cambridge University Press, Cambridge (2002)
6. Clampitt, D., Noll, T.: Modes, the height-width duality, and Handschin's tone character. Music Theor. Online **17**(1), 1–149 (2011)
7. Lothaire, M.: Combinatorics on Words. Cambridge Mathematical Library. Cambridge University Press, Cambridge (1997)

Author Index

Printed in the United States
By Bookmasters